CONFRONTING THE INTERNET'S DARK SIDE

Moral and Social Responsibility on the Free Highway

Terrorism, cyberbullying, child pornography, hate speech, cybercrime: along with unprecedented advancements in productivity and engagement, the Internet has ushered in a space for violent, hateful, and antisocial behavior. How do we, as individuals and as a society, protect against dangerous expressions online?

Confronting the Internet's Dark Side is the first book on social responsibility on the Internet. It aims to strike a balance between the free speech principle and the responsibilities of the individual, corporation, state, and the international community. This book brings a global perspective to the analysis of some of the most troubling uses of the Internet. It urges net users, ISPs, and liberal democracies to weigh freedom and security, finding the golden mean between unlimited license and moral responsibility. This judgment is necessary to uphold the very liberal democratic values that gave rise to the Internet and that are threatened by an unbridled use of technology.

Raphael Cohen-Almagor is Professor and Chair in Politics at the University of Hull, United Kingdom. He has published extensively in the fields of political science, law, ethics, and philosophy, including *The Right to Die with Dignity* (2001), *Speech, Media and Ethics* (2nd ed., 2005), and *The Scope of Tolerance* (2007). His second book of poetry, published in 2007, is entitled *Voyages*.

Confronting the Internet's Dark Side

MORAL AND SOCIAL RESPONSIBILITY
ON THE FREE HIGHWAY

RAPHAEL COHEN-ALMAGOR

Woodrow Wilson Center Press
Washington, D.C.
and
 CAMBRIDGE
UNIVERSITY PRESS

CAMBRIDGE
UNIVERSITY PRESS

32 Avenue of the Americas, New York, NY 10013-2473, USA

Cambridge University Press is part of the University of Cambridge.

It furthers the University's mission by disseminating knowledge in the pursuit of education, learning, and research at the highest international levels of excellence.

www.cambridge.org
Information on this title: www.cambridge.org/9781107513471

Woodrow Wilson Center Press
Woodrow Wilson International Center for Scholars, One Woodrow Wilson Plaza,
1300 Pennsylvania Avenue NW, Washington, DC 20004–3027
www.wilsoncenter.org

First published 2015

Printed in the United States of America

A catalog record for this publication is available from the British Library.

Library of Congress Cataloguing in Publication data
Cohen-Almagor, Raphael.
Confronting the Internet's dark side : moral and social responsibility on the
free highway / Raphael Cohen-Almagor.
     pages   cm
Includes bibliographical references and index.
ISBN 978-1-107-10559-1 (hardback) – ISBN 978-1-107-51347-1 (paperback)
1. Internet governance.   2. Internet – Moral and ethical aspects.
3. Internet – History.   I. Title.
TK5105.8854.C64   2015
174'.9025042–dc23   2015021256

ISBN 978-1-107-10559-1   Hardback
ISBN 978-1-107-51347-1   Paperback

In memory of Sarah Cohen (1930–2011), who shaped my thinking and paved the way for me

> *Not a single day passes*
> *Without seeing your faces*
> *Memories come running*
> *Different periods, different places.*

Contents

Contents

Acknowledgments

This is my fifth book in a series that started with *The Boundaries of Liberty and Tolerance* (University Press of Florida, 1994) and continued with *Speech, Media, and Ethics: The Limits of Free Expression* (Palgrave Macmillan, 2001, 2005) and *The Scope of Tolerance* (Routledge, 2006). The fourth book to address ethical boundaries to freedom of expression was *The Democratic Catch* (Maariv, 2007, in Hebrew). On completing my last book, I knew that my next project would concern the Internet, a fascinating and growing phenomenon.

This book is the result of research and thinking conducted during the past decade. I would like to thank friends and colleagues who conversed with me on pertinent questions and who read parts or the entire book. First and foremost, I am grateful to Robert Cavalier, Clifford Christians, Jack Hayward, and Steve Newman, who read and commented on the book manuscript. They provided vital suggestions and criticisms that challenged my thinking and significantly improved the quality of the book.

I communicated and exchanged ideas with Eric Barendt, Ann Bartow, Vint Cerf, Jerry Cohen, Dorothy Denning, Wilfrid Knapp, Sam Lehman-Wilzig, Jack Pole, and Mike Whine. They all provided invaluable insights. The three Oxford scholars and friends – Jerry Cohen, Wilfrid Knapp, and Jack Pole – are no longer with us.

Other people who provided useful information and helped me crystallize my thoughts are Yaman Akdeniz, Michael Bernstein, Ian Binnie, David D. Clark, Bret S. Cohen, Abraham Cooper, David Corchia, Peter Cory, Irwin Cotler, Reuven Erlich, Mark Fackler, Harold Feld, Luciano Floridi, Robert Fortner, Martin Freeman, Mark J. Freiman, Laurence Godfrey, David Goldberg, Harvey Goldberg, Wayne Hanniman, Holly Hawkins, Jayne Hitchcock, Frank Iacobucci, Steve Jones, Bonnie Jouhari, Athina Karatzogianni, Jennifer

King, Leonard Kleinrock, Marc Knobel, Marvin Kurz, Deborah M. Lauter, Peter Leitner, Jacqueline Lipton, Roderick A. Macdonald, David Matas, Giuliana Mazzoni, Bruce McFarlane, Gregory T. Nojeim, Nisha Patel, Nikolaus Peifer, Mark Pillams, Mark Potok, Joel Reidenberg, Perry Roach, Tony Rutkowski, Philippe A. Schmidt, Andrew Jay Schwartzman, Oren Segal, Steven C. Sheinberg, Dominic Sparkes, Christophe Stener, Jose Vegar, Jonathan Vick, Richard Warman, Aaron Weisburd, and Kevin Woodley. They provided important observations and clarifications that enriched my understanding of the Internet, its architecture, its merits, and its use and abuse.

I am most grateful to the interviewees for their time and willingness to share their knowledge and experience with me. I thank the following people for their kind cooperation: Ruth Allen, Robert D. Atkinson, Carolyn Atwell-Davis, Steve Balkam, Rosa Beer, Rick Boucher, Yigal Carmon, Daniel Castro, Michelle K. Collins, Robert Corn-Revere, Charles M. Firestone, Mary E. Galligan, Julie A. Gottlieb, Leslie Harris, Shawn Henry, Herb Linn, Brian Marcus, John Morris, Philip Mudd, Michael Nelson, Marc Rotenberg, Peter Swire, Adam Thierer, and Chris Wolf.

I acknowledge with gratitude the generous support of the Woodrow Wilson International Center for Scholars and the Faculty of Arts and Social Sciences, University of Hull. Both institutions provided me the opportunity to concentrate on my research for a precious period of time, allowing me to probe, think, and write. I especially cherish the year I spent at the Wilson Center, the best research institute I have ever known.

Lee Hamilton, Mike van Dusen, Joe Brinley, Lee Rawls, and George Talbot provided invaluable assistance, without which this book could not have been published. I am truly grateful to them for their support and belief in the importance of this research project.

Last but not least, I express my deep gratitude to Janet Spikes and Marco Zambotti for their superb research assistance; to Linda Lee Stringer and Shannon Granville for their excellent editorial support; and to my family for their love, understanding, support, and enduring patience. Gilad, Dana, Roei, and especially Zehavit provided me with much-needed time to complete this book.

Linda Lee Stringer and her team at Publications Professionals LLC have checked all Web pages during the month of October 2014. At that point of time, all links were viable. The nature of the Internet is such, however, that Web pages move, and are sometimes removed. But the reader is provided with ample information about sources and will be able to find information on and off the net.

An early version of Chapter 1 was published in *International Journal of Technoethics* 2, no. 2 (2011): 46–65. A very early version of Chapter 5 was coauthored with Sharon Haleva-Amir and published under the title "Bloody Wednesday in Dawson College: The Story of Kimveer Gill, or Why Should We Monitor Certain Websites to Prevent Murder," *Studies in Ethics, Law and Technology* 2, no. 3, article 1 (2008), http://works.bepress.com/raphael_cohen_almagor/1. Another version of Chapter 6 was published in Hilmi Demir, ed., *Luciano Floridi's Philosophy of Technology: Critical Reflections* (Dordrecht, Netherlands: Springer, 2012): 151–67. A different version of Chapter 8 was published in *Journal of Business Ethics* 106, no. 3 (2012): 353–65. My gratitude is granted for permissions to use the material.

The book is dedicated to my mother, Sarah Cohen, who passed away during the writing phase. My mum stood by me all her life and was also involved in the thinking and shaping of this book. She always encouraged me to study and explore social dilemmas and to work for the benefit of my community. Mum was the driving force behind me – the compass, the anchor, and the inspiration. To a large extent, I am who I am because of her love and unwavering support. My mum will live in my heart and mind until my last day.

Raphael Cohen-Almagor
Beverley

Abbreviations

ACLU	American Civil Liberties Union
ADL	Anti-Defamation League
AES	Advanced Encryption Standard
AFA	Association de Fournisseurs d'Accès et de Services Internet
AIIP	Association of Independent Information Professionals
AOL	America Online
ARPA	Advanced Research Projects Agency
ARPANET	ARPA Network
AS	Autonomous System
ASN	Autonomous System Number
AVE	Against Violent Extremism
BBN	Bolt Beranek and Newman
BBS	Bulletin Board System
BIF	Benevolence International Foundation
BPjM	Bundesprüfstelle für jugendgefährdende Medien (Federal Department for Media Harmful to Young Persons)
CCAICE	Canadian Coalition against Internet Child Exploitation
CCCP	Canadian Centre for Child Protection
ccTLD	country code top-level domain
CEM	child exploitation material
CEO	chief executive officer
CEOP	Child Exploitation and Online Protection
CETS	Child Exploitation Tracking System
CMC	computer-mediated communication
CSNET	Computer Science Network
CSR	Corporate Social Responsibility
CTIRU	Counterterrorism Internet Referral Unit (UK)
DARPA	Defense Advanced Research Projects Agency

DES	Data Encryption Standard
DNS	Domain Name System
DoS	Denial of Service
ECA	engine for content analysis
EC3	European Cybercrime Centre
ERA	engine for relationship analysis
EU	European Union
FATF	Financial Action Task Force
FBI	Federal Bureau of Investigation (US)
FEP	Firewall Enhancement Protocol
FTC	Federal Trade Commission (US)
FTP	File Transfer Protocol
gTLD	generic top-level domain
HTML	HyperText Markup Language
HTTP	Hypertext Transfer Protocol
IAB	Internet Architecture Board
IANA	Internet Assigned Numbers Authority
ICANN	Internet Corporation for Assigned Names and Numbers
ICT	information and communication technology
IDEA	International Data Encryption Algorithm
IETF	Internet Engineering Task Force
IHR	Institute for Historical Review
IM	Instant Message
IMP	Interface Message Processor
INACH	International Network against Cyber Hate
INHOPE	International Association of Internet Hotlines
IP	Internet Protocol
IPv4	Internet Protocol version 4
IPTO	Information Processing Techniques Office
IRC	Internet Relay Chat
ISOC	Internet Society
ISP	Internet service provider
ISPA	Internet Service Providers' Association (UK)
IWF	Internet Watch Foundation
Janet	Joint Academic Network
KKK	Ku Klux Klan
LICRA	Ligue Internationale Contre le Racisme et l'Antisémitisme (International League against Racism and Antisemitism)
Mbps	megabits per second
MIT	Massachusetts Institute of Technology

MRAP	Mouvement contre le Racisme et pour l'Amitié entre les Peuples (Movement against Racism and for Friendship between Peoples)
NCECC	National Child Exploitation Coordination Centre
NCMEC	National Center for Missing and Exploited Children
NCP	Network Control Protocol
NPL	National Physical Laboratory
NSA	National Security Agency (US)
NSF	National Science Foundation
NSFNet	National Science Foundation Network
NSPCC	National Society for the Prevention of Cruelty to Children
OSCE	Organization for Security and Co-operation in Europe
OTP	one-time pad (encryption technique)
PGP	Pretty Good Privacy (encryption program)
PICS	Platform for Internet Content Selection
POWDER	Protocol for Web Description Resources
P2P	peer-to-peer
RCMP	Royal Canadian Mounted Police
RMF	real-time message filter
S/MIME	Secure/Multipurpose Internet Mail Extensions
SMS	Short Message Service
SNS	Social networking site
TCP	Transmission Control Protocol
3DES	Triple DES
UCLA	University of California at Los Angeles
UEJF	Union des Étudiants Juifs de France (Union of French Jewish Students)
UGC	user-generated content
UN	United Nations
URL	uniform resource locator
VGT	Virtual Global Taskforce
VoIP	voice over Internet Protocol
WCOTC	World Church of the Creator
WHOA	Working to Halt Online Abuse
WHS	Web-hosting service
WWW	World Wide Web
ZOG	Zionist Occupation Government

Introduction

Know from whence you came. If you know whence you came, there are absolutely no limitations to where you can go

–James Baldwin

The Internet burst into our lives in the early 1990s without much preparation or planning and changed them forever. It has affected almost every aspect of society. It is a macrosystem of interconnected private and public spheres: household, literary, military, academic, business, and government networks. The Internet has produced major leaps forward in human productivity and has changed the way people work, study, and interact. The mix of open standards and diverse networks and the growing ubiquity of digital devices makes the Internet a revolutionary force that undermines traditional media, such as newspapers, broadcasting, and telephone systems, and that challenges existing regulatory institutions that are based on national boundaries.

The Internet's design and *raison d'être* are open architecture, freedom of expression, and neutral network of networks. In the prevailing Western liberal tradition, freedom of expression is perceived as a fundamental human right requiring the uninhibited free flow of information. This is especially true for the Internet. But soon enough, people began to exploit the Net's massive potential to enhance partisan interests, some of which are harmful and antisocial. Given that the Internet has been part of our lives for a relatively short time, the discussions about it concentrate on the social production and the technological, architectural, and geographic aspects of the Net. (Thinkers in this area include Yochai Benkler,[1] Manuel

---

[1] Yochai Benkler, *The Wealth of Networks: How Social Production Transforms Markets and Freedom* (New Haven, CT: Yale University Press, 2006).

Castells,[2] Aharon Kellerman,[3] Lawrence Lessig,[4] Gary P. Schneider and Jessica Evans,[5] James Slevin,[6] and Jonathan Zittrain,[7] to name a few.) The discussions about the costs and harms of such Internet content reflect on the transnational nature of the Internet. They tend to conclude that it is very difficult – some say virtually impossible – for national authorities to unilaterally implement laws and regulations that reflect national, rather than global, moral standards.[8]

Most Internet users act within the law. Thus, free speech advocates argue that the collective should not be restricted because of the few who abuse Internet freedom in order to harm others. We should not allow the abusers to dictate the rules of the game, but of course we should fight against those who abuse this freedom. The way to combat problematic speech is said to be with more speech. Organizations and associations have been set up to protect and promote freedom of expression, freedom of information, and privacy on the Internet.[9] In the United States, the land of the First Amendment,[10] emphasis is put on education (see the work of Robert D. Atkinson,[11] Robert Corn-Revere,[12]

[2] Manuel Castells, *Communication Power* (Oxford: Oxford University Press, 2009); Manuel Castells, *The Internet Galaxy: Reflections on the Internet, Business, and Society* (Oxford: Oxford University Press, 2001).

[3] Aharon Kellerman, *The Internet on Earth: A Geography of Information* (Oxford: Wiley, 2002).

[4] Lawrence Lessig, *Code and Other Laws of Cyberspace* (New York: Basic Books, 1999); Lawrence Lessig, *The Future of Ideas: The Fate of the Commons in a Connected World* (New York: Vintage, 2002); Lawrence Lessig, *Free Culture: How Big Media Uses Technology and the Law to Lock Down Culture and Control Creativity* (New York: Penguin, 2004).

[5] Gary P. Schneider and Jessica Evans, *New Perspectives on the Internet: Comprehensive*, 6th ed. (Boston: Thomson, 2007).

[6] James Slevin, *The Internet and Society* (Oxford: Polity, 2000).

[7] Jonathan L. Zittrain, *The Future of the Internet – And How to Stop It* (New Haven, CT: Yale University Press, 2008).

[8] Dick Thornburgh and Herbert S. Lin, eds., *Youth, Pornography, and the Internet* (Washington, DC: National Academies Press, 2002); National Research Council, *Global Networks and Local Values: A Comparative Look at Germany and the United States* (Washington, DC: National Academies Press, 2001). For further discussion, see Robert J. Cavalier, ed., *The Impact of the Internet on Our Moral Lives* (Albany: State University of New York Press, 2005).

[9] Among them are the Center for Democracy and Technology, http://cdt.org/ ; the Electronic Frontier Foundation, http://www.eff.org/ ; the Electronic Privacy Information Center, http://epic.org/ ; the Global Internet Liberty Campaign, http://gilc.org/ ; the Internet Society, http://www.isoc.org/ ; the Association for Progressive Communications, http://www.apc.org ; and Save the Internet, http://savetheinternet.com/ .

[10] See the text of the First Amendment and annotations at http://constitution.findlaw.com/amendment1.html .

[11] See Atkinson's posts on *The Innovation Files* (blog), http://www.innovationfiles.org/author/robatkinson/ .

[12] Robert Corn-Revere, "Caught in the Seamless Web: Does the Internet's Global Reach Justify Less Freedom of Speech?," in *Who Rules the Net? Internet Governance and Jurisdiction*, ed.

Leslie Harris,[13] Tom Head,[14] Gerson Moreno-Riaño,[15] Andrea C. Nakaya,[16] Michael R. Nelson,[17] and Adam Thierer,[18] among others[19]). "Keep the Internet free and open," reiterates Vint Cerf, Google vice president and chief evangelist.[20] These thinkers recognize the dangers of the Internet, but they commonly argue that the principle of free speech enshrined in the First Amendment shields all but the most immediately threatening expression. A strong presumption exists against speech restrictions. As Michael Nelson said, the Internet helps mitigate tensions. It conveys information, tells us about the aims and activities of terrorists and hatemongers, and shows us how poor their ideas are.[21]

These views represent the existing mode of thinking in the United States. The United States tends not to be preemptive in the sphere of freedom of expression. Among the limited boundaries to free expression on the Net are direct and specific calls for murder ("true threats"),[22] child pornography, direct

---

Adam Thierer and Clyde Wayne Crews Jr. (Washington, DC: Cato Institute, 2003), 219–38, based on amicus brief in *Yahoo!, Inc. v. La Ligue contre le Racisme et L'Antisemitisme*, Case No. 01-17424 (9th Cir.); Robert Corn-Revere, "*United States v. American Library Association*: A Missed Opportunity for the Supreme Court to Clarify Application of First Amendment Law to Publicly-Funded Expressive Institutions," in *Cato Supreme Court Review, 2002–2003*, ed. James L. Swanson (Washington, DC: Cato Institute, 2003), 105–30.

[13] Leslie Harris is president and chief executive officer of the Center for Democracy and Technology. See her biography at https://www.internetsociety.org/inet-washington-dc/speakers/ms-leslie-harris . See also Leslie Harris, "Internet Governance, or Just Governing the Internet," *Huffington Post*, July 5, 2012, http://www.huffingtonpost.com/leslie-harris/internet-governance_b_1643856.html ; Leslie Harris, "From Moment to a Movement: Sustaining Our New Power," presentation at the Personal Democracy Forum, June 13, 2012, http://personaldemocracy.com/media/moment-movement-sustaining-our-new-power .

[14] Tom Head, ed., *The Future of the Internet* (Farmington Hills, MI: Greenhaven Press, 2005).

[15] Gerson Moreno-Riaño, ed., *Tolerance in the Twenty-First Century: Prospects and Challenges* (Lanham, MD: Lexington Books, 2006).

[16] Andrea C. Nakaya, ed., *Censorship: Opposing Viewpoints* (Farmington Hill, MI: Greenhaven Press, 2005).

[17] Michael R. Nelson, "Sovereignty in the Networked World," in *Emerging Internet: Annual Review of the Institute for Information Studies* (Falls Church, VA: Institute for Information Studies, 1998).

[18] Thierer and Crews, *Who Rules the Net?*

[19] See also Mark A. Shiffrin and Avi Silberschatz, "Web of the Free," *New York Times*, October 23, 2005.

[20] Alex Fitzpatrick, "Google's Vint Cerf: Keep the Internet Free and Open," Mashable.com, December 3, 2012, http://mashable.com/2012/12/03/vint-cerf-open-internet/ ; Bonnie Tubbs, "Web Pioneer Vint Cerf Advocates a Free Internet," *iweek*, September 18, 2013, http://www.iweek.co.za/in-the-know/web-pioneer-vint-cerf-advocates-a-free-internet .

[21] Interview with Michael Nelson, former IBM director, Internet Technology and Strategy, Washington, DC, January 31, 2008.

[22] A statement is a "true threat" when a reasonable person making the statement would foresee that the statement would be interpreted by those to whom it is communicated as a serious expression of an intent to do bodily harm or assault. See *Planned Parenthood of Columbia/Willamette, Inc. v.*

calls for terrorism and spreading of electronic viruses, and material protected by copyright legislation. Threats of a general nature, hatred, bigotry, racism, and instructions on how to kill and maim and how to seduce children are all protected forms of speech under the First Amendment. Speech is afforded protection except when a life-threatening message is directed against identified individuals.[23] Blanket statements expressing hatred toward certain groups are given free sway, even if individual members of such groups are put at risk.[24] Negin Salimipour argued that government actions "limiting the spread of harmful content should be carefully designed to ensure that measures taken do not restrict hate or offensive speech on the Internet."[25] This statement may sound strange to European ears, but American courts have followed this doctrine in cyberspace, affording this form of speech broad protection. Hate is tricky because it is hard to define.

PROMISES AND CHALLENGES

The Internet contests boundaries to free expression and enlarges the scope of tolerance. With almost 40 percent of the world's population online, nearly 3 billion people,[26] the Internet has been heralded as "the best development

American Coalition of Life Activists, 290 F.3d 1058, 1080 (9th Cir. 2002). See also *Watts v. United States*, 394 U.S. 705 (1969); *United States v. Kelner*, 534 F.2d 1020 (2d Cir. 1976); Jennifer E. Rothman, "Freedom of Speech and True Threats," *Harvard Journal of Law and Public Policy* 25, no. 1 (2001): 283–367; Anna S. Andrews, "When Is a Threat 'Truly' a Threat Lacking First Amendment Protection? A Proposed True Threats Test to Safeguard Free Speech Rights in the Age of the Internet," UCLA Online Institute for Cyberspace Law and Policy, University of California, Los Angeles, May 1999; Kenneth L. Karst, "Threats and Meanings: How the Facts Govern First Amendment Doctrine," *Stanford Law Review* 58, no. 5 (2006): 1337–1412.

[23] In *Planned Parenthood of Columbia/Willamette, Inc. v. American Coalition of Life Activists*, 23 F. Supp. 2d 1182 (D. Or. 1999), an Internet site listed the names and home addresses of doctors who performed abortions. The site called for the doctors to be brought to justice for crimes against humanity. The names of doctors who had been wounded were listed in gray. Doctors who had been killed by antiabortionists had been crossed out. The court found this speech to be threatening and not protected under the First Amendment. See Richard Delgado and Jean Stefancic, *Understanding Words That Wound* (Boulder, CO: Westview, 2004), 127. Another pertinent case is *The Secretary, United States Department of Housing and Urban Development, on behalf of Bonnie Jouhari and Pilar Horton v. Ryan Wilson and ALPHA HQ*, HUDALJ 03-98-0692-8 (decided July 19, 2000).

[24] Anti-Defamation League, "Combating Extremism in Cyberspace: The Legal Issues Affecting Internet Hate Speech," Anti-Defamation League, New York, 2000); Delgado and Stefancic, *Understanding Words That Wound*, 127.

[25] Negin Salimipour, "The Challenge of Regulating Hate and Offensive Speech on the Internet," *Southwestern Journal of Law and Trade in the Americas* 8, no. 2 (2001–2002): 395.

[26] ICT Data and Statistics Division, "The World in 2014: ICT Fact and Figures," International Telecommunication Union, Geneva, 2013, http://www.itu.int/en/ITU-D/Statistics/Documents/facts/ICTFactsFigures2014-e.pdf .

in participatory democracy since universal suffrage and the most participatory form of mass speech yet developed."[27] From the highest national courts to elementary classrooms around the world, scholars, lawmakers, and adolescents alike take part in "a never-ending worldwide conversation."[28] As individual participants make connections and share information across the globe, communities form and develop. Unhindered by geographic borders, these communities create new systems of social power and exchange.[29] Collaborations never before possible blur the edges of the private and public spheres, challenging traditional constructs of self and community. Even in its infancy, the Internet as we know it has already proven a wonderful, easy-to-use mechanism to advance knowledge and learning across the world, to bridge gaps (educational, national, religious, cultural), and to promote understanding.

The impact that the rapid descent of this colossal pool of information has had on our lives and societies is nearly impossible to comprehend. The hurried acceptance of the Internet in the Western world has been accompanied by the controversial realization that no central authority sets standards for acceptable content on this network.[30] The Internet's free space is said to be subject only to obligating technical protocols and programming language rules. Orthodox liberals celebrate this lack of rules as a democratizing, publicly empowering characteristic that will promote intellectual and social progress, whereas others see it as creating a potential tinderbox of unguided lawlessness, whose messages and influence might unravel significant common values in the social framework of pluralistic societies.[31] The reasons for this situation are historical and structural: the early Internet was rooted in the United States and became global only in its recent phase. The chaotic structure of the Internet as a complex web of separate nets results in each country setting its own laws and regulations concerning Internet oversight and monitoring. These laws and regulations differ from one country to another.

Perhaps the only thing more impressive than the breadth of the Internet is its near-instantaneous arrival and restructuring of societies and lives across the

---

[27] Reid Goldsborough, "Leveraging the Internet's Marketplace of Ideas," *Community College Week*, February 16, 2004, 19.

[28] *ACLU v. Reno*, 929 F. Supp. 824 at 883 (1996).

[29] Howard Rheingold, "The Emerging Wireless Internet Will Both Improve and Degrade Human Life," in *The Future of the Internet*, ed. Tom Head (Farmington Hills, MI: Greenhaven Press, 2005), 19–32, 22.

[30] J. Michael Jaffe, "Riding the Electronic Tiger: Censorship in Global, Distributed Networks," in *Liberal Democracy and the Limits of Tolerance: Essays in Honor and Memory of Yitzhak Rabin*, ed. Raphael Cohen-Almagor (Ann Arbor: University of Michigan Press, 2000), 274–94, 275.

[31] Ibid.

globe. In historical context, the repercussions of the Internet revolution will most likely reach and surpass those of the Industrial Revolution and other comparable phenomena.[32]

Just as we are beginning to realize the seemingly infinite potential that the Internet presents for diffusion of knowledge and educational exchange, so too must we acknowledge and assess the reach the Net extends for dissemination of counterprogressive information. Freedom of expression is of utmost importance and value, but it needs to be weighed against the no less important consideration of social responsibility. The International Organization for Standardization states:

> In the wake of increasing globalisation, we have become increasingly conscious not only of what we buy, but also how the goods and services we buy have been produced.... All companies and organisations aiming at long-term profitability and credibility are starting to realise that they must act in accordance with norms of right and wrong.[33]

When I first decided to write this book, I saw clearly that I could not possibly tackle *all* the problematic information that we find on the Internet. I asked myself, "What troubles you the most, and what issues may present a compelling case for social responsibility?" I think that if I could reach some conclusions and suggestions about dealing with some highly problematic issues, possibly the discussion could then serve as a springboard to drive forward a move for Net social responsibility. After long and careful probing, I decided to concentrate attention on violent, antisocial forms of Internet expression: cyberbullying, hate speech and racism, use of the Net by terrorist organizations, crime-facilitating speech, and child pornography. Criminal expressions aimed at financial gain are outside the scope of this book. Thus, I do not address copyright violations, identity and credit theft, online piracy and counterfeiting, phishing, spamming, fraud, and other forms of financial criminal trespass. These are very important matters – so important that they deserve a separate, thorough analysis.[34] In addition, the book does not cover Internet speech designed to promote democracy and human rights in nondemocratic societies – most notably in the Arab world, Africa, and China. This important issue merits yet another, different analysis.

---

[32] Joseph S. Nye Jr., "Information Technology and Democratic Governance," in *Governance.com: Democracy in the Information Age*, ed. Elaine Ciulla Kamarck and Joseph S. Nye Jr. (Washington, DC: Brookings Institution Press, 2002), 1–16, 1–2.

[33] "What Is Social Responsibility?," *How to Become a Social Entrepreneur* (blog), http://www.imasocialentrepreneur.com/social-responsibility/ .

[34] See, for example, Hannibal Travis, ed., *Cyberspace Law: Censorship and Regulation of the Internet* (London: Routledge, 2013).

ANTI-UNIVERSALISM

The hypotheses advanced in this volume and the conclusions reached are limited to modern democracies emerging during the past century or so. *Democracy* is defined as a form of government whose power is vested in the people and exercised by them either directly or by their freely elected representatives. As Abraham Lincoln said, democracy is government of the people, by the people, for the people.[35] That is to say, one assumption of the liberal ideology that this book contests is the assumption of universalism. Clifford Christians, a renowned scholar and publicist in the area of media ethics, has emphasized that some universal ethical values withstand borders and are shared by all humans. Quoting Václav Havel, Christians writes that through human solidarity rooted in universal reverence for life, we respect ourselves and genuinely value the participation of others in a volatile age where "everything is possible and almost nothing is certain."[36] In an earlier work, Christians, John Ferré, and Mark Fackler offered mutuality as a model of community that is "universal, categorical, and normative."[37] Our membership in the human species creates the notion of universal moral obligation and a belief in shared universal values.

This belief, however, is more wishful thinking than an acknowledgment of reality. I believe that there are some basic universal needs that all people wish to secure, such as food, raiment, and shelter. I believe that sexual drives are universal and that people need to have some sleep to be able to continue functioning. I also believe that we should strive to make moral principles universal. But our ability to do so will be improved by emphasizing the differences between liberal and nonliberal values, not by blurring them and confusing the ideal and the real.

Sociologically speaking, we cannot ignore the fact that universal values do not underlie all societies.[38] Ideally, some ethical concerns should be accepted

---

[35] "A Short Definition of Democracy," Democracy Building, Lucerne, Switzerland, 2004, http://www.democracy-building.info/definition-democracy.html .

[36] Clifford G. Christians, "The Ethics of Being in a Communications Context," in *Communication Ethics and Universal Values*, ed. Clifford G. Christians and Michael Traber (Thousand Oaks, CA: Sage, 1997), 3–23, 19. See also Deni Elliott, "Universal Values and Moral Development Theories," in Christians and Traber, *Communication Ethics and Universal Values*, 68–83.

[37] Clifford G. Christians, John P. Ferré, and P. Mark Fackler, *Good News: Social Ethics and the Press* (New York: Oxford University Press, 1993), 75. See also Clifford G. Christians, "Global Ethics and the Problem of Relativism," in *Global Media Ethics: Problems and Perspectives*, ed. Stephen J. A. Ward (Hoboken, NJ: Wiley-Blackwell, 2013), 272–94.

[38] For a contrasting view, see Christians and Traber, *Communication Ethics and Universal Values*; Leonard Swidler, *For All Life: Toward a Universal Declaration of a Global Ethic* (Ashland, OR: White Cloud Press, 1999); Robert S. Fortner and P. Mark Fackler, eds., *Ethics*

by all societies, but in reality, we know this is not the case. Some countries do not adopt liberal democracy as a way of life. Instead, they adhere to other forms of government that are alien to the underpinning values of liberal democracy: liberty, equality, tolerance, and pluralism. Some societies do not accept the norms of respecting others and not harming others that form the *raison d'être* of democracy.[39]

According to Immanuel Kant, only through morality can a rational being be a law-giving member in the realm of ends, and only through morality can a rational being be an end in himself. Kant distinguishes between relative value and intrinsic value, explaining that people have intrinsic value – that is, dignity. Kant identifies dignity with moral capacity, arguing that human beings are infinitely above any price: "to compare it with, or weigh it against, things that have price would be to violate its holiness, as it were."[40] In other words, "morality, and humanity so far as it is capable of morality, are the only things that have dignity."[41] Each person has dignity and moral worth. People should be respected as human beings and should never be exploited. In this context, Stephen Darwall distinguishes between *recognition respect* and *appraisal respect*, explaining that the former includes the respect we must show to people as people, just out of recognition of their status as people, whereas the latter is the respect we show to people in virtue of their character or achievements.[42] Kant had in mind *recognition respect*. He wrote, "Such beings are not merely subjective ends whose existence as a result of our action has value for us, but are objective ends, i.e., things [*Dinge*] whose existence is an end in itself."[43]

In turn, the Millian Harm Principle holds that something is eligible for restriction only if it causes harm to others. John Stuart Mill wrote in *On Liberty*, "Acts of whatever kind, which, without justifiable cause, do harm to

and Evil in the Public Sphere: Media, Universal Values, and Global Development (Cresskill, NJ: Hampton Press, 2010).

39 On the notion of respect, see Ronald Dworkin, "Liberalism," in *A Matter of Principle* (Oxford: Clarendon Press, 1985), 181–204; Ronald Dworkin, *Taking Rights Seriously* (London: Duckworth, 1977); Raphael Cohen-Almagor, *The Boundaries of Liberty and Tolerance: The Struggle against Kahanism in Israel* (Gainesville: University Press of Florida, 1994); Raphael Cohen-Almagor, *Speech, Media, and Ethics: The Limits of Free Expression* (Houndmills, UK: Palgrave, 2005); Raphael Cohen-Almagor, *The Scope of Tolerance: Studies on the Costs of Free Expression and Freedom of the Press* (London: Routledge, 2006); Richard L. Abel, *Speaking Respect, Respecting Speech* (Chicago: University of Chicago Press, 1998).

40 Immanuel Kant, *Groundwork for the Metaphysic of Morals*, trans. Jonathan Bennett ([1865] 2008), 33, http://www.redfuzzyjesus.com/files/kant-groundwork-for-the-metaphysics-of-morals.pdf . For further discussion, see Graham Bird, ed., *A Companion to Kant* (Oxford: Blackwell, 2006).

41 Kant, *Groundwork for the Metaphysic of Morals*, 33.

42 Stephen L. Darwall, "Two Kinds of Respect," *Ethics* 88, no. 1 (1977): 36–49.

43 Kant, *Groundwork for the Metaphysic of Morals*, 29.

others, may be, and in the more important cases absolutely require to be, controlled by the unfavourable sentiments, and, when needful, by the active interference of mankind."[44] Whether an act ought to be restricted remains to be calculated. Hence, in some situations, people are culpable not because of the act that they have performed, though this act might be morally wrong, but because of its circumstances and its consequences. While Kant spoke of unqualified, imperative moral duties, Mill's philosophy is consequentialist in nature. Together the Kantian and Millian arguments make a forceful plea for moral, responsible conduct: always perceive others as ends in themselves rather than as means to something, and avoid harming others. As Ronald Dworkin suggests, the concept of dignity needs to be associated with the responsibilities each person must take for his or her own life. Dignity requires owning up to what one has done.[45]

Liberal democracies accept these ideas as the foundations of governance. In contrast, theocracy, apartheid, and forms of governance that are based on despotism, either of one person or of a small group, all deny the background rights and moral values of liberal democracy. All forms of governance, all cultures and ideologies, have a certain conception of justice, but their understanding of justice may differ from one society to another. Consequently, the specific ways in which cultures apply justice in particular situations may differ.

In *The Law of Peoples*, John Rawls drew a distinction between liberal and illiberal societies. Liberal societies are pluralistic and peaceful; they are governed by reasonable people who protect basic human rights. These rights include the right to life (to the means of subsistence and security), liberty, and personal property as well as to formal equality and self-respect as expressed by the rules of natural justice.[46] Liberal peoples are reasonable and rational. Their conduct, laws, and policies are guided by a sense of political justice.[47]

---

[44] John Stuart Mill, *Utilitarianism, Liberty, and Representative Government* (London: J. M. Dent, 1948), chapter 3 of *On Liberty*, 114–130, 114. For further discussion, see Piers Norris Turner, "'Harm' and Mill's Harm Principle," *Ethics* 124, no. 2 (2014): 299–326.

[45] Dworkin asserts that people who blame others or society for their own mistakes, or who absolve themselves of any responsibility for their conduct by blaming genetic determinism, lack dignity. "The buck stops here," says Dworkin, is an important piece of ethical wisdom. See Ronald Dworkin, *Justice for Hedgehogs* (Cambridge, MA: Harvard University Press, 2011), 210–11. For further discussion, see Jeremy Waldron, "Is Dignity the Foundation of Human Rights?," Public Law Research Paper 12-73, New York University School of Law, New York, January 3, 2013; Marcus Düwell, Jens Braarwig, Roger Brownsword, and Dietmar Mieth, eds., *The Cambridge Handbook of Human Dignity: Interdisciplinary Perspectives* (Cambridge: Cambridge University Press, 2014).

[46] John Rawls, *The Law of Peoples* (Cambridge, MA: Harvard University Press, 2002), 59–88. For further discussion, see Richard Rorty, "Justice as a Larger Loyalty," *Ethical Perspectives* 4, no. 3 (1997): 139–51.

[47] Rawls, *The Law of Peoples*, 25.

In contrast, nonliberal societies fail to treat their people as truly free and equal. Outlaw states are aggressive and dangerous,[48] while other forms of nonliberal societies might adopt skewed concepts of morality and justice based on compulsion and coercion. A nonliberal society may deem it just to cut off a thief's hand, whereas liberal societies may perceive such justice as abhorrent. Another nonliberal society may deem it just to stone a woman who is said to be an adulterer, whereas liberal societies conceive such justice as absolutely repugnant. Authoritarian societies jail their political opponents, whereas liberal societies encourage pluralism of ideas and provide avenues to empower opposition. Moral values, unfortunately, are not universally shared in all countries by all humanity. Thus my concern is with Western liberal democracies that perceive human beings as ends and that respect autonomy and variety. The arguments are relevant to other countries, but because nondemocratic countries do not accept the basic liberal principles; because their principles do not encourage autonomy, individualism, pluralism, and openness; and because their behavior is alien to the concepts of human dignity and caring, one can assume that the discussion will fall on deaf ears. Nonliberal societies based on authoritarian conceptions and principles deserve a study of their own.[49] I elaborate further and explain this argument in chapter 3.

Although I am not a relativist, I believe that history and culture do matter. Societies do not adopt a universal common denominator to define the boundaries of freedom of expression. For instance, Germany and Israel are more sensitive to Holocaust denial, and rightly so. Although the United States protects hate speech, racism, and Holocaust denial, we would be most

---

[48] Ibid., 81.

[49] For information on Internet censorship in China, see Jodie Martin, "Internet Repression in China," December 7, 2007, https://suite.io/jodie-martin/gm02sb . For information on Internet repression in Vietnam, see "Viet Nam: A Tightening Net: Web-Based Repression and Censorship," Amnesty International, October 2006, http://www.amnesty.org/en/library/info/ASA41/008/2006 . For information on Internet repression in Iran, see "Iran 'Happy' Dancers Sentenced to Jail and Flogging in Flagrant Assault on Freedom of Expression," Amnesty International, September 18, 2014, http://www.amnesty.org/en/news/iran-happy-dancers-sentenced-jail-and-flogging-flagrant-assault-freedom-expression-2014-09-18 ; Bud Simmons, "Internet Repression in Iran," *Thoughts of a Conservative Christian* (blog), September 17, 2008, http://bsimmons.wordpress.com/2008/09/18/internet-repression-in-iran/ . For information on Internet repression in Syria, see Sami Ben Gharbia, "Syria: More Victims of Internet Repression," *Global Voices Online*, October 20, 2007, http://www.menassat.com/?q=en/news-articles/1711-syria-more-victims-internet-repression . For more information on Internet repression in Ethiopia, see "Internet Repression in Ethiopia," CyberEthiopia, September 1, 2006, http://cyberethiopia.com/home/content/view/26/ ; Andrew Heavens, "Ethiopia Blocks Opposition Web Sites: Watchdog," Reuters, May 1, 2007, http://nazret.com/blog/index.php/2007/05/01/ethiopia_blocks_opposition_web_sites . See also Athina Karatzogianni, *The Politics of Cyberconflict* (London: Routledge, 2006), 121–53.

troubled if Germany were *not* to adopt restrictive measures against Internet sites that deny the Holocaust. No universally shared measure helps decide the boundaries of freedom of expression. These boundaries vary from one society to another and are influenced by historical circumstances and cultural norms. Liberal societies adhere to general liberal principles (thinly described as "human rights"), but these principles are instantiated in more detailed, specific, contextual ways (as, say, the more thickly described ways that democracies understand "freedom of expression"). Basic human rights recognize the inherent dignity of people as human beings. This broad and rather abstract idea protects the life of the person and prescribes that any form of coercion should be explained and justified. Freedom of expression is valuable and of great importance, but it might be compromised when other, no less important considerations (e.g., privacy, security, dignity of the person) come into conflict with it. In difficult or evenly balanced cases, our moral conclusions may vary. On some occasions, we may give precedence to freedom of expression; on other occasions, we may decide that the competing consideration is of utmost importance.

THE BOOK'S OBJECT AND METHODOLOGY

The object of this book is to discuss moral and social responsibility on the Internet. This issue is neglected in the new media literature. It is time to start a discussion in the realm of morality and ethics, one that supplements the many discussions in the realm of law. The book addresses the ethical problems rooted in technology in response to potential risks on the Internet. *The Internet is not the problem.* The problem arises where it is used to undermine our well-being as autonomous beings living in free societies. This book focuses on articulating possible solutions to specific problems and on providing a framework within which these problems can be identified and resolved by accentuating the concepts of moral and social responsibility. It strives to suggest an approach that is informed by the experiences of democratic societies with different norms and legal cultures and that harnesses the strengths and capabilities of the public and private sectors in offering practical solutions to pressing problems.

The research for this book involved an extensive survey of free speech literature, theories in media ethics, and theories in social responsibility; an extensive survey of problematic, violent speech on the Internet; an analysis of relevant literature, government position papers, state laws, and court cases; and a review of law enforcement measures that have been taken to combat various forms of violent speech. In addition, I conducted discussions and interviews in Canada, France, Israel, the United Kingdom, and the United States with key policy makers, public officials, elected officials, police officials, legal scholars and justices, media and Internet experts, and representatives of nongovernmental organizations devoted to human rights and free speech. I have used a similar methodology of conducting extensive surveys in the fields of philosophy, law, and communication in previous studies, benefiting from experts' experiences on topics that are not well covered in the literature. I found that interviews often highlight issues that are not discussed or not discussed enough in the literature. Conversation with experts sheds light on hidden subjects that one can easily miss when confined to libraries.

CHAPTER OUTLINE

The first three chapters lay the foundations for this book. Chapter 1 provides the historical foundations, from the Advanced Research Projects Agency (ARPA) project in the 1950s until today. Chapter 2 provides the technological foundations, explaining some of the basic innovations pertinent for the argument, and chapter 3 presents the theoretical foundations. Let me explain.

From 1960 onward, technology advanced rapidly. It has been an age of innovation in which ideas have driven the development of new applications that, in turn, have driven demand. Then we witnessed circularity. New demands yielded further innovation (mobile communication technology, cloud computing) and many more new applications – email, the World Wide Web, file sharing, social networking, blogs, Skype. These novelties were not imagined in the early stage of the Internet.

Chapter 1 outlines and analyzes milestones in the history of the Internet: its evolution from the ARPA project in 1957, its formative years (1957–84), and its current state as a global phenomenon. The early Internet was devised and implemented in American research units, universities, and telecommunication companies that had vision and interest in cutting-edge research. From 1984 to 1989, the Internet entered the commercial phase. Backbone links were upgraded, new software programs were written, and the number of interconnected international networks grew. During the 1990s, a massive expansion occurred, and the Internet became a global network. Business and personal computers with different operating systems joined the universal network, and social networking – sites that enable Netusers to share photos, private journals, hobbies, and personal and commercial interests with networks of mutual friends and colleagues – became an instant and growing success. The technology has transformed into a quotidian network for identifying, sharing, and conveying information and ideas and exchanging graphics, videos, sounds, and animation among hundreds of millions of Netusers around the world.

Chapter 2 is designed to explain technological aspects and concepts essential to understanding how the Internet works and how it can be abused. New media technology offers many desirable benefits: velocity, scalability, standardization, and cheap cost. In the foci of analysis are the basic characteristics of the Net, its most prevalent modes of communication, the concept of file sharing, the work of search engines, and the tools we have to increase security and privacy: filtering, monitoring, and encryption. Promoting privacy via encryption may contribute to Web users' security, but it might also undermine their security.

Chapter 3 introduces the theoretical principles of the discussion. Relying on Aristotle and contemporary philosophers, I distinguish between legal, moral, and social responsibility and present the notion of Internet trust. *Legal responsibility* refers to addressing the issue by agencies of state power. *Moral responsibility* concerns the personal responsibility of the agent to conscience. *Social responsibility* relates to the societal implications of a given conduct. It concerns the responsibility of individuals and customers, of governments and law enforcement agencies, of business and Internet intermediaries, and of the public at large.

I also introduce two pertinent theories: The democratic catch and moral panics. The democratic catch is my attempt to find the Golden Mean for the sustained working of democracies. A delicate balance should be maintained between measures taken to protect democracy while adhering to the underpinning liberal values.

The next six chapters, 4 to 9, discuss social and moral responsibility of different agents and actors: responsibility of Netusers who upload information to the Internet, of readers who encounter information on the Web, of Internet service providers (ISPs) and Web-hosting services (WHSs), of the state, and of the international community. These chapters were enriched by fieldwork in Canada, Israel, the United States, and the United Kingdom. Chapter 4 focuses on the tragedy of Megan Meier, a teenage girl who committed suicide after she was harassed on the Internet. Then I discuss the antisocial problem of cyberbullying, which exemplifies lack of responsibility by Netusers. I highlight the need for obtaining Net education and caring for the consequences of one's actions. *Cyberbullying* means use of the Internet, cell phones, or other devices to send or post text or images intended to hurt or embarrass another person. The need for Netusers' responsibility is apparent considering the limited ability and will of governments to police the Internet. We cannot expect others – administrators, governments, members of the international community – to be responsible while we, as Netusers, shake off any notion of responsibility.

As the Internet continues to grow, the responsibility of the reader is especially important in identifying new websites that serve as a vehicle for the expression of murderous thoughts that potentially lead to murderous action. What is the responsibility of readers when they encounter violent expressions on the Net? Do readers of websites have any moral and social responsibility to warn against potentially harmful uses of the Net that might be translated into real, practical harms? To address these questions, chapter 5 focuses on the Kimveer Gill story. Gill, a person full of hatred and rage, vented his hostilities on the Net prior to embarking on a shooting spree at Dawson College in Montreal. None of his readers alerted the police. Since this murder, we have witnessed a growing phenomenon of mass murders in the United States and in other parts of the world. In the United States, mass shootings have risen markedly from an annual average of 6.4 a year between 2000 and 2006 to 16.4 a year between 2007 and 2013. They resulted in 1,043 casualties, including 486 deaths.[50] In quite a few incidents, the murderers announced their intention to kill on the Internet, yet not enough was done to stop them.

The responsibility of ISPs and host companies is arguably the most intriguing and complex issue. With the advancement of technology at large and specifically the Internet, responsibility for gaining and maintaining trust in the Net increasingly falls on the companies that operate the Net – namely ISPs and WHSs. Some of these companies act responsibly, making an effort to provide a safe environment for their Netusers, thinking that this policy is beneficial to their reputation and business. Other companies uphold Internet neutrality and conduct their business in accordance with direct monetary consequences.

In chapters 6 and 7, I elaborate on and explore this issue in detail from the ethical and social perspectives. The main question is whether Internet intermediaries should be proactive – that is, not only cooperate on receiving information from various sources but also scrutinize their sphere for problematic, antisocial, and potentially harmful material to promote trust among their subscribers. Here I discuss the concepts of net neutrality, perfectionism, and discrimination. I distinguish between three different meanings of *neutrality*: (a) net neutrality as nonexclusionary business practice, highlighting the economic principle that the Internet should be

---

[50] Michael McCarthy, "Mass Shootings on Rise in U.S., Says FBI Report," *British Medical Journal* 349 (2014): g5895.

open to all business transactions; (b) net neutrality as an engineering principle, enabling the Internet to carry the traffic uploaded to the platform; and (c) net neutrality as content nondiscrimination, accentuating the free speech principle. I call the last *content net neutrality*. While endorsing the first two meanings of net neutrality, I argue that Internet gatekeepers should adhere to the promotional approach rather than to neutrality. The promotional approach accentuates ethics and social responsibility, holding that ISPs and WHSs should promote the basic ideas of showing respect for others and not harming others. They should scrutinize content and discriminate not only against illegal content (child pornography, terrorism) but also against content that is morally repugnant and hateful. Here the concept of responsibility comes into play. I argue that some value screening of content may be valuable and that the implications from affording the Internet the widest possible scope can be very harmful. Being cognizant of the possibility that "morally repugnant" might open wide the gate to further restrictions, I emphasize that only cyberbullying and hate speech feature in this category.

The concluding two chapters concern state responsibility and the responsibility of the international community. In chapter 8, a clash is seen between (a) the view that upholds cross-boundary freedom of information around the globe and (b) the right of states to assert their jurisdiction on the Net. The first view holds that because the Internet knows no frontiers, data must have no limitations and states should not erect them, whereas the second view holds that the Internet is no different from any other medium of information. Because the state regulates in one way or another all forms of communication and sees that they abide by the law, so the Internet should abide by state law. The Internet's distinct architecture does not make it aloof from the law.

It is argued that in the late 1990s, the Internet seemed a perfect medium for business: people could be anywhere and make investments anywhere without any regulatory limitations. I discuss in detail the contested Yahoo! saga in which the French authorities wished to assert their laws over the conduct of Yahoo!, thus preventing the company from posting on its auction sites Nazi artifacts, which are illegal in France under the country's hate laws. Further appeals in American courts did not yield the right result for Yahoo!. This case, among others, demonstrates that ISPs have to respect domestic state legislation to avoid legal risks. The Internet is international in character, but it cannot be abused to override law. We do not have one law for people and another for the Internet. The Internet is made by the people, for the people, and it needs to abide by the laws of the people.

Chapter 9 supplements the previous chapter. It reflects on the responsibility of the international community. Does the international community have a responsibility to unite to combat antisocial activities? Because the Internet is an international medium, a need exists for transnational coordination and cooperation to respond to global concerns. Indeed, the international community has legal, social, and moral responsibilities. Hate, terrorism, and child pornography are decentralized and diffused, lack a coherent global system, and are organized in cells with clear agendas and sophisticated means of communication. The Internet is an obvious force in allowing their operation. Addressing those challenges requires international coordination. In this context, I discuss the Council of Europe Convention on Cybercrime and modes of cooperation that are and can be used to promote Net security. I also suggest further mechanisms that can be implemented to promote international cybersecurity. I argue that cross-country challenges require cross-country cooperation.

My research shows a pattern of closely linked virtual threats and violent conduct. The ascending frequencies with which these events happen are more of a reason to act on the international level. The nature of the Internet is such that it serves a certain function for would-be killers. Usually, people do not just snap. A psychological process takes place, a mental journey that killers experience from the inception of thoughts to the actual action. The process begins with bitterness and degenerates into anger and rage. Lacking mitigating circumstances, the wrath might end with a brawling explosion. People need to vent their hostility, acrimony, and anger. They provide traceable signs and hints. They find it difficult to contain all these boiling emotions inside. In the Internet age, it is convenient to vent into the virtual world. The global Internet, where people adopt different personas and have a perceived sense of anonymity, is becoming a vital component of this crystallizing process. As the Internet continues to grow, the responsibility of the reader who encounters murderous thoughts, of the ISP that hosts those thoughts, and of law enforcement agencies that cooperate across continents to protect the lives of innocent people are all important in the identification of websites that serve as a vehicle for the crystallizing process of potential murderers.

I close by proposing to establish a new browser for liberal democracies called CleaNet. Through mechanisms of deliberative democracy, Netusers would agree on what constitutes illegitimate expression to be excluded from the browser. CleaNet would facilitate safer and more responsible surfing of the Internet.

1

Historical Framework

History consists of a series of accumulated imaginative inventions.

–Voltaire

The aim of this chapter is to outline the milestones that led to the establishment of the Internet as we know it today, from its inception as an idea in the 1950s until the early 21st century.[1] The varied and complex social and technological transformations we witness today have their roots in the way the Internet was conceived. This survey shows that the Internet was developed through research grants from the US Department of Defense's Advanced Research Projects Agency. Scientists wanted to establish a system for maintaining communication links between distant locations in the event that the electrical route was destroyed. The early Internet was devised and implemented in American research units, universities, and telecommunication companies that had vision and interest in cutting-edge research. The program grew in the 1960s and 1970s, becoming a network of computers designed to transmit information by packet switching. In 1983, Ithiel de Sola Pool wrote in what has become a classic text:

> The broadband digital channel coming into the home can be multiplexed to bring in different streams at different data rates simultaneously. Subscribers can talk on the phone, have their utilities metered, watch a video picture on their television, and receive their electronic mail, all at once without interference. The loop is likely to be an optical fiber rather than a copper wire because the fiber has the needed bandwidth at lower cost.[2]

[1] Vinton G. Cerf, in an email message to the author on January 16, 2012, commented, "While some of the ideas that are incorporated into the Internet can be traced to the 1950s, the Internet design was first written in 1973 by Bob Kahn and me."
[2] Ithiel de Sola Pool, *Technologies of Freedom: Our Free Speech in an Electronic Age* (Cambridge, MA: Harvard University Press, 1983), 177.

The network of computers was, from the start, an open, diffused multi-platform. Up until the 1990s, the network developed in the United States[3] and then, within a few years, expanded globally at remarkable pace and with no less impressive technological innovations, the end of which we are yet to witness. In its postwar incarnation, the Internet has transformed into a global system for sharing ideas and data and for exchanging text, graphics, software, music, and video clips to almost 2 billion Netusers in nearly 200 countries.

This chapter will reflect on the history of the Internet. Some of the technological aspects of the Internet will be explained in chapter 2.

THE FORMATIVE YEARS

The history of the Internet started in the United States in the early 1960s. This was the Cold War period, when the world was bipolar: the United States and the Soviet Union were competing in expanding their influence in the world, viewing each other with great caution and suspicion.

On October 4, 1957, the Soviet Union launched the first space satellite, Sputnik. The Sputnik success necessitated American reaction. It was a question of pride and leadership. The US Department of Defense responded by establishing the Advanced Research Projects Agency (ARPA),[4] designed to promote research that would ensure that the United States could compete with and excel the Soviet Union in any technological race. ARPA's mission was to produce innovative research ideas, to provide meaningful technological influence that went far beyond the conventional evolutionary developmental approaches, and to act on those ideas by developing prototype systems.[5] Although ARPA's first priority was to get the United States into space, one of its offices, the Information Processing Techniques Office (IPTO), funded research in computer science designed to mobilize American universities and research laboratories so they could build up a strategic communication network (Command and Control Research) that would make messaging capabilities available to the government.[6]

---

[3] Vinton Cerf had people from Japan, France, and Norway working on the program at Stanford University as early as 1974.

[4] During its lifetime, this agency has used two acronyms, ARPA and DARPA, or Defense Advanced Research Projects Agency.

[5] DARPA Fact Sheets, http://www.darpa.mil/About.aspx .

[6] James Curran and Jean Seaton, *Power without Responsibility: How Congress Abuses People through Delegation* (London: Routledge, 2009), 257; Kathleen Conn, *The Internet and the Law: What Educators Need to Know* (Alexandria, VA: Association for Supervision and Curriculum Development, 2002), xiii.

A popular myth holds that the Department of Defense scientists thought that if the Soviets were capable of launching satellites, they might also be capable of launching long-distance nuclear missiles. Because networks at the time relied on a single, central control function, so the myth goes, the main concern was networks' vulnerability to attack: once the network's central control point ceased to function, the entire network would become unusable. The scientists wanted to diffuse the network so that it could be sustained after an attack on one or more of its communication centers.[7] They had in mind a "decentralized repository for defense-related secrets" during wartime.[8] However, the pioneers of the ARPA Network (ARPANET) project argue that ARPANET was not related to building a network resistant to nuclear war:

> This was never true of the ARPANET, only the unrelated RAND study on secure voice considered nuclear war. However, the later work on Internetting did emphasize robustness and survivability, including the capability to withstand losses of large portions of the underlying networks.[9]

Leonard Kleinrock, the father of modern data networking and one of the pioneers of digital network communications who helped to build ARPANET, explained that the reason ARPA wanted to deploy a network was to allow its researchers to share each others' specialized resources (hardware, software, services, and applications). The object was not to protect against a military attack.[10] And David D. Clark, a senior research scientist at the Massachusetts Institute of Technology (MIT) Laboratory for Computer Science who worked on the ARPANET project in the early 1970s, said he never heard of nuclear survivability and that there is no mention of this idea in the ARPA records from the 1960s. In a personal communication, Clark wrote:

> I have asked some of the folks who pushed for the ARPAnet: Larry Roberts and Bob Kahn. They both assert that nobody had nuclear survivability on their mind. I was there from about '73, and I never heard it once. There might have been somebody who had the idea in the back of his mind, but 1) if so, he held it real close, and 2) I cannot figure out who it might have been. We know who more or less all the important actors were. (Sadly, [J. C. R.] Licklider has

---

7 Gary P. Schneider and Jessica Evans, *New Perspectives on the Internet: Comprehensive*, 6th ed. (Boston: Thomson, 2007), APP 12–13; William Stewart, "DARPA/ARPA: Defense/Advanced Research Project Agency," http://www.livingInternet.com/i/ii_darpa.htm ; Scott Griffin, "Internet Pioneers," http://www.ibiblio.org/pioneers/ .

8 Conn, *Internet and the Law*, xiii.

9 Barry M. Leiner, Vinton G. Cerf, David D. Clark, Robert E. Kahn, Leonard Kleinrock, Daniel C. Lynch, Jon Postel, Lawrence G. Roberts, and Stephen Wolff, "A Brief History of the Internet," Internet Society, 2003, http://www.isoc.org/internet/history/brief.shtml .

10 Leonard Kleinrock, email message to the author, July 19, 2010.

died, but I think I did ask him when he was still alive. I wish I had better notes.) So I am very confident that [Paul] Baran's objective did not survive to drive the ARPA effort. It was resource sharing, human interaction … and command and control.[11]

In 1962, J. C. R. (Joseph Carl Robnett) Licklider became the first director of the IPTO. His role was to interconnect the Department of Defense's main computers through a global, dispersed network. Licklider articulated the vision of a "galactic" computer network – a globally interconnected set of processing nodes through which anyone anywhere could access data and programs.[12] In August 1962, Licklider and Weldon Clark published the first paper about the concept of the Internet. Titled "On-line Man–Computer Communication,"[13] the paper projected the communication network as a tool for scientific collaboration. Here the seeds for what would later become the Internet were planted.

Paul Baran of the RAND Corporation deserves particular attention for many innovative ideas, including the research project that created the myth of ARPANET's involvement in the development of a robust decentralized network that would enable the United States to have a second-strike capability.[14] Baran had been commissioned by the US Air Force to study how the military could maintain control over its missiles and bombers in the aftermath of a nuclear attack. In 1964, Baran proposed a resilient command and control voice network based on digitized voice packets he called *message blocks*. This distributed scheme for US telecommunications infrastructure that had no central command or control point would survive a "first strike" attack. In the event of such an attack on any one point, all surviving points would be able to reestablish contact with each other.[15] Note that Baran's research project came about six years after ARPA was established. Lawrence G. Roberts, the principal architect

[11] David D. Clark, email message to the author, July 19, 2010. See also Leonard Kleinrock, "An Early History of the Internet," *IEEE Communications Magazine* 48, no. 8 (2010): 26–36.

[12] "Timeline," National Academy of Engineering, 2014, http://www.greatachievements.org/ Default.aspx?id=2984 ; Mitch Waldropt, "DARPA and the Internet Revolution," in *DARPA: 50 Years of Bridging the Gap*, 78–85 (Tampa, FL: Faircount, 2008), http://www.darpa.mil/ WorkArea/DownloadAsset.aspx?id=2554 .

[13] J. C. R. Licklider and Weldon E. Clark, "On-line Man–Computer Communication," in *Proceedings 1962: Spring Joint Computer Conference* (Palo Alto, CA: National Press, 1962), 113–28 .

[14] "Internet Pioneer Paul Baran Passes Away," *BBC News*, March 28, 2011, http://www.bbc.co.uk/ news/technology-12879908 .

[15] "On Distributed Communications Series," RAND, http://www.rand.org/about/history/baran-list. html ; "Paul Baran and the Origins of the Internet," RAND, http://www.rand.org/about/history/ baran.html ; James Slevin, *The Internet and Society* (Oxford: Polity, 2000), 29–30. See also the list of reports written by Baran for RAND at http://www.rand.org/pubs/authors/b/baran_paul.html .

of ARPANET, wrote that the RAND work had no significant impact on the ARPANET plans and Internet history.[16]

In 1965, Donald Davies of the British National Physical Laboratory (NPL) began thinking about packet networks and coined the term *packet*. In fact, at that time, three scientists in three different locations were thinking independently about that same technology: Leonard Kleinrock developed the first mathematical models of the behavior of packet switching at the MIT labs. He called them *message switching*.[17] At RAND, Baran formulated the idea of standard-size addressed message blocks and adaptive alternate routing procedures with distributed control. And Davies thought similarly that to achieve communication between computers, a fast message-switching communication service was needed, in which long messages were split into chunks that were sent separately to minimize the risk of congestion.[18] The chunks he called *packets*, and the technique became known as *packet switching*. Davies's network design was received by the ARPA scientists. The ARPANET and the NPL local network became the first two computer networks in the world using the technique.[19]

In 1968, Licklider and Robert Taylor described their vision of interactive communities that would consist of geographically separated individuals and teams. Although these communities would exist across different locations, a common interest would support a comprehensive system of field-oriented programs and data. Licklider and Taylor thought people would cease to send letters or telegrams and would seldom make phone calls. Instead, people

---

[16] Lawrence G. Roberts, "Internet Chronology," December 15, 2007, http://www.packet.cc/internet.html .

[17] Leonard Kleinrock, "Information Flow in Large Communication Nets," PhD thesis proposal, Massachusetts Institute of Technology, Cambridge, MA, May 31, 1961, http://www.lk.cs.ucla.edu/data/files/Kleinrock/Information%20Flow%20in%20Large%20Communication%20Nets.pdf ; Leonard Kleinrock, *Communication Nets: Stochastic Message Flow and Design* (New York: Dover, 1973); Leonard Kleinrock, "Creating a Mathematical Theory of Computer Networks," *Operations Research* 50, no. 1 (2002): 125–31. See also Kleinrock's biography at http://www.lk.cs.ucla.edu/ . Vinton Cerf commented, "[I]t is important to note that there were military message systems called AUTODIN that used message switching independent of Kleinrock's brilliant queuing theoretic analysis of the performance of message (packet) switching." Vinton G. Cerf, email message to the author, January 16, 2012.

[18] Vinton Cerf commented, "I think Davies was more concerned about latency than congestion although he did try out an idea he called 'isarithmic' networks in which only a fixed number of packets could be in transit at any time. The idea didn't work out." Vinton G. Cerf, email message to the author, January 16, 2012.

[19] "Donald W. Davies CBE, FRS," The History of Computing Project, March 8, 2013, http://www.thocp.net/biographies/davies_donald.htm ; Leonard Kleinrock, "History of the Internet and Its Flexible Future," *IEEE Wireless Communications* 15, no. 1 (2008): 8–18, 11.

would send electronic files, specifying the level of urgency.[20] Their vision failed to predict the extent to which people would become producers of data and consumers of technology. They thought that new forms of communication would replace old ones. They were correct only in part, and to date they were absolutely wrong regarding the place of the telephone in modern life.

The ARPANET was launched by Bolt Beranek and Newman (BBN) at the end of 1969.[21] BBN was commissioned to design four Interface Message Processors (IMPs). These machines would create open communication between four different computers running on four different operating systems, thus creating the first long-haul computer network and connecting users at the University of California at Los Angeles (UCLA); the Stanford Research Institute in Menlo Park, California; the University of California at Santa Barbara; and the University of Utah. These four nodes together comprised the Network Working Group.[22] A fifth ARPANET node was installed at BBN's headquarters. Each node consisted of an IMP, which performed the store-and-forward packet-switching functions. Packet switching was a new and radical idea in the 1960s. Through ARPANET's Network Control Protocol (NCP), developed by Steve Crocker and originally called Network Control Program, users were able to access and use computers and printers in other locations and transport files between computers. NCP was an investigational project that explored the most favorable way of building a network that could function as a trustworthy communications medium. The main hurdle to overcome was to develop an agreed-upon set of signals between different computers that would open up communication channels, thereby enabling data to pass from one point to another. These agreed-upon signals were called *protocols*.

The network then expanded to other institutions, including Harvard, MIT, Carnegie Mellon, Case Western Reserve, and the University of Illinois at Urbana. Within 16 months, more than 10 sites had been established, with an estimated 2,000 users and at least 2 routes between any 2 sites for the transmission of information packets.[23] ARPANET was the world's first advanced computer network using packet switching. Leonard Kleinrock wanted to develop a design methodology that would scale to very large networks, and

[20] J. C. R. Licklider and Robert W. Taylor, "The Computer as a Communication Device," *Science and Technology* 76, no. 2 (1968): 20–41, 38.

[21] In 1948, two MIT professors, Richard Bolt and Leo Beranek, established a small acoustics consulting firm and soon added a former student of Bolt's, Robert Newman. See the firm's website at http://www.bbn.com/about/timeline/ .

[22] Jonathan Strickland, "How ARPANET Works," HowStuffWorks, http://www.howstuffworks.com/arpanet.htm/printable ; David Beckett, "Internet Technology," in *Internet Ethics*, ed. Duncan Langford (New York: St. Martin's Press, 2000), 13–46, 15.

[23] Slevin, *Internet and Society*, 31; Conn, *Internet and the Law*, xiii–xiv.

he thought that the only way to do so was to introduce the concept of *distributed control*. With distributed control, the responsibility for controlling the network routing would be shared among all the nodes, and therefore, no node would be unduly tasked.[24] This idea resulted in robust networks.

One of the major characteristics of the network is innovation. One development quickly leads to another. In the early 1970s, scientists tried to overcome new problems. The communication ideas, the experiments, the testing, the tentative designs – all brought about an endless stream of networks that were ultimately interlinked to become the Internet. Someone had to record all the protocols, identifiers, networks and addresses, and names of the things in the networked universe. And someone had to keep track of all the information that stemmed from the discussions. That someone was Jonathan B. Postel, a young computer scientist who worked at that time on the ARPA project at UCLA.[25] Postel devoted himself to building and running the Internet's naming and numbering structure. He proposed the top-level domains dot-com, dot-edu, and dot-net.[26] In those pioneering, unstructured, and building years, Postel became known as the Internet Assigned Numbers Authority (IANA). Postel was not elected to the position of responsibility he held in the Internet community; he was simply, in the words of the White House's Internet policy adviser, Ira Magaziner, "the guy they trust."[27]

A second characteristic of the ARPANET is that it succeeded in connecting the computers used in different time-sharing systems. Now the ARPANET scientists wished to connect the packet-switching network of the ARPANET with a satellite packet-switching network and a packet radio packet-switching network. In July 1970, the first packet radio, ALOHANET, based on the concept of random packet transmission, was developed at the University of Hawaii by Norman Abramson and became operational. ALOHANET linked the University of Hawaii's seven campuses to each other and to the ARPANET, which sponsored the ALOHANET project. On the basis of this model, ARPA built its own packet radio network, which

[24] Bob Kahn, Dave Walder, and William Crowther, among others, were major players in the system architecture and detailed protocol design. Leonard Kleinrock, email message to the author, July 19, 2010; Vinton G. Cerf, email message to the author, January 16, 2012.

[25] Vinton G. Cerf, "I Remember IANA," Request for Comments 2468, Network Working Group, Internet Society, October 17, 1998, http://www.rfc-editor.org/rfc/rfc2468.txt .

[26] Katie Hafner and Matthew Lyon, *Where Wizards Stay Up Late: The Origins of the Internet* (New York: Simon & Schuster, 1996), 252–53. See also Danny Cohen, "Working with Jon," *Remembering Jonathan B. Postel* (website), November 2, 1998, http://www.postel.org/remembrances/cohen-story.html .

[27] "'God of the Internet' Is Dead," BBC News, October 19, 1998, http://news.bbc.co.uk/1/hi/sci/tech/196487.stm .

was called PRNET.[28] During that same period, ARPA also developed a satellite network, called SATNET.

In early 1973, the network had grown to 35 nodes and was connected to 38 host computers.[29] That year, Norway and England were added to the network, and traffic expanded significantly. ARPANET grew into the Internet out of the idea that there would be multiple independent networks of rather arbitrary design.[30] The term *internetworking* was first used by Vint Cerf and Robert Kahn in their 1974 article about the Transmission Control Protocol.[31] The term *Internet* was first used by Cerf, Yogen Dalal, and Carl Sunshine in Request for Comments 675 in December 1974.[32]

During that period, Vint Cerf and Robert Kahn developed a set of protocols that implemented the open architecture philosophy.[33] These protocols were the Transmission Control Protocol (TCP), which was developed in 1973, and the Internet Protocol (IP), which was defined and split from TCP in 1977. TCP organizes data into packages, puts the packages into the right order on arrival at their destination, and checks them for errors. TCP is a higher-layer protocol embedded in IP packets. IP, in turn, standardizes the packet formats, but it does not have any rules for routing. The routing protocols are distinct from the basic IP protocol. Each IP packet includes a header that specifies source, destination, and other information about the data as well as the message data itself. TCP and IP are fully symmetric and support peer-to-peer usage.[34] The importance of TCP and IP in the history of the Internet is so great that many people consider Cerf to be the father of the

[28] Johnny Ryan, "The Essence of the Internet," in *A History of the Internet and the Digital Future* (London: Reaktion Books, 2011). An adapted version of the chapter is available at http://arstech nica.com/tech-policy/news/2011/03/the-essence-of-the-net.ars/ .

[29] Ira Rubinstein, "Anonymity Reconsidered," presented at the Second Annual Privacy Law Scholars Conference, hosted by the Berkeley Law School, Berkeley, CA, June 4–6, 2009.

[30] Barry M. Leiner, Vinton G. Cerf, David D. Clark, Robert E. Kahn, Leonard Kleinrock, Daniel C. Lynch, Jon Postel, Lawrence G. Roberts, and Stephen S. Wolff, "The Past and Future History of the Internet," *Communication of the ACM* 40, no. 2 (1997): 102–8, 103.

[31] Vinton G. Cerf and Robert E. Kahn, "A Protocol for Packet Network Interconnection," *IEEE Transactions on Communications* 22, no. 5 (1974): 637–48.

[32] Vinton G. Cerf, Yogen Dalal, and Carl Sunshine, "Specification of Internet Transmission Control Program," Request for Comments 675, Network Working Group, Internet Society, December 1974, http://tools.ietf.org/rfc/rfc675.txt .

[33] The idea was originally introduced by Kahn in 1972 as part of the packet radio program. Cerf and Kahn first presented their work to the International Network Working Group – a group that Cerf established in 1972.

[34] Vinton G. Cerf, email message to the author, January 16, 2012. See also Charles M. Kozierok, "TCP Functions: What TCP Does," in *The TCP/IP Guide* (San Francisco: No Starch Press, 2005), http://www.tcpipguide.com/free/t_TCPFunctionsWhatTCPDoes.htm ; Bradley Mitchell, "IP - Internet Protocol," About.com, http://compnetworking.about.com/od/networkprotocolsip/g/ ip_protocol.htm .

Internet. A number of TCP/IP-based networks – independent of the ARPANET – were created in the late 1970s and early 1980s. The National Science Foundation (NSF) funded the Computer Science Network (CSNET) for educational and research institutions that did not have access to the ARPANET.[35]

In 1983, 500 computer hosts were connected to the Internet. In 1984, the number of hosts increased to more than 1,000.[36] As more researchers connected their computers and computer networks to the ARPANET, interest in the network grew in the academic community. One reason for increased interest in the project was its adherence to an open architecture philosophy. Each network could continue using its own protocols and data transmission methods internally. There was no need for special accommodations to be connected to the Internet, there was no global control over the network, and anyone could join. This open architecture philosophy was revolutionary at the time. Most companies used to make their networks distinct and incompatible with other networks. They feared competition and strove to make their products inaccessible to competitors. The shift to an open architecture approach is one of the most celebrated features of the Internet.

THE MASSIVE EXPANSION

By the late 1980s, a significant number of people (mostly professionals) were using email, but the Internet was not in the public eye. I was a student at Oxford University at that time and can testify that using the Internet was a most frustrating experience. Most websites were not accessible. Navigating between sites was anything but seamless. It was easier to retrieve information from the library in the good, old-fashioned way.

But things were about to change. During the 1990s, we witnessed a massive expansion of the Net. The Internet's accessibility, its multiple applications, and its decentralized nature were instrumental to this rapid growth. Business and personal computers with different operating systems could join the universal network. The Internet became a global phenomenon, more countries and people joined, and groundbreaking minds expanded the horizons of the platform with imaginative innovations. In 1990, the ARPANET project was officially over when it handed over control of the public Internet backbone to the NSF.[37] In 1991, Croatia (HR), Hong Kong (HK), Hungary (HU), Poland

---

[35] Schneider and Evans, *New Perspectives on the Internet: Comprehensive*, 15.

[36] "Time Line of the History of the Internet," http://www.the-history-of-the-internet.com/time-line.html .

[37] Curran and Seaton, *Power without Responsibility*, 263; Slevin, *Internet and Society*, 33.

(PL), Portugal (PT), Singapore (SG), South Africa (ZA), Taiwan (TW), and Tunisia (TN) joined the NSFNet (National Science Foundation Network), whose backbone was upgraded to DS-3 (44.736 megabits per second [Mbps]) as the traffic passed to 1 trillion bytes and 10 billion packets per month. That year, 1991, saw another milestone when Philip Zimmermann released the popular encryption program PGP (Pretty Good Privacy).[38]

In 1992, the Internet Society was formed, and the number of Internet hosts broke 1 million.[39] In 1993, there were 623 websites worldwide.[40] The United Nations came online, and the NSFNet expanded internationally as Bulgaria (BG), Costa Rica (CR), Egypt (EG), Fiji (FJ), Ghana (GH), Guam (GU), Indonesia (ID), Kazakhstan (KZ), Kenya (KE), Liechtenstein (LI), Peru (PE), Romania (RO), Russian Federation (RU), Turkey (TR), Ukraine (UA), the United Arab Emirates (AE), and the US Virgin Islands (VI) joined the network. The World Wide Web proliferated at a 341,634 percent annual growth rate of service traffic.[41] By the end of 1993, there were 2.1 million hosts.[42] The phenomenal growth and success of the Internet were the result of technological creativity, flexibility, and decentralization as well as the healthy curiosity of people who wanted to be part of the scene.

In 1995, Cerf argued that the "Internet has gone from near-invisibility to near-ubiquity."[43] The growth of the Internet, its expanding international character, and awareness of its effective features led more and more businesses to believe in the innovation and to invest in it. Shopping malls arrived on the Internet. First Virtual, the first cyberbank, opened for business. Two Stanford PhD students, Jerry Yang and David Filo, started a website called "Jerry and David's Guide to the World Wide Web." This guide swiftly expanded and later changed its name to one word, Yahoo! (Yet Another Hierarchical Officious Oracle).[44] More countries joined the network, including Algeria

[38] Philip R. Zimmermann, *The Official PGP User's Guide* (Cambridge, MA: MIT Press, 1995); "Recap the Internet History," VAC Media, http://www.broadbandsuppliers.co.uk/uk-isp/recap-the-history-of-Internet/ ; Robert H. Zakon, "Hobbes' Internet Timeline 11," Zakon.org, January 30, 2014, http://www.zakon.org/robert/Internet/timeline/ .

[39] For more information about the Information Society, visit the organization's website at http://www.internetsociety.org/ . See also "Recap the Internet History."

[40] Matthew Gray, "Web Growth Summary," Massachusetts Institute of Technology, Boston, 1996, https://stuff.mit.edu/people/mkgray/net/web-growth-summary.html .

[41] Zakon, "Hobbes' Internet Timeline 11."

[42] Matthew Gray, "Internet Growth: Raw Data," Massachusetts Institute of Technology, Boston, 1996, https://stuff.mit.edu/people/mkgray/net/internet-growth-raw-data.html .

[43] Vinton G. Cerf, "Computer Networking: Global Infrastructure for the 21st Century," Computing Research Association, Washington, DC, 1995, http://www.cs.washington.edu/homes/lazowska/cra/networks.html .

[44] "The History of Yahoo!: How It All Started . . .," Yahoo!, 2005, https://sites.google.com/a/hawaii.edu/the-yahooligans/Home/history-of-yahoo-com .

(DZ), Armenia (AM), Bermuda (BM), Burkina Faso (BF), China (CN), Colombia (CO), Jamaica (JM), Jordan (JO), Lebanon (LB), Lithuania (LT), Macao (MO), Morocco (MA), New Caledonia (NC), Nicaragua (NI), Niger (NE), Panama (PA), the Philippines (PH), Senegal (SN), Sri Lanka (LK), Swaziland (SZ), Uruguay (UY), and Uzbekistan (UZ). The number of Internet hosts increased to 3 million. This growth necessitated technological accommodation and, indeed, the same year, the NSFNet backbone was upgraded to OC-3 (155 Mbps) links and the volume of traffic increased to 10 trillion bytes per month. To navigate between the growing number of sites, the first version of the popular Netscape Web browser was released by Mosaic Communications Corporation.[45] Mosaic made using the Internet as easy as pointing a mouse and clicking on icons and words.[46] By then, the birth pangs of the global network were over, and information retrieval became more efficient and effective.

In 1995, major carriers such as British Telecom, France Telecom, Deutsche Telekom, Swedish Telecom, Norwegian Telecom, and Telecom Finland, among many others, announced Internet services. An estimated 300 service providers were in operation, ranging from very small resellers to large telecom carriers. More than 30,000 websites were in operation, and the number was doubling every two months.[47] The growing importance of commercial traffic and commercial networks was discussed at a series of conferences initiated by the National Science Foundation on the commercialization and privatization of the Internet. The NSF awarded a contract to Merit Network, in partnership with IBM and MCI Communications, to manage and modernize the Internet backbone. In 1995, NSFNet was shut down completely, and the American core Internet backbone was privatized.[48] PSINET, UUNET, and CERFNET were joined by MCINET as major backbone carriers.

The result was that the number of hosts more than doubled in one year, reaching 6.6 million.[49] The mid 1990s were the years when the Internet established itself as the focal point for communication, information, and business. A number of Net-related companies went public, with Netscape leading the pack with the third-largest-ever NASDAQ initial public offering share value.[50] At the same time, many people began creating their own

[45] Mosaic Communications Corporation, "Who Are We," http://home.mcom.com/MCOM/ mcom_docs/backgrounder_docs/mission.html ; "Recap the Internet History"; Zakon, "Hobbes' Internet Timeline 11."
[46] Hafner and Lyon, *Where Wizards Stay Up Late*, 257–58.
[47] Cerf, "Computer Networking."
[48] Curran and Seaton, *Power without Responsibility*, 253.
[49] Gray, "Internet Growth: Raw Data."
[50] Zakon, "Hobbes' Internet Timeline 11."

personal Web areas. Homepages and bookmarks were introduced to allow Netusers (about 16 million in 1995)[51] to organize their personal documents and to keep track of useful information. The Internet was growing at a rapid pace, attracting more and more people, who grew to depend on it to meet their daily needs for information, research, business, commerce, entertainment, travel arrangements, and so forth. For each and every need, there came an entrepreneur who seized the opportunity and opened a website addressing that need.

In 1996, the number of Netusers more than doubled, from 16 million to 36 million.[52] From the mid 1990s, the development of the Internet took a turn as a growing number of large and medium-size organizations started running TCP/IP on their internal organizational communication networks, called *intranets*. For security purposes, intranets used firewalls to shield themselves from the outside world. These protection systems often allow for the exchange of information with the Internet through specified "gateways." These private networks are called *extranets* and allow organizations to exchange data with each other. By 1997, the market for intranets and extranets was growing annually at a rate of 40 percent worldwide.[53] The number of Netusers was estimated at 70 million by the end of the year.[54]

At that time, the number of hosts was about 10 million, with an untold number of links between them.[55] Finding information on the Web became, yet again, a tricky issue, but for different reasons. Connectivity was no longer the problem; rather, navigating and finding the information you needed in the growing maze had become difficult. Addressing this challenge, two Stanford graduate students, Larry Page and Sergey Brin, started to work on a search engine, which they called BackRub because they designed it to analyze a "back link" on the Web. Later they renamed their search engine Google, after *googol*, the term for the numeral one followed by 100 zeroes. They released the first version of Google on the Stanford website in August 1996.[56] In a few years, Google became the most popular search engine on campuses and in businesses in the United States. Following its national thriving penetration, Google

[51] "Internet Growth Statistics," Miniwatts Marketing Group, August 12, 2014, http://www.Internet worldstats.com/emarketing.htm .
[52] Ibid.
[53] Slevin, *Internet and Society*, 34.
[54] "Internet Growth Statistics."
[55] Gray, "Internet Growth: Raw Data."
[56] John Battelle, "The Birth of Google," *Wired*, August 2005, http://www.wired.com/wired/archive/13.08/battelle.html?pg=2&topic=battelle&topic_set= ; Richard T. Griffiths, "Search Engines," in *History of the Internet, Internet for Historians (and Just about Everyone Else)* (Leiden, Netherlands: Universiteit Leiden, 2002), http://www.let.leidenuniv.nl/history/ivh/chap4.htm .

began to develop internationally. Having achieved success in organizing electronic information and making it freely and universally accessible to all Netusers, Google decided to make as many books as possible accessible on Google Books and to provide helpful geographic information through Google Maps, Google Earth, and Google Street View.

Large corporations became more aware of the massive potential of the Internet. America Online (AOL), Microsoft, Sun Microsystems, Inktomi, Yahoo!, and Cisco caught the attention of Wall Street. AOL alone saw its stock rise 50,000 percent.[57] In 1998, AOL acquired Netscape Communications Corporation for a stock transaction valued at $4.2 billion. Microsoft bought Hotmail for $400 million. In 1999, online retailers reported $5.3 billion in sales.[58]

Not only legitimate businesses realized the potential of the Internet. Criminals were also quick to abuse the Internet for profit. The Council of Europe finalized its international Convention on Cybercrime on June 22, 2001, and adopted it on November 9, 2001.[59] It was the first treaty addressing criminal offenses committed over the Internet (see chapter 9). The same year, the Firewall Enhancement Protocol (FEP) was proposed, and Jimmy Wales and Larry Sanger launched Wikipedia, the Web-based free encyclopedia. The term *wiki* means "quick" in Hawaiian. This collaborative, multilingual project is supported by the nonprofit Wikimedia Foundation. In 2014, there were 287 language editions of Wikipedia. The English edition contains 4.6 million articles. The wiki entries are written by volunteers around the world, and almost all of the articles can be edited by anyone with access to the site.[60] Wikipedia has become the largest and most popular general reference resource on the Internet. More and more organizations and individuals have sought their own wiki entry, believing that if you do not exist on Wikipedia, then you do not exist.

SOCIAL NETWORKING

Social media are Internet applications that enable the sharing of content: status updates, graphics, blogs, voice, games, photos, and audio and video files.

---

[57] Harry McCracken, "A History of AOL, as Told in Its Own Old Press Releases," *Technologizer*, May 24, 2010, http://technologizer.com/2010/05/24/aol-anniversary/ ; Charles Dubow, "The Internet: An Overview," in *Does the Internet Benefit Society?*, ed. Cindy Mur (Detroit, MI: Greenhaven Press, 2005), 7–13, 11.

[58] "Recap the Internet History."

[59] The full text of the convention is available on the Council of Europe's website at http://conventions.coe.int/Treaty/en/Treaties/Html/185.htm .

[60] See the "Wikipedia" entry on the Wikipedia website at http://en.wikipedia.org/wiki/Wikipedia .

Social media are defined by a culmination of two World Wide Web principles: Web 2.0 and user-generated content (UGC).[61] Web 2.0 provides the platform for UGC. Social media enable Netusers to manifest their imagination and creativity by adding and modifying data.

Early in the 21st century, social network sites were launched. These sites enable Netusers to share information, photos, private journals, hobbies, and interests with networks of mutual friends. They provide friends with the ability to send email and chat online, connect with classmates and study partners, and connect with friends of friends. They provide forums where business people and co-workers can network and interact, where single people can meet other singles, where matchmakers can connect their friends with other friends, and where families can map their family trees.

Facebook.com was founded on February 4, 2004, by Mark Zuckerberg, Eduardo Saverin, Dustin Moskovitz, and Chris Hughes.[62] Facebook started as a social network for American universities, but in September 2006, the network was extended beyond educational institutions to anyone with a registered email address. The site remains free to join and makes a profit through advertising revenue.

In 2005, there were 1,018 million Netusers.[63] That year, three former employees of PayPal – Chad Hurley, Steve Chen, and Jawed Karim – created a video file–sharing website called YouTube. The official debut was December 15, 2005. Two days later, *Saturday Night Live* aired "Lazy Sunday." The digital clip attracted nearly 2 million views in a week. In April 2006, the venture capital firm Sequoia injected $8 million in funding, and Judson Laiply uploaded "The Evolution of Dance," a six-minute mashup of 50 years of dance crazes, which enjoyed immense popularity and led the top YouTube list for a number of years.[64] The success of this video platform attracted the attention of big investors. On October 9, 2006, Google bought YouTube for $1.65 billion.[65] In 2012, "Gangnam Style" was uploaded to YouTube, and in a span of two years it has attracted more than 2 billion viewers.

Also in 2006, Jack Dorsey started the free social networking site Twitter. Essentially, Twitter combines SMS (Short Message Service) with a way to

[61] Andreas M. Kaplan and Michael Haenlein, "Users of the World, Unite! The Challenges and Opportunities of Social Media," *Business Horizons* 53, no. 1 (2010): 59–68, 60.

[62] Nicholas Carlson, "At Last – The Full Story of How Facebook Was Founded," *Business Insider*, March 5, 2010, http://www.businessinsider.com/how-facebook-was-founded-2010-3#we-can-talk-about-that-after-i-get-all-the-basic-functionality-up-tomorrow-night-1 .

[63] "Internet Growth Statistics."

[64] Judson Laiply, "Evolution of Dance," http://www.youtube.com/watch?v=dMHobHeiRNg .

[65] David Lidsky, "The Brief but Impactful History of YouTube," *Fast Company*, February 1, 2010, http://www.fastcompany.com/magazine/142/it-had-to-be-you.html .

create social groups. Netusers can send information to their followers and receive information from individuals or organizations they have chosen to follow.[66] Tweets are instantly broadcast to followers and posted on their news feed. The retweet function allows a Netuser to transfer another user's tweet to his or her followers. A Netuser can remove a post but not a retweet. There are more than 271 million active registered Twitter users.[67]

The number of Netusers continued to grow from 1,319 million in 2007, to 1,574 million in 2008, to 1,802 million in 2009, to 1,971 million in 2010, to 2,267 million in 2011, to 2,497 million in 2012, to 2,802 million in 2013, to an estimated 2,937 million in March 2014.[68] As of October 2014, the Indexed Web contains at least 4.62 billion pages.[69] Further usage statistics are given in table 1.1.

CONCLUSION

The Internet and its architecture have grown in an evolutionary fashion from modest beginnings rather than from a Grand Plan.[70] The ingenuity of the Internet as it was developed in the 1960s by the ARPA scientists lies in its layered protocol architecture and in the packet-switching technology. Until ARPANET was built, most communications experts claimed that packet switching would never work.[71] In 1965, when the first network experiment took place, and for the first time packets were used to communicate between computers, scientists did not imagine the multiple uses of this technology for society. Leonard Kleinrock explicitly wrote that he did not foresee the

---

[66] Erick Enge, "Jack Dorsey and Eric Enge talk about Twitter," Stone Temple Consulting, Boston, October 15, 2007, http://www.stonetemple.com/articles/interview-jack-dorsey.shtml ; Om Malik, "A Brief History of Twitter," Gigaom, February 1, 2009, http://gigaom.com/2009/02/01/a-brief-history-of-twitter/ .

[67] See the Twitter website at https://about.twitter.com/company . See also Dhiraj Murthy, *Twitter: Social Communication in the Twitter Age* (Cambridge, UK: Polity, 2013); Jeffrey Rosen, "The Web Means the End of Forgetting," *New York Times*, July 19, 2010, http://www.nytimes.com/2010/07/25/magazine/25privacy-t2.html .

[68] "Internet Growth Statistics."

[69] For the size of the World Wide Web, see http://www.worldwidewebsize.com/ .

[70] Brian E. Carpenter, "The Architectural Principles of the Internet," Request for Comments 1958, Network Working Group, June 1996, http://www.ietf.org/rfc/rfc1958.txt .

[71] See Roberts, "Internet Chronology." According to Roberts, "Packet switching was new and radical in the 1960's. In order to plan to spend millions of dollars and stake my reputation, I needed to understand that it would work. Without Kleinrock's work of Networks and Queuing Theory, I could never have taken such a radical step. All the communications community argued that it couldn't work. This book was critical to my standing up to them and betting that it would work." See James Gillies and Robert Cailliau, *How the Web Was Born: The Story of the World Wide Web* (Oxford: Oxford University Press, 2000), 26.

TABLE 1.1. *World Internet Usage and Population Statistics*
June 30, 2014 – Mid-Year Update

World Regions	Population (2014 Est.)	Internet Users Dec. 31, 2000	Internet Users Latest Data	Penetration (% Population)	Growth 2000–2014	Users % of Table
Africa	1,125,721,038	4,514,400	297,885,898	26.5%	6,498.6%	9.8%
Asia	3,996,408,007	114,304,000	1,386,188,112	34.7%	1,112.7%	45.7%
Europe	825,824,883	105,096,093	582,441,059	70.5%	454.2%	19.2%
Middle East	231,588,580	3,284,800	111,809,510	48.3%	3,303.8%	3.7%
North America	353,860,227	108,096,800	310,322,257	87.7%	187.1%	10.2%
Latin America / Caribbean	612,279,181	18,068,919	320,312,562	52.3%	1,672.7%	10.5%
Oceania / Australia	36,724,649	7,620,480	26,789,942	72.9%	251.6%	0.9%
World Total	7,182,406,565	360,985,492	3,035,749,340	42.3%	741.0%	100.0%

Notes: (1) World Internet Usage and Population Statistics are for June 30, 2014. (2) Demographic (Population) numbers are based on data from the US Census Bureau and local census agencies. (3) Internet usage information comes from data published by Nielsen Online, by the International Telecommunications Union, by GfK, local ICT Regulators, and other reliable sources. (4) For definitions, disclaimers, navigation help, and methodology, please refer to the Site Surfing Guide at http://www.internetworldstats.com/surfing.htm .

powerful community side of the Internet and its influence on every aspect of society.[72] The diffusiveness of the Net and its focus on flexibility, decentralization, and collaboration brought about the Internet as we know it today. In the initial stages, the Internet was promoted and funded, but not designed, by the US government. Allowing the original research and education network to evolve freely and openly without any restrictions, selecting TCP/IP for the NSFNet and other backbone networks, and subsequently privatizing the NSFNet backbone were the most critical decisions for the Internet's evolution.

The fact that the Internet has no central management or coordination and that routing computers do not retain copies of the packets they handle accentuates the free-spirited environment of the Net. Multiple governments and proprietary companies own pieces that make up the Internet. The leading principle is liberty: we are all at liberty to exercise our autonomy by expressing ourselves and by posting our ideas on the Web.

[72] Kleinrock, "History of the Internet and Its Flexible Future," 12.

In the second decade of the 21st century, the Internet embraces some 300,000 networks stretching across the planet. Its communications travel on optical fibers, cable television lines, and radio waves, as well as on telephone lines. The traffic continues to grow at a rapid pace. Mobile phones and other communication devices are joining computers in the vast network. Some data are now being tagged in ways that allow websites to interact.[73] Mobile phones are extensively in use to access the Internet.[74] The growth of cloud computing is providing powerful ways to easily build and support new software. *Cloud computing* is a generation of computing that uses distant servers for data storage and management, thereby allowing the device to use smaller and more efficient chips that consume less energy than standard computers. The responsibility of data storage and control is left to the provider. Large companies such as Microsoft, Google, Yahoo!, AT&T, Amazon, and Salesforce[75] offer cloud service.[76] Because companies and individuals can "rent" computing power and storage from services such as the Amazon Elastic Compute Cloud, turning a good idea into an online service has become much faster. This power is leading to an explosion in uses for the Internet and a corresponding explosion in the amount of traffic flowing across the Internet.[77] The result is the most impressive web of communications in the history of humanity. Millions of people around the globe cannot describe their lives and function as they wish without the Internet.

The next chapter supplements this historical overview with a detailed analysis of the technological aspects of the Internet.

---

[73] "Internet History Part 5: Revolution," National Academy of Engineering, 2014, http://www.great achievements.org/?id=3747 .

[74] "Mobile Internet Use Nearing 50%," *BBC News*, August 31, 2011, http://www.bbc.co.uk/news/ technology-14731757 .

[75] For more information about Salesforce, see the company's website at https://www.salesforce. com/uk/form/sem/landing/service-cloud.jsp .

[76] For further discussion, see Jon Brodkin, "10 Cloud Computing Companies to Watch," *NetworkWorld*, May 18, 2009, http://www.networkworld.com/supp/2009/ndc3/051809-cloud-companies-to-watch.html?page=1 .

[77] Eric Knorr and Galen Gruman, "What Cloud Computing Really Means," *InfoWorld*, http://www.infoworld.com/d/cloud-computing/what-cloud-computing-really-means-031?page=0,0 ; Michael R. Nelson, "A Response to Responsibility of and Trust in ISPs by Raphael Cohen-Almagor," *Knowledge, Technology, and Policy* 23, no. 3 (2010): 403–7. Further information on cloud computing can be found through the Google search engine.

2

Technological Framework

We're changing the world with technology.

–Bill Gates

Never before in history has innovation offered promise of so much to so many in so short a time.

–Bill Gates

The aim of this chapter is to shed light on the technological aspects of the Internet. The chapter serves as a technological resource for the entire book. This resource bank is designed to make technological quandaries accessible to interested laypeople.

When visionaries such as J. C. R. Licklider,[1] Robert W. Taylor,[2] Vint Cerf,[3] Larry Roberts,[4] Robert E. Kahn,[5] Ted Nelson,[6] Leonard

---

[1] J. C. R. Licklider, "Man–Computer Symbiosis," *IRE Transactions on Human Factors in Electronics* HFE-1 (March 1960): 4–11, http://groups.csail.mit.edu/medg/people/psz/Licklider.html .

[2] J. C. R. Licklider and Robert W. Taylor, "The Computer as a Communication Device," *Science and Technology* 76, no. 2 (1968): 20–41, 38.

[3] Vinton G. Cerf, "Computer Networking: Global Infrastructure for the 21st Century," Computing Research Association, Washington, DC, 1995, http://www.cs.washington.edu/homes/lazowska/cra/networks.html .

[4] "Larry Roberts," Internet Pioneers, http://www.ibiblio.org/pioneers/roberts.html ; Cade Matz, "Larry Roberts Calls Himself the Founder of the Internet. Who Are You to Argue?" *Wired*, September 24, 2012, http://www.wired.com/wiredenterprise/2012/09/larry-roberts/ . See also Roberts's website at http://www.packet.cc/ .

[5] William Stewart, "Robert Kahn: TCP/IP Co-Designer," http://www.livinginternet.com/i/ii_kahn.htm ; Robert E. Kahn, "Internet Essay," National Academy of Engineering, 2014, http://www.greatachievements.org/?id=3749 .

[6] William Stewart, "Ted Nelson Discovers Hypertext," http://www.livinginternet.com/w/wi_nelson.htm .

Kleinrock,[7] and Douglas C. Engelbart[8] first conceived of the Internet, they could not have imagined the fascinating state of the Internet today. Its rich and diversified nature and its wide circulation have benefited millions of users around the world. The Net serves as a communication medium comprising all other media. It is an infrastructure for digital commercial activities, an arena for a wide array of public debates and social networks, and a mega-size information resource.

The Internet presents a mode of technology that breaks the monopoly of conventional media. As Johannes Gutenberg's print machine broke the monopoly of the monks over books and brought about new ideas, so the Internet has changed how people think, communicate, and learn. This interactive medium is different from the print and electronic media. Its construction enables anybody and everybody, people who have access to conventional media as well as those who are denied access, to voice their opinions, even the vilest and most disturbing opinions.

First, the Internet is about communication. It comprises the basic components of communication: a sender, a message, a channel, and a receiver. Generally speaking, the Internet also provides opportunity for feedback. One Netuser can receive data from and send data to one or many other Netusers.

Second, the Internet is based on packet switching – technology that uses connectionless protocols for host-to-host resource sharing. As explained in chapter 1, the packet-switching system was designed by Bolt Beranek and Newman under contract to Advanced Research Projects Agency (ARPA) in an effort to extend the usability of the researchers' computing machinery beyond those located locally. The ARPA project scientists wanted the capability to copy files and have remote access to the research laboratories' computers.[9]

A packet is a short message, usually part of a larger message containing a source and destination address, control information, and a given amount of data with an identifying numerical address. Packet switching is based on the ability to place content in digital form, which means that content can be

7 Leonard Kleinrock, "Information Flow in Large Communication Nets," PhD thesis proposal, Massachusetts Institute of Technology, Cambridge, MA, May 31, 1961, http://www.lk.cs.ucla.edu/data/files/Kleinrock/Information%20Flow%20in%20Large%20Communication%20Nets.pdf .

8 John Markoff, "Computer Visionary Who Invented the Mouse," *New York Times*, July 3, 2013, http://www.nytimes.com/2013/07/04/technology/douglas-c-engelbart-inventor-of-the-computer-mouse-dies-at-88.html?nl=todaysheadlines&emc=edit_th_20130704&_r=0 ; Om Malik and Mathew Ingram, "Doug Engelbart, American Inventor and Computing Legend, Has Passed Away," *Gigaom*, July 3, 2013, http://gigaom.com/2013/07/03/doug-engelbart-american-inventor-computing-legend-passes-away/ .

9 Duncan Langford, ed., *Internet Ethics* (New York: St. Martin's Press, 2000), 13–14.

coded into binary numbers, or bits (1 or 0), which is how computers store information.[10] Anything that can be digitized can be sent as a packet. Intermediary computers called *routers* direct each packet along the best outgoing link to its final destination. Routers serve as traffic wardens. They are able to process millions of packets a minute. Routers open the Internet Protocol (IP) packets of data to read the destination address, find the best route, and then send the packet toward its destination. Information usually passes through more than one router before reaching its final destination. If one router does not work properly, the packet seeks an alternative one. Once all of the packets arrive, the receiving machine reassembles the information in accordance with the sequence numbers in the headers of each packet. It is possible to send an unlimited number of packets with different addresses over the same circuit.[11]

Third, because information on the Internet is in digital format, it can be easily manipulated. Images can change color and form, text can be modified, music and films can be edited and altered, and virtual people and other entities can be created. Human imagination found new frontiers to explore.

Fourth, the volume, scope, and variety of data that the sender is able to transmit over the Internet are massive. Unlike a physical book or photograph or an analog audio recording, a digital information object can be copied an infinite number of times, often without losing fidelity or quality.[12] The Internet presents information in textual, audible, graphical, and video formats, evoking the appearance and function of print mass media, radio, and television combined. People can speak, chat, send email, buy merchandise, gamble, post various forms of data, and share and transfer files. Because information is easily copied, it is also easy to distribute to many Netusers in diverse and vastly remote physical locations. The Internet is a network of networks that consists of millions of private, public, academic, business, military, and government networks of local to global scope that are linked by copper wires, fiber-optic cables, wireless connections, and other technologies.

Fifth, the lack of centralized control means that those users who are determined to abuse the Internet for their own purposes can easily succeed in their plans.

[10] Luciano Floridi, *Information: A Very Short Introduction* (Oxford: Oxford University Press, 2010), 27–28.

[11] John Mathiason, *Internet Governance: The New Frontier of Global Institutions* (Abingdon, UK: Routledge, 2009), 7.

[12] Dick Thornburgh and Herbert S. Lin, eds., *Youth, Pornography, and the Internet* (Washington, DC: National Academy Press, 2002), 32.

Sixth, the Internet is borderless. The technology is of global connectivity that encompasses continents. It is no longer solely American: the Internet has spread to all parts of the world. In 1996, the United States accounted for 66 percent of the world's Netusers. With the rapid and astonishing growth of other markets, especially in Asia, by 2008 the American market was reduced to 21.0 percent,[13] in 2009 to 14.6 percent,[14] in April 2010 to 14.4 percent, in March 2011 to 13.0 percent, and in December 2013 to 10.7 percent.[15] Still, the value of the American market continues to be of major significance.

Seventh, the portability of the Internet led to its adoption in low-income regions, where people access the Net through mobile phones rather than computers. Portability enables Netusers to capture and share real-time events, both positive (family events, public demonstrations for social justice, uprisings against dictators) and negative (road accidents, criminal offenses, natural catastrophes).

Lastly, the flexible, multipurpose nature of the Internet, with its potential to operate as a set of interpersonal, group, or mass media channels, is a unique communication system. The Internet provides many more avenues in comparison with any other form of communication to preserve anonymity and privacy with little or no cost involved. For the cost of a subscription to an Internet service provider or to a Web-hosting service, Netusers are able to interact through instant messages and email, and for a relatively low cost, any Netuser can become a Net publisher by establishing some presence on the Net. Net publishing will be covered later in this book.

The discussion opens with an explanation of the basic characteristics of the Net. Then comes a discussion of file sharing. The chapter concludes with the intricate issues of filtering, monitoring, and encryption.

BASIC CHARACTERISTICS

On October 24, 1995, the Federal Networking Council agreed on the definition of *Internet*:

> "Internet" refers to the global information system that – (i) is logically linked together by a globally unique address space based on the Internet Protocol (IP) or its subsequent extensions/follow-ons; (ii) is able to support communications using the Transmission Control Protocol/Internet Protocol

---

[13] "Worldwide Distribution of Internet Users," Tech Crunchies, March 19, 2008, http://techcrunchies.com/worldwide-distribution-of-internet-users/ .

[14] *Internet World Stats News*, no. 51, April 4, 2010.

[15] "Internet Usage Statistics," Miniwatts Marketing Group, http://www.internetworldstats.com/stats.htm .

(TCP/IP) suite or its subsequent extensions/follow-ons, and/or other IP-compatible protocols; and (iii) provides, uses or makes accessible, either publicly or privately, high level services layered on the communications and related infrastructure described herein.[16]

The TCP/IP suite enables the social media, peer-to-peer networking, and communication transmissions that constitute the Internet.

The great thing about TCP/IP is its generality. It can accommodate different devices and different types of networks of varying sizes and purposes. Data are transferred by means of the TCP/IP network technology, which allows for complete interoperability on the Internet so that computers can communicate with one another even if they have different operating systems or application software. TCP/IP consists of two pieces. TCP, or Transmission Control Protocol, manages network communication over the Internet. The data are broken up into packets, with the first part of each packet containing the address where it should go. IP, or Internet Protocol, enables network communication. IP is a means of labeling data so that it can be sent to the proper destination in the most efficient way possible. It establishes a unique numeric address (a series of four numbers, ranging from 0 to 255, separated by decimal points, which looks like this: 87.102.64.135) for each system connected to the Internet. This labeling system is known as *Internet Protocol version 4* (IPv4). When the Internet Assigned Numbers Authority gave out the last IPv4 addresses in 2011, the Internet shifted to IPv6 protocol, which is formatted and notated differently than IPv4, with more than 34 undecillion addresses (340 trillion groups of 1 trillion networks each) available.[17] TCP/IP makes the network virtually transparent to Netusers regardless of which system they are using; it enables decentralized communication, and it allows the Internet to function as a single, unified network.[18]

Internet service providers (ISPs) are providers of content or technical services for the use or operation of content and services on the Internet. Control over the Internet's operation is decentralized. As explained in chapter 1, the underlying

[16]　Barry M. Leiner, Vinton G. Cerf, David D. Clark, Robert E. Kahn, Leonard Kleinrock, Daniel C. Lynch, Jon Postel, Lawrence G. Roberts, and Stephen Wolff, "A Brief History of the Internet," Internet Society, 2003, http://www.isoc.org/internet/history/brief.shtml .

[17]　Hayley Tsukayama, "What Is IPv6, and Why Does It Matter?," *Washington Post*, June 6, 2012; Fahmida Y. Rashid, "IPv4 Address Depletion Adds Momentum to IPv6 Transition," e-Week.com, February 3, 2011, http://www.eweek.com/c/a/IT-Infrastructure/IPv4-Address-Depletion-Adds-Momentum-to-IPv6-Transition-875751/ .

[18]　Richard A. Spinello, *Cyberethics: Morality and Law in Cyberspace* (Sudbury, MA: Jones & Bartlett, 2000), 26; Johnny Ryan, "The Essence of the 'Net: A History of the Protocols That Hold the Network Together," Ars Technica, March 8, 2011, http://arstechnica.com/tech-policy/news/2011/03/the-essence-of-the-net.ars/ .

philosophy of the Internet since its inception has been to push management decisions to as decentralized a level as possible. No single government controls the entire cyberspace, but each has the capacity to regulate ISPs within its borders. As long as the ISP has network access points within the geographic space of any given nation, the nation can impose limitations on the information that flows through those points. The international nature of the Internet makes it difficult for one governing board to gain the consensus necessary to impose policy across the global Net. The complexity of this issue is discussed in chapters 8 and 9.

When many people use the word *Internet*, they are really talking about the World Wide Web.[19] Surfing the Net requires a host (that is, a computer with a distinct Internet address); server software that runs on the host; and a browser, such as Firefox, Chrome, or Explorer, to find data. The host manages the communications protocols and houses the pages and the related software required to create a website on the Internet. The host uses an operating system, such as Windows, Unix, Linux, or Macintosh, which has the TCP/IP protocols built in. The browser locates the information and displays the results in hyperlink format. With a click of a mouse, the Internet surfer can access vast amounts of information.

The Internet supports a high degree of interactivity. Thus, a Netuser who is searching for content receives content customized to his or her own needs. In this regard, the Internet is similar to a library, in which the Netuser can search for books and other forms of information. Chapter 6 elaborates on this analogy.

FILE SHARING

Peer-to-peer (P2P) file sharing represents a drastic change in the way Netusers traditionally find and exchange information. It is a process by which devices controlled by end users (i.e., "peers") interact directly with each other to transfer usually bulky files (small files can be transferred via email), rather than interacting through a specific server. File sharing has triggered major policy and legal disputes related to infringement of copyright.

File-sharing programs came to worldwide prominence with the release of Napster in 1999 by Shawn Fanning. Napster allowed Netusers to share mp3 (MPEG Audio Layer 3) music files. Since then, a host of other file-sharing programs have been released, which allow large numbers of Netusers to

[19] The Web is not the entire Internet, because many services operate using protocols other than HTTP (hypertext transfer protocol). However, in this book I am using the terms *Internet*, *Web*, and *Net* interchangeably. In common parlance, the term *World Wide Web* is rather quaint.

download and share not only mp3 music files but also software files, images, movies, and other data files.[20] TeamViewer is another innovation that facilitates file sharing. Once this free software is downloaded to Netusers' computers, it enables people to log in to another Netuser's computer and, with permission, download a file or folder, no matter how large, directly from that computer without leaving a digital trace over the Internet.[21] More than 200 million Netusers use TeamViewer.

FILTERING

The World Wide Web contains vast amounts of information. When one uses a search engine to find data on a theme of interest, one might be exposed to unwanted information, some of which might be offensive or could potentially result in harmful conduct. For instance, pedophiles luring children is a major concern.[22] To address such concerns, Netusers and technology specialists use programs to interfere with Internet transmissions and to filter or block offensive content.

Filters are generally divided into two categories: client side and server side. *Client-side* software is installed locally on the Netuser's computer and is maintained by the user. Its effectiveness depends on the user's installation, configuration, regular maintenance, and use of the software. Client-side filtering tools are very popular because they are relatively straightforward to implement and offer parents and guardians an easy way to provide a safer Internet environment. Indeed, a common personal use is a client-side filter installed on a home computer by a parent wishing to protect children from inappropriate material. Client-side filtering is also feasible in an environment in which only some access points in a local area network must be filtered – for example, in a library attempting to segregate children's areas from areas for all patrons.[23] In the *server-side* approaches, filtering of inappropriate content is performed before the content reaches a Netuser's computer and is restricted

---

[20] "File Sharing Programs: Online File Sharing," InternetGuide, http://www.internet-guide.co.uk/file-sharing.html ; Amy E. White, *Virtually Obscene: The Case for an Uncensored Internet* (Jefferson, NC: McFarland, 2006), 17.

[21] See the software's website at http://www.teamviewer.com/en/ .

[22] Vinton Cerf commented, "It is a concern, but I think evidence is that the Internet is less used for this purpose than some think. The luring takes place in other ways." Vinton G. Cerf, email message to the author, January 16, 2012.

[23] Thornburgh and Lin, *Youth, Pornography, and the Internet*, 271. Vint Cerf commented that client-side filtering may not be entirely effective. Vinton G. Cerf, email message to the author, January 16, 2012.

by the standards of the website or service platform. Server-side filtering often is used to refer to content filtering at the ISP level.[24] ISPs can block access to all websites associated with a blacklisted IP address or filter content that matches a list of blacklisted keywords.

The blacklisting can be done in one of two ways. The first, called *packet-level filtering*, is a mechanism that requires programming router computers to compare the IP address of the packet sender with predetermined blacklist content. The second is called *application-level filtering*, by which reception of specific uniform resource locators (URLs) is denied.

Some software programs use the World Wide Web Consortium's Platform for Internet Content Selection (PICS), which was later superseded by the Protocol for Web Description Resources (POWDER).[25] The architecture of PICS divides the problem of filtering into two parts: labeling (rating content) and then filtering the content according to those labels. Hence, websites first self-rate their content. They, for instance, could attach labels indicating whether content contains nudity or violence. This system is very similar to the movie ratings system. Second, PICS supports the establishment of third-party labeling bureaus to filter content.[26] The labels identify the content of sites by measuring their level of decency through a rating system, by implementing privacy vocabulary that describes a website's information practices, or by using similar techniques.[27]

Filters can be installed at institutions (e.g., schools, universities, workplaces) and at homes. Server-side filtering is useful in institutional settings, where Netusers at all access points within the institution's purview must conform to the institution's access policy. Thus, for instance, a school district that provides Internet service to all schools in the district or a library system that provides Internet service to all libraries in the system may use server-side filtering. Ideally, when no funding is available to individuals and institutions, human rights organizations may sponsor these filters so the cost does not prohibit their use.

---

[24] Internet Safety Technical Task Force, "Appendix D: Technology Advisory Report," in *Enhancing Child Safety and Online Technologies: Final Report of the Internet Safety Technical Task Force* (Boston: Berkman Center for Internet and Society, 2008); interviews with Internet experts, Washington, DC, May 9–15, 2008.

[25] See "Platform for Internet Content Selection (PICS)," World Wide Web Consortium, Cambridge, MA, http://www.w3.org/PICS/ ; Lawrence Lessig, *Code and Other Laws of Cyberspace* (New York: Basic Books, 1999), 177.

[26] "PICS HOWTO: Using PICS Headers in HTML," Vancouver Webpages, http://vancouver-webpages.com/PICS/HOWTO.html .

[27] Edgar Burch, "Comment: Censoring Hate Speech in Cyberspace – A New Debate in a New America," *North Carolina Journal of Law and Technology* 3, no. 1 (2001): 175–92, 191. Vint Cerf thinks that PICS has not been implemented well. Vinton G. Cerf, email message to the author, January 16, 2012.

Among the growing number of fairly effective filtering devices are Net Nanny and CyberPatrol.[28] These devices empower parents wanting to prevent children – and employers wanting to prevent employees – from browsing Internet sites with objectionable messages. The software scans Web pages for specific words or graphic designs and then restricts user access to them. HateFilter, developed by the Anti-Defamation League (ADL), specifically targeted several hundred sites identified as hate sites (i.e., sites carrying hateful messages in any form of textual, visual, or audio-based rhetoric). When it was activated, HateFilter denied access to Internet sites advocating hatred, bigotry, or violence against Jews, minorities, and homosexuals.[29] ADL abandoned HateFilter because it was a master-list, exclusion-based software package that became impossible to manage with the amount and variety of hate online. That in itself says volumes about the problem of Net bigotry and hate.

MONITORING

In some situations, Netusers want to be advised of dangerous content on the Internet but they either cannot filter or would prefer to be cognizant about the vile material. Reports have tied MySpace to a variety of societal ills, such as harassment, cyberbullying, cyberstalking, alcohol and drug abuse, hate crimes, planned or executed bombings, school shootings, suicide, and murder. MySpace has also been vilified as a haven for predators and pedophiles who may troll the site for unsuspecting victims.[30] Interviews conducted with security agents reveal that they often prefer to monitor problematic sites rather than filter their content or shut them down altogether. Various programs allow Netusers to monitor the activities of other users, such as the websites they visit, the materials they download, or the keystrokes typed.[31]

Monitoring technology has been in use for many years. As a matter of fact, computer network administrators have been using so-called packet sniffers for years to monitor their networks and perform diagnostic tests or troubleshoot problems. Essentially, a packet sniffer is a program that can see all of the

---

[28] More information about Net Nanny and CyberPatrol is available on their websites, http://www. netnanny.com/ and http://www.cyberpatrol.com/cpparentalcontrols.asp , respectively.

[29] Communication with Brian Marcus, former director of ADL Internet Monitoring, Washington, DC, June 5, 2008. Although HateFilter is no longer available, information about it can be found at http://www.internet-filters.net/hatefilter.html .

[30] Sameer Hinduja and Justin W. Patchin, "Changes in Adolescent Online Social Networking Behaviors from 2006 to 2009," *Computers in Human Behavior* 26, no. 6 (2010): 1818–21; Justin W. Patchin and Sameer Hinduja, "Trends in Online Social Networking: Adolescent Use of MySpace over Time," *New Media and Society* 12, no. 2 (2010): 197–216.

[31] See, for instance, the product webpage for Spector 360 Recon at http://www.spectorsoft.com/ .

information passing over the network to which it is connected. As data travel on the Net, the program looks at, or "sniffs," each packet. A packet sniffer usually can be set up in unfiltered or filtered modes. The *unfiltered* mode captures all of the packets, whereas the *filtered* mode captures only those packets containing specific data elements.[32]

CRYPTOGRAPHY OR DATA ENCRYPTION

For each technology, there often is a contra form of technology. Most Internet communications – including email, messaging, discussion groups, and Web browsing – easily reveal the IP address and the likely physical location of Netusers unless they are taking precautions to hide their identity or communication. Encryption is a method that protects the communications channel from sniffers. Based on the science of cryptography (secret writing), encryption has been used for many centuries, particularly for military purposes. The Greek historian Plutarch wrote about Spartan generals who communicated by using a *scytale*, a thin cylinder made of wood. The general would wrap a piece of parchment around the scytale and write his message along its length. When an unauthorized person removed the paper from the cylinder, the writing appeared unintelligible. But the intended recipient had a similar scytale around which he could wrap the paper and easily read the intended message. The Greeks were also the first to use ciphers, specific codes that involved substitutions or transpositions of letters and numbers.[33] Similarly, the key used by the Roman conqueror Julius Caesar was the replacement of a letter by the letter that was three places ahead of it in the alphabet (thus the letter *d* would be replaced by the letter *g*). With the aid of this key, recipients could easily decrypt and render messages understandable.[34]

Electronic varieties of encryption arose not long after the development of telegraphs and telephones. Following Samuel Morse's introduction of the telegraph in 1844, a commercial coding device was developed[35] as a result of the growing concern with keeping for-profit transactions private. Computer cryptography, or encryption, became increasingly widespread in the 1960s. Numerous encryption algorithms have been developed, each more advanced than the previous one. For some time, the most popular commercial

---

[32] Jeff Tyson, "How Carnivore Worked," HowStuffWorks, http://computer.howstuffworks.com/carnivore2.htm .

[33] Jeff Tyson, "How Encryption Works," HowStuffWorks, http://computer.howstuffworks.com/encryption1.htm .

[34] Spinello, *Cyberethics*, 139.

[35] Jonathan Wallace and Mark Mangan, *Sex, Laws, and Cyberspace* (New York: Henry Holt, 1996), 42.

encryption algorithm was DES (Data Encryption Standard). DES was designed by IBM in the early 1970s under the name Lucifer. It was adopted as a standard by the US government in November 1976 after being modified by the National Security Agency. DES is a symmetric private key cryptography system: the same secret binary key (the specific set of instructions that will be used to apply the algorithm to a specific message) is used for both encryption and decryption. For this system to work properly, both parties, the sender and the receiver of the data, must have access to this key. DES has two major weaknesses. With the advance of technology, modern computers can decode the 72 quintillion possible combinations of the encryption algorithm. And as is the case with all kinds of encryption, the decoding key must be communicated in a secure fashion. Both the sender and the receiver of the encrypted data must take precautions that the key will not be intercepted by a third party and fall into the wrong hands.[36]

With the advancement of technology and decryption capabilities, DES ceased to be a US government standard, although Triple DES (3DES) is still used. Triple DES has a longer key length that eliminates many of the shortcut attacks that can be used to reduce the amount of time it takes to break DES.[37] Three 56-bit keys are used, instead of one, for an overall key length of 168 bits. The main government encryption standard is now the Advanced Encryption Standard (AES), which was developed because of the vulnerability to decryption of DES and because 3DES is too slow. AES supports key sizes of 128, 192, and 256 bits; uses 128-bit blocks to encrypt data; and is efficient in both software and hardware.[38]

[36] "What Is DES?," Next Wave Software, http://www.thenextwave.com/page19.html ; Joseph Migga Kizza, *Ethical and Social Issues in the Information Age* (London: Springer, 2007), 105. In the early 1990s, two Swiss researchers, Xuejia Lai and James Massey, developed an encryption system similar to DES but using a longer key length: 128 bits compared with 56. The effort required to decrypt increases exponentially with the length of the key: the longer the key, the less the likelihood of breaking the code. Lai and Massey called the system International Data Encryption Algorithm (IDEA). IDEA was also a symmetric encryption: the encoder and decoder use the same key. See Sheila Robinson and Lamar Stonecypher, "An Overview of the International Data Encryption Algorithm (IDEA)," Bright Hub, June 22, 2010, http://www.bright hub.com/computing/smb-security/articles/74996.aspx ; Martin C. Golumbic, *Fighting Terror Online: The Convergence of Security, Technology, and the Law* (New York: Springer, 2008), 141.

[37] Triple DES is described by D.I. Management Services, maker of cryptography software, at http://www.cryptosys.net/3des.html . See also "Triple DES Encryption," Tropical Software, http://www.tropsoft.com/strongenc/des3.htm . Vint Cerf commented that the effective key length of 3DES is only $2 \times 56 = 112$ bits. Vinton G. Cerf, email message to the author, January 16, 2012.

[38] For an overview of the AES development, see "AES," National Institute of Standards and Technology, December 4, 2001, http://csrc.nist.gov/archive/aes/ ; see also Margaret Rouse, "Advanced Encryption Standard (AES)," TechTarget, http://searchsecurity.techtarget.com/definition/Advanced-Encryption-Standard .

Careful abusers of the Internet, such as child pornographers, use steganography to conceal sensitive files. *Steganography* is a technique that creates virtually invisible data by concealing such data behind legitimate data. Seemingly harmless text, audio, and video files may contain illegal information. The special software needed for steganography is freely available on the Net.[39] Although hiding information is quite easy, finding it requires resources and effort.

Other milestones occurred in the development of modern encryption. In 1976, two researchers in the Stanford University Electrical Engineering Department, Whitfield Diffie and Martin Hellman, developed the first feasible way to encrypt and exchange secure data over insecure channels.[40] A year later, in 1977, three MIT scientists – Ron Rivest, Adi Shamir, and Leonard Adelman – developed a public key cryptosystem for both encryption and authentication, which they called RSA, the initials of their surnames. RSA is an encryption algorithm that uses very large prime numbers to generate the public key and the private key. Unlike DES, which works on the principle of symmetric encryption, this method works on the principle of asymmetric encryption: the sender and the recipient use different keys. In this method, every user has two keys: a public key, used to send messages to the recipient, and a private key, used by the recipient to decipher the messages.[41] To encrypt a message from sender A to recipient B, both A and B need to create their own pairs of keys. Then A and B exchange their public keys, which anybody can acquire. Although the public key is published and available to anyone, the message cannot be read without the private key. When A sends a message to B, A uses B's public key to encrypt the message. On receipt of the message, B then uses the private key to decrypt it.[42]

In June 1991, Philip Zimmermann completed a complex and elaborate encryption program, which he called Pretty Good Privacy (PGP). This is a modest name for powerful, sophisticated software that enables fast encryption with advanced security. PGP encrypts the message content with a symmetric cipher. (Microsoft Outlook provides a built-in, standardized alternative to PGP called S/MIME, or Secure/Multipurpose Internet Mail Extensions.) PGP allows Netusers to encrypt messages so that unauthorized individuals, including law enforcement authorities, cannot decipher them.

[39] Vangie Beal, "Steganography," Webopedia, http://www.webopedia.com/TERM/S/stegano graphy.html .

[40] Wallace and Mangan, *Sex, Laws, and Cyberspace*, 45.

[41] Microsoft Support, "Description of Symmetric and Asymmetric Encryption," Article 246071, October 26, 2007, http://support2.microsoft.com/kb/246071 . See also Tyson, "How Encryption Works."

[42] Kizza, *Ethical and Social Issues in the Information Age*, 106.

Zimmermann handed PGP over to an unidentified friend in the summer of 1991. That individual subsequently placed the program on a bulletin board system on the Internet for anyone to access, with no fees to pay, no registration forms to fill out, and no questions to answer.[43] Zimmermann, like the Internet's founding fathers, espoused the free-spirited, open culture of networking and communication. Whether this conduct was socially responsible is questionable. PGP became a widely used encryption program in all parts of the world by all kinds of people, users and abusers.

In 2013, Edward Snowden leaked top-secret information indicating that American and British intelligence agencies have successfully cracked much of the online encryption software. According to the Snowden documents, the US National Security Agency (NSA) inserted secret vulnerabilities, known as *backdoors*, into the existing encryption software. The NSA also allegedly hacked into target computers to capture messages before they were encrypted. Sometimes, the documents revealed, the NSA coerced companies to hand over their master encryption keys or to build backdoors. The NSA has allegedly invested more than $250 million a year to develop groundbreaking capabilities against encrypted Web chats and phone calls.[44]

Whether security agencies are able to decrypt *every* form of encryption and to unveil *every* effort to hide one's privacy is unclear. Sites such as Anonymouse.com promote proxy server, anonymous browser, anonymous web surfing proxy, and anonymous email for Netusers to surf the Net anonymously.[45] Anonymous proxy masks the IP address of Netusers. No logs or other identifying information is kept. Another platform is GoTrusted, which provides a fast, easy way to secure Internet data on a personal computer while protecting one's privacy. GoTrusted makes all Web surfing anonymous and secures Web, email, video, instant messaging, and P2P file sharing. GoTrusted uses industrial strength encryption and hides one's IP.[46]

[43] Philip R. Zimmermann, *The Official PGP User's Guide* (Cambridge, MA: MIT Press, 1995); Simson L. Garfinkel, *PGP: Pretty Good Privacy* (Sebastopol, CA: O'Reilly, 1995). For more about Philip Zimmermann, see his website at http://www.philzimmermann.com/EN/background/.

[44] James Ball, Julian Borger, and Glenn Greenwald, "Revealed: How U.S. and U.K. Spy Agencies Defeat Internet Privacy and Security," *Guardian*, September 5, 2013; Jeff Larson, Nicole Perlroth and Scott Shane, "Revealed: The NSA's Secret Campaign to Crack, Undermine Internet Security," *ProPublica*, September 5, 2013, http://www.propublica.org/article/the-nsas-secret-campaign-to-crack-undermine-internet-encryption ; Matt Buchanan, "How the N.S.A. Cracked the Web," *New Yorker*, September 6, 2013, http://www.newyorker.com/tech/elements/how-the-n-s-a-cracked-the-web .

[45] See the company's website at http://anonymouse.com/ .

[46] For more information, visit the GoTrusted website at http://www.gotrusted.com/ .

A further tool to protect privacy is Tor. Tor was originally designed by the US Naval Research Laboratory to protect sensitive government communications.[47] Later Tor was adopted by journalists, human rights activists, hackers, law enforcement officers, and the public. It is free software and an open network that helps Netusers defend against a form of network surveillance that threatens personal freedom and privacy, confidential business activities and relationships, and state security. Tor comprises software that can be downloaded for free and a volunteer network of computers that enables the operation of the software.[48] The program employs cryptography in a multilayered manner so that Netusers can communicate secretly through different routers. Tor protects the identity of Netusers by bouncing their communications around a distributed network of relays. It prevents anyone who is watching the Internet connection of Netusers from learning the sites they visit, protects their physical location, and hides their real IP address.

At present, Tor is the most popular onion router. It is called an *onion router* because it builds a new chain of servers and encryption keys for each Netuser at frequent intervals. When a Netuser connects to a website in Tor, that Netuser encrypts the information. Only the final Tor server in the chain has the capability to decrypt and read the message before forwarding it to its final destination. Hundreds of thousands of people around the world – journalists and bloggers, human rights workers and activists, protestors against authoritarian governments and for democracy, law enforcement officers, soldiers, corporations, citizens of repressive regimes, and ordinary citizens – use Tor for a wide variety of reasons.[49] Unfortunately, Net abusers also use encryption programs and onion routers. Because these tools are freely available and powerful, criminals use encryption programs and onion routers to hide their Net identity. Hence, these tools are double-edged swords that may enhance privacy and anonymity but may also undermine security.

---

[47] "Tor: Overview," Tor Project, Cambridge, MA, https://www.torproject.org/about/contact.html.en .

[48] "What Is Tor?," Electronic Frontier Foundation, San Francisco, https://www.eff.org/torchallenge/what-is-tor .

[49] For more information about Tor, see the Tor Project's website at http://www.torproject.org/ . See also Linux Reviews' wiki page on Tor at http://en.linuxreviews.org/Tor ; Roger Dingledine, Nick Mathewson, and Paul Syverson, "Tor: The Second-Generation Onion Router," in *Proceedings of the 13th Conference on USENIX Security Symposium* (Berkeley, CA: USENIX Association, 2004), 21; Kim Zetter, "Rogue Nodes Turn Tor Anonymizer into Eavesdropper's Paradise," *Wired*, September 10, 2007; John Alan Farmer, "The Specter of Crypto-Anarchy: Regulating Anonymity-Protecting Peer-to-Peer Networks," *Fordham Law Review* 72, no. 3 (2003): 725–84. See also the Tor Challenge website at http://www.eff.org/torchallenge .

CONCLUSION

The Internet's design was unprecedented because it was conceived as a decentralized, open, and neutral network of networks. The open architecture of the Internet allows free access to protocols from anywhere in the world and is accessible via almost any kind of computer or network. The choice of individual network technology is not dictated by particular network architecture but rather may be selected freely by a provider and made to work with other networks through a meta-level internetworking architecture.[50] This open architecture encourages the development of more net applications. And the Internet is neutral regarding different applications of text, audio, and video, thus allowing constant development of applications to evolve. Wireless and mobile communications enable accessibility to the Internet almost anywhere, at any time. In May 2012, Virgin Atlantic announced that it would provide passengers in-flight access to email, text messages, and the Internet.[51]

The various technological tools have inherent advantages as well as disadvantages. Those designed to enhance one's privacy may harm security and vice versa. They can be put to good use (filtering pornography), but they also facilitate crime (encrypting child porn images).[52] Encryption promotes privacy and anonymity on the Net, but anonymity does not contribute to cultivating a sense of Net responsibility or trust.

In the next chapter, I will discuss the theoretical foundations underpinning the discussion of this book. I will then explain the meaning of the democratic catch, moral panics, trust, moral responsibility, and social responsibility. Finally, I will make a distinction between *Netusers* and *Netcitizens*.

[50] Leiner et al., "Brief History of the Internet."

[51] Mark Molloy, "Virgin Atlantic Become First Airline to Offer In-Flight Mobile Calls," *Metro*, May 14, 2012, http://www.metro.co.uk/tech/899035-virgin-atlantic-become-first-airline-to-offer-in-flight-mobile-calls .

[52] Vint Cerf commented, "Fixation on child porno as the archetype of abuse seems to me overblown and diverts from the significant other abuses including industrial espionage." Vinton G. Cerf, email message to the author, January 16, 2012. For further discussion, see P. W. Singer and Allan Friedman, *Cybersecurity and Cyberwar: What Everyone Needs to Know* (New York: Oxford University Press, 2014).

3

Theoretical Framework

We do not act rightly because we have virtue or excellence, but we rather have those because we have acted rightly.

–Aristotle

First, Do No Harm.

–Vinton G. Cerf

The Internet provides cheap, virtually untraceable, instantaneous, anonymous, uncensored distribution that can be easily downloaded and posted in multiple places. The transnational nature of the World Wide Web, its vast content, the lack of central management or coordination, and the fact that routing computers do not retain copies of the packets they handle provide ample opportunities for people to exploit the Net's massive potential to enhance partisan interests, some of which are harmful and antisocial. Although a relatively small number of people use the Net to harm others, they undermine people's sense of trust in the Net.

Luciano Floridi argues that we are now experiencing the fourth scientific revolution. The first was that of Nicolaus Copernicus (1473–1543), the astronomer who formulated a scientifically based, heliocentric cosmology that displaced the Earth – and hence, humanity – from the center of the universe. The second was that of Charles Darwin (1809–1882), who showed that all species of life have evolved over time from common ancestors through natural selection, thus displacing humanity from the center of the biological kingdom. The third was that of Sigmund Freud (1856–1939), who recognized that the mind is also unconscious and subject to the defense mechanism of repression; thus, we are far from having Cartesian minds entirely transparent to ourselves. And now, in the information revolution, we are in the process of dislocation and reexamination of humanity's primary nature and role in the universe.[1]

Floridi argues that while technology keeps growing, we must start digging deeper to expand and reinforce our conceptual understanding of the information age – its nature, its less visible implications, and its impact on human and environmental welfare. Doing so will enable us to anticipate difficulties; identify opportunities; and resolve challenges, conflicts, and dilemmas.[2]

Floridi has made many contributions in his attempts to "dig deeper." In this chapter, I would like to follow some of Floridi's ideas on information ethics, which he describes as the study of the moral issues arising from the availability, accessibility, and accuracy of informational resources, independent of their format, type, and physical support. He further clarifies that information ethics, understood as information-as-a-product ethics, may cover moral issues arising, for example, in the context of accountability, liability, testimony, propaganda, and misinformation.[3] I will add to this list answerability and responsibility and will focus on these two concepts, as well as on accountability.

The concept of *answerability* implies responsiveness to the views of all, with a legitimate interest in what is conducted, whether as individuals who are affected or on behalf of society. It includes a willingness to explain and justify actions of publication or omission. The outcomes of answerability express and reaffirm various norms relevant to the wider responsibilities of an organization in society. The emphasis is on the quality of performance.[4]

Answerability is closely related to *accountability*. The former accentuates more the need to respond to external claims, pressures, and demands, thereby providing explanation for one's conduct. The accompanying concept of accountability refers to people or organizations that are able to answer for their conduct and obligations. When we speak of *social responsibility*, we refer to the responsibility of individuals, groups, corporations, and governments to

[1] Luciano Floridi, *The Fourth Revolution: How the Infosphere Is Reshaping Human Reality* (Oxford: Oxford University Press, 2014); Luciano Floridi, "The Information Society and Its Philosophy: Introduction to the Special Issue on 'The Philosophy of Information, Its Nature, and Future Developments,'" *Information Society* 25, no. 3 (May–June 2009): 153–58, http://www.philosophyofinformation.net/publications/pdf/tisip.pdf . See also Luciano Floridi, *Information: A Very Short Introduction* (Oxford: Oxford University Press, 2010), 11.

[2] Floridi, "Information Society and Its Philosophy." For further discussion, see Alistair S. Duff, *A Normative Theory of the Information Society* (New York: Routledge, 2012).

[3] Luciano Floridi, "Information Ethics: Its Nature and Scope," in *Moral Philosophy and Information Technology*, ed. Jeroen van den Hoven and John Weckert (Cambridge: Cambridge University Press, 2008), 40–65, http://www.philosophyofinformation.net/publications/pdf/ieinas.pdf .

[4] Denis McQuail, *Media Accountability and Freedom of Publication* (New York: Oxford University Press, 2003), 204.

society. The difference between responsibility, on the one hand, and answerability and accountability, on the other, is that the first connotes a more voluntary and self-directed character. Responsibilities are typically accepted, not imposed by another person or entity, although they can be contracted and attributed. In contrast, answerability and accountability have a more external character, although they also can be voluntary. The more voluntary, the more compatible conduct is with freedom and even coterminous with responsibility. The accountable person or organization is also answerable.

In other words, responsibility, answerability, and accountability complement each other, one being an extension of the others.[5] They are designed to improve the quality of the service or product, to promote trust of those who are using the service or product, and to protect the interests of all parties concerned, including the business at hand. Businesses known to be responsible, answerable, and accountable for their services or products enjoy a solid reputation and may attract more customers.

Accountability involves three key elements:[6]

1. *A set of outcome measures that reliably and objectively evaluates performance.* Every profession has a minimum set of measures that every professional must meet. These measures vary according to the profession and the individual activity to be performed by the professional. For example, in high-tech professions, the measures include the ability to sell the product and to attract more customers.

2. *A set of performance standards defined in terms of these outcome measures.* Like outcome measures, performance standards must be carefully chosen and attainable. Each profession has a set of common performance standards for all its members for every type of service or product provided by that profession.[7] For the high-tech profession, the standard of output measures may be the ability to accommodate the product to technological developments and to maintain relevance and competitiveness.

3. *Incentives for meeting the given standards and penalties for failing to meet them.* Incentives should also be carefully chosen, maintaining a fine balance between the organization's interests and customers' benefits. Sometimes professionals may sacrifice customers' interests to attain their company's standards of success. Monetary incentives may push professionals to put profit ahead of maintaining a good service. As

---

[5] Ibid., 306. See also Herman T. Tavani, *Ethics and Technology: Controversies, Questions, and Strategies for Ethical Computing* (Hoboken, NJ: Wiley, 2011), 119–123.

[6] Joseph Migga Kizza, *Ethical and Social Issues in the Information Age* (London: Springer, 2007), 73–74.

[7] Ibid. See also Silvio Waisbord, *Reinventing Professionalism* (Cambridge: Polity, 2013).

deterrence, the prescribed penalties should be clear, transparent, and commensurate with one's level of responsibility. Thus, every profession prescribes certain standards that professionals must meet to stay on the job. Professionals know what they gain when they overperform and what they might lose if they underperform or act irresponsibly. High-tech professionals may receive company shares for excellent conduct, and they might be demoted, transferred, or fired for failure. Penalties for irresponsible behavior vary from one profession to another, bearing in mind the consequences of the professional's irresponsibility. The penalty for a pilot's irresponsible behavior is far weightier than the penalty for a teacher's irresponsible behavior.

In addition, people have wider moral and social responsibilities to their community, which are dictated by social norms and by their conscience. Residents in an apartment building are expected to keep the elevator tidy, especially when they are carrying garbage. Someone who notices a little child wandering alone on the street is morally obligated to inquire of the where-abouts of the child's guardian.

Some things are not to be done. Common standards of civility and decency compel us to keep some activities private. One knows that one should not cook when traveling by public transportation. Such a restriction is not written anywhere, but consideration for others – as well as health and safety reasons – makes people realize that such conduct is unacceptable. Autonomous people who are responsible for their actions normally do not urinate in public or leave banana peels on the pavement. Considerate people do not knock on their neighbor's door at 3:00 a.m. to ask for a glass of milk. Considerate drivers do not park their cars in the middle of the road, not even for only five minutes. People live in a community and understand that actions have consequences. Most of our conduct regards others in one way or another, affecting the lives of other people. Acting responsibly means acting with foresight – seeing that offensive and harmful consequences of one's conduct that can be avoided are, indeed, avoided.

Responsibility and accountability are important because sometimes people and organizations seek independence from their responsibilities. Ambrose Bierce, an American journalist and satirist, described responsibility as a "detach-able burden easily shifted to the shoulders of God, Fate, Fortune, Luck, or one's neighbor. In the days of astrology it was customary to unload it upon a star."[8] In the Internet age, Netusers unload it upon cyberspace. Here an interesting

---

[8] Ambrose Bierce, *The Devil's Dictionary* (New York: Neale, 1911), http://www.alcyone.com/max/lit/devils/ .

phenomenon has emerged that confuses the concept of moral and social responsibility. In the offline, real world, people know that they are responsible for the consequences of their conduct – speech as well as action. In the online, cyberworld, we witness responsibility shake-off. One can assume a dream identity, and then anything goes. The Internet has a disinhibition effect. The freedom allows language one would dread to use in real life, words one need not abide by, and imagination that trumps conventional norms and standards. It is high time to bring to the fore discussion about morality and responsibility.

In the focus of my discussion are the neglected concepts of trust and social responsibility as they apply to the Internet realm. I will discuss and explain the concepts and their implications. Fostering trust requires cultivating social responsibility on the Internet. The concept of social responsibility should guide Netusers, Net readers, Internet service providers, states, and the international community at large. But before I turn to a discussion of trust and responsibility, I will introduce the underpinning premises of this book and then present two important theories that are pertinent to my discussion. The first is the democratic catch, a theory that I developed in my previous work. The second is Stanley Cohen's theory of moral panics. A delicate balance should be maintained between these theories. Although some boundaries to liberty and tolerance must be established to maintain a responsible Internet, we should be careful not to awaken moral panics by either exaggerating the challenges or instilling chilling effects that might silence Netusers and organizations.

UNDERPINNING PREMISES

First, as explained in the introduction, the hypotheses advanced in this book and the conclusions reached are limited to Western modern democracies emerging during the past century or so. My concern is with liberal democracies, which respect autonomy and variety. The notion of autonomy involves the ability to reflect on one's beliefs and actions and the ability to form an idea regarding them, so as to decide the way to lead one's life. By deciding between conflicting considerations, people consolidate their opinions more fully and review the ranking of values for themselves with a clear frame of mind.

Second, free expression is a fundamental right and value in democracies. Individuals are free to realize themselves and to form a worldview and an opinion by giving flight to their spirit. Freedom of the individual and the community brings truth to light through a struggle between truth and falsity. The underlying assumption is that truth will prevail in a free and open encounter with falsehood. Furthermore, freedom of expression is necessary to keep beliefs vital. It is the freedom to exchange opinions and views in a spirit of tolerance, with respect to the autonomy of every individual, and to persuade one another in a way that develops, strengthens, and secures the democratic regime. Freedom of expression is crucial to identify the causes of discontent, the presence of cleavages, and possible future conflicts.[9] It is essential for the development and cultivation of societal norms.

The third premise holds that – generally speaking – a balance must be found between the right to freedom of expression and the harms that might result from such expression. Proponents of this premise argue that the right to exercise free expression does not include the right to do unjustifiable harm to others.[10] I believe free expression is a basic tenet of liberal democracy; take away free expression, and democracy becomes a dead dogma. At the same time, I am not an absolutist. I do not believe free expression is without boundaries. Take away boundaries, and freedom might turn into a jungle. We should always examine circumstances and consequences of speech and action. When free expression comes into conflict with other important principles, such as the security of a person, we need to carefully weigh the price we are required to pay for free expression and decide whether we can afford it.

Fourth, democracy and free communication industries exist and act under certain basic tenets of liberty and tolerance, from which they draw their strength and vitality and preserve their independence. I accentuate (see the Introduction) that two of the most fundamental rights underlying every democracy are showing respect for others and not harming others.[11] Those responsibilities should not be held secondary to considerations of profit and personal prestige of anyone, including Internet service providers, reporters, and bloggers. We should perceive one another as ends and not as means – a Kantian deontological approach[12] – and we should avoid knowingly harming others. John Stuart Mill believed that (a) it is possible to evaluate the rightness

---

9 Aharon Barak, "Freedom of Expression and Its Limitations," in *Challenges to Democracy: Essays in Honour and Memory of Professor Sir Isaiah Berlin*, ed. Raphael Cohen-Almagor (Aldershot, UK: Ashgate, 2000), 167–88, 168; Raphael Cohen-Almagor, *The Boundaries of Liberty and Tolerance: The Struggle against Kahanism in Israel* (Gainesville: University Press of Florida, 1994), 89–93; Thomas I. Emerson, *Toward a General Theory of the First Amendment* (New York: Random House, 1966), 5–15.

10 *Canadian Charter of Rights and Freedoms*, section 1; *R. v. Keegstra*, [1990] 3 S.C.R. 697; *Canadian Human Rights Commission et al. v. Taylor et al.*, [1990] 75 D.L.R. (4th); *R. v. Butler*, [1992] 1 S.C.R. 452.

11 John Stuart Mill, *Utilitarianism, Liberty, and Representative Government* (London: J. M. Dent, 1948); R. S. Downie and Elizabeth Telfer, *Respect for Persons* (London: Allen & Unwin, 1969).

12 Immanuel Kant, *Foundations of the Metaphysics of Morals*, ed. Robert Paul Wolff (Indianapolis, IN: Bobbs-Merrill, 1969). For further discussion, see Pepita Haezrahi, "The

and wrongness of an action by considering its consequences, and (b) the morality of an action depends on the consequences it is likely to produce. Because we are to judge before acting, then we must weigh the probable results of what we do, given the conditions of the situation.

Fifth, we respect others as autonomous human beings who exercise self-determination to live according to their life plans; we respect people as self-developing beings who are able to expand their inherent faculties as they choose – that is, to develop the capability they want to cultivate, not every capability with which they are blessed. We respect people's desire to realize their potential and achieve their goals. We perceive each individual as a bearer of rights and responsibilities in relation to other persons. To treat others with respect is to respect their right to make decisions, regardless of our opinions of those decisions. We accept that each of us believes that our own course of life has intrinsic value, at least for the individual, and we respect the individual's reasoning – *as long as he or she does not harm others*. We respect the individual's rights as a person even if we have no respect for that person's specific decisions and choices.

I construe the respect-for-others argument as grounded in Kantian and liberal ethics. Kantian ethics is based on reflexive self-consciousness. It speaks of respecting people as rational beings and of autonomy in terms of self-legislation. Immanuel Kant calls the ability to be motivated by reason alone the *autonomy of the will*, which he contrasts with the heteronomy of the will that is subject to external causes. An *autonomous agent* is someone who is able to overcome the promptings of all heteronomous counsels, such as those of self-interest, emotion, and desire, should they be in conflict with reason. Only an autonomous being perceives genuine ends of actions (as opposed to mere objects of desire), and only such a being deserves our esteem, espouses Kant, as the embodiment of rational choice. The autonomy of the will, Kant argues, "is the sole principle of all moral laws, and of all duties which conform to them; on the other hand, heteronomy of the will not only cannot be the basis of any obligation, but is, on the contrary, opposed to the principle thereof, and to the morality of will."[13]

The notion of obligation instructs us how to behave. According to Kant, an action has moral worth only if it is performed from a sense of duty. Duty, rather

Concept of Man as End-in-Himself," in *Kant: A Collection of Critical Essays*, ed., Robert Paul Wolff (London: Macmillan, 1968), 291–313.

[13] Immanuel Kant, *Critique of Practical Reason*, trans. Thomas Kingsmill Abbott (London: Longmans, Green, & Co., [1788] 1898), 122, http://philosophy.eserver.org/kant/critique-of-practical-reaso.txt . For further deliberation, see Lawrence Jost and Julian Wuerth, eds., *Perfecting Virtue: New Essays on Kantian Ethics and Virtue Ethics* (Cambridge: Cambridge University Press, 2011).

than purpose, is the fundamental concept of ethics. It is the practical, uncon-
ditional necessity of action, and therefore, it applies to all rational beings.
Thus it can be a law for all human wills. The moment one sets up a categorical
imperative for oneself ("Act on maxims that can at the same time have
themselves as universal laws of nature as their object. That gives us the formula
for an absolutely good will"[14]) and submits to it, one is then governed by
reason; when reason becomes the master of one's desires, one is capable of
imposing certain limitations on oneself. Duty commands us to accept moral
codes because they are just, regardless of others' attitudes toward them. This
deontological ethic proscribes a set of actions, which constrains our range of
options – not because the results will be useful but because this set of actions is
incompatible with the concept of justice.[15] In transgressing the rights of others,
one tends to treat them merely as means to an end, without considering that, as
rational beings, they must always be esteemed as highly as the ends.

Sixth, the Internet is certainly a wonderful addition to our lives. The
Internet is here to stay – and to grow. To enable its growth in positive
directions that may contribute to society and enrich it, we urgently need to
address the abuses of the Net. The problem, I emphasize, is not technology;
technology is merely an instrument. Instead, the problem lies with antisocial
elements that use the Internet to undermine the basic tenets of democracy.

Seventh, in addressing these concerns we need to acknowledge the
democratic catch, avoid moral panics, and espouse the notions of moral and
social responsibility (as discussed below). People need to be aware of the
consequences of their actions. We are expected to think and reflect on possible
outcomes of choices we make, life projects we pursue, and conceptions of the
good that we bring into practice. Responsibility is not restricted to one specific
segment of society; no one is exempted.

Eighth, the Internet is a complex, multifaceted form of communication,
but like other modes of communication, it is not above the law. Harmful,
unlawful speech in the offline world is equally unlawful in the online world.
There is not one law for all forms of expression and another for the Internet.
Morally speaking, electronic communication is no different from other forms
of communication. People should be held accountable for what they publish

---

[14] Immanuel Kant, *Groundwork for the Metaphysic of Morals*, trans. Jonathan Bennett
([1865] 2008), 35, http://www.redfuzzyjesus.com/files/kant-groundwork-for-the-metaphysics-of-
morals.pdf . For discussion of the categorical imperative, see Wolff, *Kant: A Collection of
Critical Essays*, 211–336; Paul Guyer, *Kant's System of Nature and Freedom* (Oxford:
Clarendon Press, 2005), 146–68.

[15] Kant, *Foundations of the Metaphysics of Morals*, 54–55. See also Paul Dietrichson, "What
Does Kant Mean by 'Acting from Duty'?," in Wolff, *Kant: A Collection of Critical Essays*,
314–30.

on the Net, even when the content of their expression is *prima facie* lawful. The concern, of course, is with dangerous speech, the consequences of which could be harmful to society.

Ninth, the open architecture of the Internet is such that we *all* shape the Net by our actions. We therefore all have a vested interest in ensuring that it facilitates the positive elements of society. The future of world communications, with vast social and political implications, will depend on people who use the Internet in a responsible fashion, on businesses that see beyond profit, on governments that take active steps to achieve safety on the free information highway, and on responsible international cooperation to fight international manifestations of antisocial activities on the Net.

Lastly, we should seek the essence of democratic legitimacy by fostering the ability of all citizens to collectively engage in authentic deliberation about their conduct and the future of the Internet. Public deliberation enhances understanding of complicated issues, facilitates learning, and creates a vital and inclusive pluralistic democracy in which citizens believe that they can make a difference by shaping and reshaping the decision-making processes. The only meaningful democracy is participatory democracy, and in addressing developing technologies that affect our lives, deliberative democracy may serve as a guiding model. Deliberative democracy evokes ideals of rational legislation, of participatory politics, and of civic self-governance and autonomy. It presents an ideal of political autonomy that is based on practical reasoning and expressed in an open and accountable discourse, leading to an agreed judgment on substantive policy issues concerning the common good. Deliberative discourse is uncoerced, pluralistic, inclusive, reasoned, and equal debate that shapes the common interests of participating citizens. Habermas notes that the success of deliberative democracy depends on the institutionalization of the corresponding procedures and conditions of communication and on the interplay of deliberative processes and informed public opinions.[16]

---

[16] Jürgen Habermas, *Between Facts and Norms: Contributions to a Discourse Theory of Law and Democracy* (Cambridge: Polity, 1996), 298. See also Jürgen Habermas, *Moral Consciousness and Communicative Action* (Cambridge, MA: MIT Press, 1990); David Miller, "Deliberative Democracy and Social Choice," *Political Studies* 40, suppl. 1 (1992): 54–67; Joseph Bessette, *The Mild Voice of Reason: Deliberative Democracy and American National Government* (Chicago: University of Chicago Press, 1994); Carlos Santiago Nino, *The Constitution of Deliberative Democracy* (New Haven, CT: Yale University Press, 1996); Jon Elster, ed., *Deliberative Democracy* (Cambridge: Cambridge University Press, 1998); Amy Gutmann and Dennis F. Thompson, *Why Deliberative Democracy?* (Princeton, NJ: Princeton University Press, 2004); Stephen Macedo, *Deliberative Politics: Essays on Democracy and Disagreement* (New York: Oxford University Press, 1999); James S. Fishkin and Peter Laslett, *Debating Deliberative Democracy* (Oxford: Wiley-Blackwell, 2003); John S. Dryzek, *Deliberative*

Let me now explain the meaning of the democratic catch and the concept of moral panics.

THE DEMOCRATIC CATCH

Democracy in its modern, liberal form is a young phenomenon. It was crystallized only after the First World War. Viscount James Bryce wrote the following in 1924: "Seventy years ago . . . the approaching rise of the masses to power was regarded by the educated classes of Europe as a menace to order and prosperity. Then the word Democracy awakened dislike or fear. Now it is a word of praise."[7] The idea that governments would be elected through popular vote alarmed and frightened the 19th-century decision makers. Now we are so accustomed to the idea of democracy that we tend to forget how young and fragile it is.

People living in a democratic society hold a prevailing perception that all forms of government but democracy contain the seeds of the society's self-destruction. The principles of authoritarian, totalitarian, theocratic, and other forms of government that are coercive in nature deny human autonomy and wish to dictate to people how they should lead their lives. Because people wish to live as free human beings, they will rebel against oppression when the right opportunity presents itself. You can do many things with bayonets, but you cannot sit on them for long. Basing a regime on violence undermines that very regime. The bleeding that results from sitting on bayonets will eventually cripple the system.

The situation, however, seems to be different in a democracy. In a democracy, so the argument goes, people are free to pursue their conceptions of the good as long as they do not harm others. People, as moral agents, have their conceptions of a moral life and determine accordingly what they deem to be the most valuable or best form of life worth leading. A conception of the good involves a mixture of moral, philosophical, ideological, and religious notions, together with personal values that contain some picture of a worthy life. One's conception of the good does not have to be compatible with moral excellence. It does not mean a conception of justice. The assumption is that a conception of the good constitutes a basic part of our overall moral scheme and that it is

Democracy and Beyond: Liberals, Critics, Contestations (Oxford: Oxford University Press, 2002); John S. Dryzek, *Foundations and Frontiers of Deliberative Governance* (Oxford: Oxford University Press, 2012); John Parkinson and Jane Mansbridge, eds., *Deliberative Systems: Deliberative Democracy at the Large Scale* (Cambridge: Cambridge University Press, 2012); Maurizio Passerin d'Entrèves, *Democracy as Public Deliberation: New Perspectives* (Piscataway, NJ: Transaction, 2006).

[17] James Bryce, *Modern Democracies*, vol. 1 (New York: Macmillan, 1924), 4.

public insofar as it is something we advance as good for others as well as for ourselves. Consequently, we want others to hold this conception of the good for their sake. We refrain from coercing others, though. When our desire for other people to share our view of the good is based on coercion, it cannot be said to be moral, because they are no longer autonomous to decide on their way of life.[18] They are then forced to follow a scheme that they may not consider to be a conception of the good life.

Thus, it seems that the avenues opened for individuals to pursue their conception of the good make democracy different from other forms of government. Its principles of liberty, tolerance, pluralism, participation, and representation constitute a solid regime. I contest this argument, asserting that democracy is no different from other forms of government in having self-government capabilities that contain the seeds of its destruction. The very principles of democracy might undermine it. Limitless liberty might lead to anarchy. Tolerating the intolerant might lead to coercion and violence. Respecting *all* conceptions of the good might allow the harming of the more vulnerable people in society, such as women and children. Excessive participation might lead to "flooding" of the system and to the inability of government to function. And no democracy aims to secure representation for each and every idea in society.

Moreover, because democracy is a relatively young phenomenon, it lacks experience in dealing with pitfalls involved in the working of the system. This is what I call the "catch" of democracy.[19] The freedoms we enjoy are respected as long as they do not imperil the basic values that underlie democracy. Freedom of speech, for instance, is a fundamental right, an important anchor of democracy, but it should not be used without boundaries. Although we dread censorship, there is a room to consider time and space regulations. For instance, during a lecture, the audience should be silent. When the time comes for the audience to ask questions, the speaker should listen. If a human rights organization arranges a demonstration in the city square, another organization may hold its own gathering in the city square on another day. Under no circumstances, however, will people be allowed to hold a picket on the main highway leading to the city. Liberty and tolerance are not prescriptions for lawlessness and violent anarchy.

[18] Raphael Cohen-Almagor, "On Compromise and Coercion," *Ratio Juris* 19, no. 4 (2006): 434–55.
[19] Cohen-Almagor, *Boundaries of Liberty and Tolerance*. See also Raphael Cohen-Almagor, *Speech, Media, and Ethics: The Limits of Free Expression* (Houndmills, UK: Palgrave, 2005); Raphael Cohen-Almagor, *The Scope of Tolerance: Studies on the Costs of Free Expression and Freedom of the Press* (London: Routledge, 2006); Raphael Cohen-Almagor, *The Democratic Catch: Free Speech and Its Limits* [in Hebrew] (Tel Aviv: Maariv, 2007).

Like every young phenomenon, democracy needs to develop gradually, with great caution and care. Because democracies lack experience, they are uncertain how to fight against explicit antidemocratic and illiberal practices. Abundant literature exists about the advantages of democracy, the value of liberty, and the virtue of tolerance.[20] Liberal thinkers want to promote liberty and tolerance; urge governments not to apply partisan considerations that affirm principally their own interests and conceptions; and seek ways to accommodate different conceptions of the good, to reach compromises by which democracy will respect variety and pluralism. In comparison to these concepts, much less has been written in the field of political theory about the intricate issue of the appropriate scope of tolerance.

Even less is written about the limits of free expression on the Internet, an even younger phenomenon than democracy. We are in the early learning stages of understanding the effect of the Internet on our moral lives. Liberals are willing to restrict freedom of expression when a clear and direct link is established between harmful speech and harmful action. Pursuing this reasoning on the Internet is difficult because sites often are open to all. And even if certain content is deemed to be very harmful, when should we restrict it? When it leads to, say, one murder, although thousands of people watched the same content? Ten murders? One hundred murders?

The film *Natural Born Killers* is said to have triggered a few killings. Some of its viewers were "inspired" to go on killing sprees,[21] yet liberals do not support censorship. The film was watched by millions of people all over the world, and most of them did not go out to kill. Therefore, we cannot hold a film responsible for violence. We hold the killers responsible for their

[20] See, for instance, Susan Mendus, ed., *Justifying Toleration: Conceptual and Historical Perspectives* (Cambridge: Cambridge University Press, 2009); David Heyd, ed., *Toleration: An Elusive Virtue* (Princeton, NJ: Princeton University Press, 1996); John Horton and Susan Mendus, eds., *Toleration, Identity, and Difference* (Houndmills, UK: Macmillan, 1999); Susan Mendus, ed., *The Politics of Toleration: Tolerance and Intolerance in Modern Life* (Edinburgh, UK: Edinburgh University Press, 1999); T. M. Scanlon, *The Difficulty of Tolerance: Essays in Political Philosophy* (Cambridge: Cambridge University Press, 2003); Catriona McKinnon and Dario Castiglione, eds., *The Culture of Toleration in Diverse Societies: Reasonable Tolerance* (New York: Manchester University Press, 2003); Andrew G. Fiala, *Tolerance and the Ethical Life* (New York: Continuum, 2005); David Hackett Fischer, *Liberty and Freedom: A Visual History of America's Founding Ideas* (Oxford: Oxford University Press, 2005); Bican Sahin, *Toleration: The Liberal Virtue* (Lanham, MD: Lexington Books, 2010); Preston King, *Value of Tolerance* (London: Routledge, forthcoming).

[21] Karina Wilson, "*Natural Born Killers* and the Media Violence Debate," Mediaknowall, http://www.mediaknowall.com/as_alevel/mediaviolence/violence.php?pageID=nbk ; "Natural Born Killers," http://www.manics.nl/site/movies/naturalbornkillers.htm ; Katherine Ramsland, "Movies Made Me Murder," http://www.crimelibrary.com/criminal_mind/psychology/movies_made_me_kill/5.html .

wrongdoing. The same argument can be made even more forcefully with regard to the Internet. Websites may have many millions of hits. They may contain antisocial content, but most readers abide by the law. We should punish those who inflict harm on others, not censor the Net.

Hatemongers who exploit the Internet to advance bigotry present a particularly hard case. In the United States, the American Civil Liberties Union (ACLU), among other organizations, has supported the rights of racist and antisemitic organizations – most notoriously, the Ku Klux Klan (KKK) and the American National Socialist Party – to organize, to speak, to demonstrate, and to march.[22] In their defense of radical political groups, the ACLU and others have not claimed that the words, pictures, and symbols of such groups have no negative consequences.[23] The constitutional protection accorded to freedom of speech is not based on a naive belief that speech can do no harm but on the confidence that the benefits society reaps from the free flow and exchange of ideas outweigh the costs society endures by allowing reprehensible and even dangerous ideas. Free speech activists acknowledge that the racist and antisemitic images and discourse of these groups might offend and harm the targeted individuals, might corrupt the level and nature of civic discourse, and might increase the probability of hate crimes. Yet the admission of speech's causal propensities and harmful consequences has not lessened the strength of the free speech principle. The Nazis and the KKK have free speech rights not because what they say is harmless but despite their harmful expressions.[24] American society has been willing to pay a substantial price for allowing

---

[22] See *Collin v. Smith*, 578 F.2d 1197 (7th Cir.), cert. denied 439 U.S. 915 (1978); *Village of Skokie v. National Socialist Party of America*, 69 Ill. 2d 605, 373 N.E.2d 21 (1978). For further deliberation, see Aryeh Neier, *Defending My Enemy: American Nazis, the Skokie Case, and the Risks of Freedom* (New York: Dutton, 1979).

[23] On the free speech rights of the KKK and other radical organizations, see *Brandenburg v. Ohio*, 395 U.S. 444 (1969); *Forsyth County v. Nationalist Movement*, 505 U.S. 123 (1992); *R.A.V. v. City of St. Paul*, 505 U.S. 377 (1992). See also *United States v. Hayward and Krause* 6 F. 3d 1241 (7th Cir. 1993) and *United States v. Juveniles J.H.H., L.M.J, and R.A.V.* 22 F. 3d 821 (8th Cir. 1994), in which the court held that some forms of expression that are used to intimidate (e.g., cross burning) are harmful and damaging to others and, as such, do not enjoy the protecting cover of speech in the constitutional sense. However, see *Virginia v. Black*, 123 S. Ct. 1536 (2003), which declared the Virginia cross-burning statute unconstitutional because it discriminated on the basis of content and viewpoint. For a critique of this highly controversial opinion, see Steven G. Gey, "A Few Questions about Cross Burning, Intimidation, and Free Speech," Public Law Research Paper 106, Florida State University College of Law, Tallahassee, February 2004. For a broad discussion, see Ivan Hare and James Weinstein, eds., *Extreme Speech and Democracy* (Oxford: Oxford University Press, 2009).

[24] Frederick Schauer, "The Cost of Communicative Tolerance," in *Liberal Democracy and the Limits of Tolerance*, ed. Raphael Cohen-Almagor (Ann Arbor: University of Michigan Press, 2000), 28–42, 29. For further discussion, see Stephen L. Newman, "American and Canadian Perspectives on Hate Speech and the Limits of Free Expression," in *Constitutional Politics in*

hatemongers to spread their racist ideology on the streets as well as on the Internet. American liberals believe it is better to have democracy that allows hate speech than a society that does not. Rather than panicking or being afraid of such vile ideas, we need to expose the falsity of hatred and to educate people to embrace tolerance and equal liberties for all.

This liberal viewpoint is prevalent in America; it has been criticized elsewhere. Not all countries are able and willing to pay the price that the United States is paying for its stance on freedom of expression. Not all liberal democracies are as strong and viable as the United States and can endure the consequences of wide freedom of expression. The American stance gives precedence to the rights of the *speaker*, while its critics argue that the rights of the speaker are surely important but need to be weighed and balanced against the rights of others – the audience, the target group, the bystanders, and the society at large. Indeed, this is my view. I argue that democracies have a responsibility to counter hate propaganda. Responsibility should evoke concern about the potential contagion of hate, the psychological harm likely to be suffered by any minority group targeted by hatemongers, and the threat to society's stability and tranquility should hate propaganda succeed in spreading its malice within the majority. Countering hate is important for its symbolic value – for the message it conveys, especially to targeted minorities, that they are not neglected and abandoned if so attacked and that their dignity and equal standing in society are important.[25]

Culture and norms are also important. In an environment that includes groups of people who have very little respect for one another and very little tolerance, a message of hate might ignite and lead to violence. In such an atmosphere, upholding education as a cure for all such evil might be unrealistic. Because education is a long-term process and can rarely resolve pressing concerns, a case can be made for proactive action here and now to address urgent tensions that could erupt into violence.

Participants in a liberal democracy ought to weigh the costs of allowing hate speech, as well as the risks involved, and balance these against the costs and

---

Canada and the United States, ed. Stephen L. Newman (Albany: State University of New York Press, 2004), 153–73.

[25] Jeremy Waldron forcefully makes the point that hate speech undermines the dignity of the person. He argues that a person's dignity is not just some Kantian aura. It is the person's social standing, the fundamentals of basic reputation that entitle people to be treated as equals in the ordinary operations of society. Hate speech aims "to besmirch the basics of their reputation, by associating ascriptive characteristics like ethnicity, or race, or religion with conduct or attributes that should disqualify someone from being treated as a member of society in good standing." See Jeremy Waldron, *The Harm in Hate Speech* (Cambridge, MA: Harvard University Press, 2012), 5; see also 165–72.

risks to demòcracy and free speech associated with censorship. Supporters of free expression may insist on a direct link being proven between the harmful expression and the resulting harmful action: the government must establish a nexus of harm linking the proscribed utterance to some grave and imminent threat of tangible injury. This approach would require that the government perform a contextual analysis drawing on empirical data. Who was harmed? How were they harmed? It is similar to what we demand of the plaintiff in a libel case. And if the argument also references society's right of self-defense, then we should seek evidence of a real threat to individuals or social stability. I will explain and elaborate on these ideas in chapters 6 and 7. I share the concern of American liberals about exaggerated, ill-focused reaction. Indeed, evoking moral panics and resorting to excessive limitations at the expense of freedom of speech are unnecessary measures. Responsible response to hate should be measured and balanced.

MORAL PANICS

Moral panics is a sociological term coined by Stanley Cohen.[26] It refers to the reaction of a group of people based on a false or exaggerated perception that a cultural phenomenon, behavior, or group (mostly minority groups or subcultures) is dangerously deviant and poses a threat to society. In other words, moral panics do not refer to situations in which nothing gives rise to fear but to situations that evoke a heightened level of disproportionate concern over a certain conduct or phenomenon. The scope and scale of the threat have been exaggerated and have come to be perceived as a menace to consensual societal values, morals, or interests.[27] The challenge evokes calls to strengthen the social control apparatus of society by tougher rules, intense condemnation, more legislation, and more law enforcement aimed to restore stability and peace (both societal peace and peace of mind).

An important factor in moral panics is the *deviancy amplification spiral* – an increasing cycle of media reports on undesirable events or behaviors. These reports induce moral panics in society and can lead to legislation designed to further penalize those established as the threatening deviants who are the

---

[26] Stanley Cohen, *Folk Devils and Moral Panics* (London: Routledge, 1987); Stanley Cohen and Jock Young, *The Manufacture of News: Deviance, Social Problems, and the Mass Media* (Thousand Oaks, CA: Sage, 1981).

[27] Stuart Hall, Chas Critcher, Tony Jefferson, John Clarke, and Brian Roberts, *Policing the Crisis: Mugging, the State, and Law and Order* (London: Macmillan, 1978); Sean P. Hier, ed., *Moral Panic and the Politics of Anxiety* (London: Routledge, 2011); Erich Goode and Nachman Ben-Yehuda, *Moral Panics: The Social Construction of Deviance* (Chichester, UK: Wiley-Blackwell, 2009), 22–30.

source of the panic.[28] In other words, when the reaction to a person, a group, or a certain phenomenon is out of proportion to the actual threat, when the perceived threat is magnified beyond realistic appraisal, then we may be facing a moral panic.

Moral panics are sociological phenomena and social constructs. The combined effect of the media coverage of a phenomenon, public opinion, and the reaction of the authorities can have the spiral-like effect of creating moral panics. For moral panics to evolve, a belief must exist that something revered as sacred by or fundamental to society is being threatened.[29]

In the Internet age, mass media images of e-criminality and e-terror provoke anxiety, insecurity, and fear. Media representation of Internet threats focuses mainly on terrorism and child pornography. Other popular concerns are cybercrime and cyberbullying. All are certainly valid and important concerns, but to fully appreciate these phenomena and the scope of the threats they present, we must put them in perspective. We need to understand the sociology and politics of fear, who the agents are, and what their motivations and interests are for evoking fear. Being realistic is certainly warranted; overblown exaggeration, however, is counterproductive and undermines the construction of adequate policies designed to address the challenges.

I should say at the outset that I do not see the Internet as "a threat to societal values and interests."[30] The Internet is a useful platform that has changed daily life forever and is here to stay, but we must devise ways to deal with its less positive aspects. To fault or fear technology as the source of problems is critically shortsighted. People tend to blame new technology and innovation for social ills. They did so when the radio was invented, when comics became prevalent, when the electronic box that now may occupy a cinematic wall but is still called television was introduced, when rock 'n' roll became popular, and when video games became a part of life. It is always easier to blame entities than to blame people.

The challenge, however, comes from individuals who abuse technology to advance criminal agendas. My intention is to evoke social and moral

---

[28] Cohen, *Folk Devils and Moral Panics*; Chas Critcher, *Moral Panics and the Media: Issues in Cultural and Media Studies* (Buckingham, UK: Open University Press, 2003); Charles Krinsky, ed., *Moral Panics over Contemporary Children and Youth* (Farnham, UK: Ashgate, 2008); Karina Wilson, "Moral Panics and Violence in the Media," Mediaknowall, http://www.mediaknowall. com/as_alevel/mediaviolence/violence.php?pageID=moral .

[29] Suzanne Ost, *Child Pornography and Sexual Grooming: Legal and Societal Responses* (Cambridge: Cambridge University Press, 2009), 15–16.

[30] Cohen, *Folk Devils and Moral Panics*, 9. For further discussion, see Sonia Livingstone and Magdalena Bober, "U.K. Children Go Online: Final Report of Key Project Findings," London School of Economics and Political Science, London, 2005, 7, http://eprints.lse.ac.uk/archive/ 00000399 .

responsibility on the part of people who upload material on the Internet, readers of Internet publications, Internet service providers, governments, law enforcement agencies, and the international community at large. All need to address the urgent need to devise ways to counter the challenges that Internet abusers are posing to free societies. My plea is for self-regulation, for respect of the law of the state, and for international cooperation by all segments of society. My plea is based not on existing legal obligations but rather on the moral and social obligations that cross borders and cultures regarding the two foundations of liberal democracies: respect for others and not harming others. These obligations are of urgency when human lives are at stake.

The Internet business sector (Internet service providers, website administrators, and website owners) bears heavy responsibility because the moral obligations imposed on it may, in due course, become a legal obligation. This was the case regarding child pornography and cybercrime.[31] Of concern are potentially problematic websites – that is, websites that attract criminals to post their criminal ideas and criminal intentions. Monitoring such sites and fighting their antisocial content will increase trust on the Internet and promote the notion of responsibility on the free highway.

TRUST

There cannot be meaningful relationships between people without trust. Trust is an essential component of friendship, of marriage, or of a working relationship. The Oxford Dictionary defines trust as "confidence, strong belief in the goodness, strength, reliability of something or somebody."[32] Trust is a social phenomenon that requires acquaintance and some knowledge on the part of the person who trusts and the person or thing that is trusted. It is about transparency and expectations of the trusted partner. The more transparent the relationships are, the more expectations people possess regarding the level of

[31] See the Council of Europe's Convention on Cybercrime, which is available at http://conven tions.coe.int/Treaty/en/Treaties/Html/185.htm . See also U. S. Department of Justice, Computer Crimes and Intellectual Property Section, "International Aspects of Computer Crime," US Department of Justice, http://www.irational.org/APD/CCIPS/intl.html and the CCIPS website at http://www.justice.gov/criminal/cybercrime/ . I will elaborate on this issue in chapters 6 and 7.

[32] Albert Sydney Hornby, Jack Windsor Lewis, and Anthony Paul Cowie, eds., *Oxford Advanced Learner's Dictionary of Current English*, 3rd edition, 4th impression, s.v. "trust" (London: Oxford University Press, 1975). For further discussion, see Francis Fukuyama, *Trust: The Social Virtues of and the Creation of Prosperity* (New York: Free Press, 1995); Ian Richards, "The Dilemma of Trust," in *The Handbook of Global Communication and Media Ethics*, ed. Robert S. Fortner and P. Mark Fackler (Oxford: Wiley-Blackwell, 2011), 247–62.

trust they can have in the relationship. When transparency is dubious, trust is difficult to gain and maintain.

To trust one another, people need to understand one another. Here, language is crucial. People are social beings. We communicate with one another, converse, and exchange ideas and different points of view through language. Language constructs, affects, and changes reality; it facilitates communication, promotes understanding, and helps to erect bridges between people and cultures.

Every profession has its concepts, phrases, and keywords that are important to help categorize phenomena, save time, and provide a framework for working together. The Internet is no exception. Here, too, we find some important concepts and terms that call for probing and analysis. In response to the changing reality – a reality that is very much influenced by advances in technology – people have adopted concepts and developed a new set of terminology to handle the new technological challenges. I have highlighted some of the main concepts and terms in chapters 1 and 2. It is crucial to have a common understanding of the technology at hand and how it came about if we are to promote trust and responsibility on the Internet. We need to understand the meaning and significance of Net anonymity, social networking, and e-friendship and how and why information is disseminated.

Close friends trust one another. They share personal stories, difficulties, and secrets. The more they prove to each other that they are trustworthy, the more intimacy they will feel with each other, be ready to reveal themselves, and be committed to their friendship. Building trust requires time and an investment in each other. It is a long process involving an exchange of gestures – a mutual give-and-take that slowly builds a viable relationship that can withstand challenges and obstacles. Gradually people construct an evidence record of their relationship over a period of time. The experience builds a certain expectation level between people that is analogous to trust. As long as friends sustain and nourish their friendship and as long as they resolve misunderstandings and disagreements with mutual respect, the trust will be maintained.

In business, time can be a luxury. Partners to some deals may require closing a transaction in a short period of time without really coming to know one another. To gain trust, they sign a contract with clear obligations and sanctions in case one of the partners breaches the agreement. The relationship is built and secured by clear and transparent guarantees and safeguards that make the contract meaningful and sustainable to all parties.

Although trust may require a meaningful relationship, it need not require goodwill or warm sentiments. I may trust the bank to keep my money safe, but I do not imagine for a moment that the bank is warmly disposed toward me. We clearly have shared interests, and our relationship will continue only as long as we share the interests. If I accumulate debt, the bank will quickly penalize me for my irresponsible spending. If I decide to transfer my money to another bank, my relationship with my former bank will normally end. Similarly, when I seek a physician's advice, I may trust the physician's professional expertise and integrity even if the cold, detached professional seems to be indifferent to me as a person.[33] In both cases, the trust is limited in scope, relevant only to the professional relationship we develop.

Considerable uncertainty abounds about how trust in the offline world transfers to cyberspace and about the trustworthiness of the components of the cyberspace system. On the one hand, trust is closely connected with a greater level of certainty or confidence in the reliability and security of the Internet. Thus, trust will likely be enhanced as a person learns more about the technology. As technology becomes familiar, it becomes less intimidating. People who know more about the Internet may become more trusting of the medium. On the other hand, information can create uncertainty and contribute to an increased perception of risk.[34] Similarly, knowledge of medicine or of law does not necessarily increase our confidence and trust in these professions.

As explained in chapter 1, the culture of the ARPANET (precursor to the Internet) community was one of open research, free exchange of ideas, no overbearing control structure, and mutual trust. The founders of the Internet were a close-knit group. They trusted each other and viewed their network as a trustworthy environment. They therefore considered access controls unnecessary and extraneous. This is not to say that the network lacked accountability. The ARPANET relied on the informal social norms characteristic of close-knit communities.[35] But as the network experienced geometric growth in the decades that followed and as security problems emerged, the picture significantly changed, undermining the initial assumption of trust.

The basic features of the Internet may come into contradiction. On the one hand, the Internet is a highly convenient medium. It is accessible to all

---

[33] Marek Kohn, *Trust: Self-Interest and the Common Good* (New York: Oxford University Press, 2008), 12. See also Matthew Harding, "Manifesting Trust," *Oxford Journal of Legal Studies* 29, no. 2 (2009): 245–65; Matthew Harding, "Responding to Trust," *Ratio Juris* 24, no. 1 (2011): 75–87.

[34] William H. Dutton and Adrian Shepherd, "Confidence and Risk on the Internet," in *Trust and Crime in Information Societies*, ed. Robin Mansell and Brian S. Collins (Cheltenham, UK: Edward Elgar, 2005), 207–44, 211. See also William H. Dutton and Adrian Shepherd, "Trust in the Internet as an Experience Technology," *Information, Communication, and Society* 9, no. 4 (2006): 433–51.

[35] Leonard Kleinrock, "History of the Internet and Its Flexible Future," *IEEE Wireless Communications* 15, no. 1 (2008): 8–18, 12.

Netusers; its ever-developing technology enables storage of a vast amount of information; the advanced research crawlers facilitate rapid location of information. No wonder many people use the Internet as their first-resort information resource.

On the other hand, easily accessible information does not mean reliable information. Because nearly anyone can put information onto the Internet, the appropriateness, utility, and authenticity of information are generally uncertified and hence unverified. With important exceptions (generally associated with institutions that care about their reputation), the Internet is a "buyer beware" information marketplace, and the unwary Netuser can be misinformed, tricked, deceived, seduced, or led astray.[36] We all have certain agendas, and we do not always reveal them or admit to them. Being aware of such hidden agendas is difficult when we know very little about the authors of information and even more so when the authors come from a different culture, do not communicate in their native language, hide their identity, or lead us to believe that they are someone else.

Helen Nissenbaum notes that specific features of the online world bear on the formation and sustenance of trust, which conceal many of the aspects of character and personality, nature of relationship, and setting that normally function as triggers of trust or as reasons for deciding to trust. First, the medium allows agents to cloak or obscure their identity. Anonymity reduces the number of mutual cues on which trust may develop. It may further obscure the nature of mutual relations and suggest a diminished sense of responsibility to one another. Second, online we are separated from others in time and space. Personal characteristics such as gender, age, race, socioeconomic status, or occupation can be hidden from us. Thus we lack cues that may give evidence of similarity, familiarity, or shared value systems – cues that mediate in the construction of trust. A 67-year-old man may present himself as a 14-year-old girl. Anyone can assume the looks and characteristics of another in an effort to try to gain the attention of a person he or she wishes to attract. Third, the settings of an online environment frequently are inscrutable in ways that affect our readiness or inclination to trust. Clear social and professional role definitions often are lacking, and duties and responsibilities are less well defined and understood. As a result, we cannot rely on traditional mechanisms for cultivating trust.[37] In addition, because we seldom see other Netusers in person, we miss bodily clues that often help us to consolidate opinions about

---

[36] Dick Thornburgh and Herbert S. Lin, eds., *Youth, Pornography, and the Internet* (Washington, DC: National Academies Press, 2002), 35.

[37] Helen Nissenbaum, "Securing Trust Online: Wisdom or Oxymoron," *Boston University Law Review* 81, no. 3 (2001): 635–64, http://www.nyu.edu/projects/nissenbaum/papers/securingtrust.pdf.

others: facial expressions, body movement, gestures, and our general impression of people. Tools such as Skype may help eliminate this difficulty.

These characteristics are obvious obstacles to developing personal relationships online. As I have already stated, to secure some level of trust takes time and human investment, and this is the optimistic view. Pessimists such as Paul De Laat argue that establishing online trustworthiness is next to impossible and that any gesture of trust tends to be seen as an "opportunistic and ludicrous move which will only solicit ridicule instead of esteem."[38] The Internet is an environment in which shape and form are hidden; thus, knowing who is who often is difficult. As the famous cartoon says, "On the Internet nobody knows you're a dog."[39] Neil Barrett poses rhetorically, "Ask yourself quite how much you trust the Internet, a system built on invisible, unknown components, performing unknown functions in an unknown manner ... still happy to shop, bank and flirt online?"[40] The fact that sites fluctuate and that you cannot be sure that they will be there the next time you want to access them does not increase trust in the medium.

One study suggests that the social connections that people make on the Internet do not promote trust – indeed, some evidence indicates that chat rooms may bring together mistrusting people.[41] Evidence also shows, however, that close and meaningful relationships do form online, and many of those relationships involve some degree of trust.[42] Tracking past conduct through online reputation systems increases trust.[43] The more people are familiar with certain websites and Netusers, the more trustful people become of those websites and Netusers.[44] Familiarity is crucial. Indeed, it is little surprise

---

[38] Paul B. De Laat, "Trusting Virtual Trust," *Ethics and Information Technology* 7, no. 3 (2005): 167–80, 170. See also Philip Pettit, "Trust, Reliance, and the Internet," *Analyse and Kritik* 26, no. 1 (2004): 108–21.

[39] See http://knowyourmeme.com/memes/on-the-internet-nobody-knows-youre-a-dog . The problem of hidden identity has spawned a number of identity verification services. These services provide a verification-chain framework to both parties, while protecting sensitive information.

[40] Neil Barrett, "Criminal IT: Should You Trust the Internet?," ZDNet, January 27, 2005, http://www.zdnet.com/news/criminal-it-should-you-trust-the-internet/140965 . See also Information Resources Management Association, *Cyber Crime: Concepts, Methodologies, Tools and Applications* (Hershey, PA: IGI Global, 2012).

[41] Eric M. Uslaner, "Trust, Civic Engagement, and the Internet," *Political Communication* 21, no. 2 (2004): 223–42.

[42] John Weckert, "Trust in Cyberspace," in *The Impact of the Internet on Our Moral Lives*, ed. Robert J. Cavalier (Albany: State University of New York Press, 2005), 95–120, 112.

[43] Thomas W. Simpson, "e-Trust and Reputation," *Ethics and Information Technology* 13, no. 1 (2011): 29–38, 29.

[44] "eCommerce Trust Study," Cheskin Research and Studio Archetype/Sapient, San Francisco, January 1999, http://experiencecurve.com/articles/ecommerceTrustCheskin.pdf ; Gregory Ciotti, "10 Proven Ways to Make Your Tweets More Trustworthy," *Buffer* (blog), May 2, 2012, http://blog.bufferapp.com/tweets-more-trustworthy ; Christian Holst, "16 Ways to Make Your Website Seem

that people are more likely to trust brand names with which they are familiar and consider reputable than to take a risk with unknown brand names. Similarly, people are unlikely to trust an unknown, obscure website that they have never visited.

When people want to obtain information about any phenomenon in the universe, they often go first to the Web. The Web's easy accessibility makes it an attractive choice. In their approach to the Internet, people exhibit thick trust, thin trust, or no trust depending on their knowledge, common sense, and experience. For instance, I exhibit *thick trust* when I use *The Stanford Encyclopedia of Philosophy* as a major resource for my studies, and I refer my students to the Stanford site.[45] This trust was initially based on the reputation of Stanford University and was substantiated by my reading of the material in the encyclopedia.[46] In turn, *thin trust* is a qualified, circumscribed, cautious trust. People usually exhibit this kind of trust in financial affairs, typically when dealing with strangers. Companies and services try to convince us that we should do business with them, or through them, by fostering a sense of trust. They do so by building a good reputation, upholding certain norms, and providing signals that may gain our assurance. Reputation systems enable Netusers to learn and benefit from the experience of others. Such systems promote trust between people who find themselves in similar situations by providing cues about people's honesty. For instance, eBay developed a system of rated sellers with compliments that are designed to develop trust. Sellers who consistently receive the highest buyers' ratings and who ship items quickly have earned a track record of excellent service.[47] The feedback forum allows Netusers to police themselves and has been a key factor in eBay's success.[48] Such reputation systems should be done in a bona fide manner and be as transparent as possible; otherwise they will backfire, the reputation of the company will be undermined, and customers' trust in the company will fade. Thus, if a company establishes its reputation with a clear bias toward positive feedback, it may improve its sales in the short run, but its reputation will be depleted

More Trustworthy," Baymard Institute, Copenhagen, November 1, 2011, http://baymard.com/blog/ways-to-instill-trust .

[45] The encyclopedia can be accessed at http://plato.stanford.edu .

[46] I cannot say the same about Wikipedia, especially on reading at one point my own entry, which said that I am "an Islamic extremist educator, researcher, and human rights activist." Wikipedia needs to devise rigorous ways to stop abusers from inserting false, misleading, and even malicious information into its entries.

[47] You can access the website at http://www.ebay.com .

[48] "Queen of the Online Flea Market," *Economist*, January 3, 2004, http://www.economist.com/node/2312964 .

in the long run.[49] To paraphrase a popular saying, you can fool some of the people some of the time, but you cannot fool all of the people all of the time.

Search engines come to our assistance in establishing trust. They may influence Netusers to increase or decrease their level of trust in websites and blogs. Because anyone may upload information on the Net, we need to establish the credibility of the source before we rely on it. The more clues authors reveal about their identities, the easier it is for us to conduct a Google search about them and make up our minds about whether a site is a credible resource. For instance, quite a few people are named Robin Wright. If the author signs an article about revolutions in the Middle East without providing any clues about her (or possibly his) identity, we may exhibit thin trust in the piece. Suppose we know nothing about the subject matter, and we are just starting to research the topic for a seminar paper. We lack the ability to appreciate the content of the piece. For all we know, it might be the famous actress Robin Wright who wrote the piece, and, with all due respect to Ms. Wright's acting abilities, we may not regard her as an authority on the Middle East. However, if the author identifies herself as a journalist, author, and foreign policy analyst, a quick Google search reveals her notable credentials and provides us with the needed reassurance that we may rely on the information in her article and use it for our research.

One of my students used material posted on the Institute for Historical Review (IHR) website[50] in a seminar paper he wrote about Sir Oswald Mosley, the notorious British fascist during World War II. The student, who exhibited thick trust in the IHR publications, was unaware of the clear bias and agenda of that revisionist organization that was set up to deny the Holocaust and to vindicate Adolf Hitler and his Nazi regime. The student should have verified the credentials of this pseudoacademic organization before relying on its predisposed publications for academic research concerning a British fascist. Thus, the guise of an "institute" is, in itself, insufficient to warrant trust. Researchers must always do their homework to verify credibility. Relying on IHR publications for the study of the Holocaust, Nazism, fascism, or the Israeli–Palestinian conflict is a mockery of good academic practice; however, IHR publications are a good resource for studying Holocaust denial and anti-Zionism.

[49] In what its marketing director called "a poorly written e-mail" by "an overzealous employee," Elsevier apparently offered a $25 voucher to academics who rated their publications with five stars. The publisher withdrew the offer when it became public knowledge. See Thomas W. Simpson, "e-Trust and Reputation," *Ethics and Information Technology* 13, no. 1 (2011): 29–38. For further discussion, see Mariarosaria Taddeo and Luciano Floridi, "The Case for E-trust," *Ethics and Information Technology* 13, no. 1 (2011): 1–3.

[50] You can access the IHR website at http://www.ihr.org/ .

Finally, most of us do not trust strangers who notify us out of the blue that we have won, for example, a $2 million prize in a lottery in which we never took part. Fraudulent e-commerce schemes, such as Nigerian letter fraud, are prevalent on the Internet.[51]

Like trust, e-trust requires effort to be sustained. People develop a certain level of expectations and, when their expectations are not met, thick trust might transform into thin trust or evaporate altogether. When expectations are met and pleasant surprises occur along the way, thick trust will not likely transform into thin trust, and thin trust, in turn, may become thick. Co-workers develop different levels of trust because they depend on each other. Over a period of time, the trust level tends to fluctuate as co-workers pleasantly surprise or disappoint one another.

Breaking of trust might lead to breaking of relationships. Some breaches of trust, both in the offline and the online worlds, are unforgivable. A person uses a credit card to buy a phone card on the Internet. A few days later, he or she discovers that personal details were compromised and someone else was using the card to make phone calls to foreign countries. When the person complains, the company absolves itself of responsibility and is unwilling to provide compensation. Trust is hence broken beyond repair. We can assume that the person will cease to do business with the irresponsible company.

One way to increase integrity and trust in e-commerce is by using digital signatures. As in asymmetrical encryption, the technology uses private and public keys. Customers first register by submitting their public key to a certification or registration authority. That entity verifies the subscriber's identity and then issues a digital certificate. The certificate includes the subscriber's public key and identity information and is digitally signed by the entity. When a customer sends a digitally signed message to a merchant, the customer's private key is used to generate the digital signature, and the customer's public key is used by merchants to verify the signature. The digital signature ensures that the customer issued the communication and that content was not tampered with from the time the signature was made.[52] The transaction can be verified in a transparent way via email. The technology thus provides the four key ingredients for trust in e-commerce: confidentiality, authentication, integrity, and nonrepudiation, meaning that the signer

[51] See "Common Fraud Schemes," http://www.fbi.gov/scams-safety/fraud , for information from the Federal Bureau of Investigation (FBI) on various types of fraud, including Nigerian letter fraud, as well as tips on avoiding being a victim of such scams.
[52] For further discussion, see the American Institute of CPAs' website at http://www.aicpa.org/InterestAreas/InformationTechnology/Resources/TrustServices/DownloadableDocuments/tpafile7-8-03fortheweb.doc .

cannot deny having signed the message. To ensure that appropriate proce-dures are followed in e-commerce activities, public key infrastructure, and cryptography, a special program called the WebTrust Program for Certification Authorities was launched. It ensures that specific procedures are implemented.[53]

Given the number of people posting information on the Net, the percentage of us who exhibit thick trust when we obtain information from the Net, perceiving it as fairly reliable, is reassuring. A 2009 poll that asked what people consider the most reliable source of information shows that the Internet comes first with 37 percent, before television (17 percent), newspapers (16 percent), and radio (13 percent).[54] Although much information on the Net is reliable and most users trust it, those who abuse the Net undermine users' confidence. Problems and perceived dangers may be seen as failures either of the technical systems or of the system designers and users. We expect responsible business professionals to take steps to prevent abuse or to reduce the vulnerabilities in the system.[55] However, technical measures alone do not suffice to foster trust in cyberspace. Transparency and trust will not garner significant economic benefits for a corporation unless they are backed by a genuine concern for stakeholder interests.[56] A growing sense of urgency has arisen to foster social responsibility on the Net by increasing awareness of abuse and by developing transparent ways to fight it. This kind of transparency is likely to maintain trust over time.

MORAL RESPONSIBILITY

We need to distinguish between legal, moral, and social responsibility. *Legal responsibility* refers to the addressing of issues by agencies of the state. Through

---

[53] See "WebTrust Program for Certification Authorities," CPA Canada, Toronto, ON, http://www.webtrust.org/homepage-documents/item27839.aspx ; "What is WebTrust for CAs (Certification Authorities)?," SSL Shopper, http://www.sslshopper.com/article-what-is-webtrust-for-cas-certification-authorities.html .

[54] John Zogby, "Why Do People Trust the Internet More?" *Forbes*, June 18, 2009.

[55] Mansell and Collins, *Trust and Crime in Information Societies*, 4. Internet expert Brian Marcus argues that the Internet is based on trust. Most of its users act within the law. Therefore, the United States imposes far more protections on free expression than restrictions on speech. Marcus contends that we should not punish most users because of the small numbers who exploit the Internet to violate the law, and we should not allow a small number of abusers to dictate the rules of the game. Personal communica-tion with Brian Marcus, Washington, DC, June 5, 2008. FBI senior official Philip Mudd has voiced a similar argument. Interview with Philip Mudd (associate executive assistant director, National Security Branch, Federal Bureau of Investigation), Woodrow Wilson Center, Washington, DC, March 25, 2008.

[56] John Elia, "Transparency Rights, Technology, and Trust," *Ethics and Information Technology* 11, no. 2 (2009): 145–53, 150.

its various institutions, the state sees fit to provide and administer certain services, thereby removing these responsibilities from citizens. For example, the state is responsible for securing its borders against external attacks and for providing security for citizens inside its borders. For those purposes, the state ensures the existence of army and police forces, acting in accordance with legally binding decrees that clarify what is allowed in the administration of security. A further example concerns the administration of justice: the state is responsible for establishing courts to settle disputes between individuals and grievances between citizens and agencies of the state.

Moral responsibility is the personal responsibility of the agent to demonstrate conscience, with appeals to moral consideration. We assume that a causal connection exists between the agent and the action or the consequences of the action and that the action was intentional. When people perform a morally significant act, we may believe that they deserve praise. When they fail to perform a morally significant act, we may blame them for their omission. For example, a person who witnesses a child drowning is worthy of praise for saving the child. Alternatively, a person who lacks the ability to save the child and does nothing may still be regarded as worthy of blame for not summoning the lifeguard for help. To regard such agents as worthy of praise or blame is to ascribe moral responsibility to them on the basis of what they have done or failed to do.[57]

Aristotle was the first to construct a theory of moral responsibility. In *Nicomachean Ethics*, Aristotle explored the underpinnings of human virtues and their corresponding vices. He stated that it is sometimes appropriate to respond to an agent with praise or blame on the basis of his or her actions or dispositional traits of character. Of course, people who are being coerced cannot be held responsible for their deeds. People are responsible when they are informed and aware of what they are doing.[58] Only a certain kind of agent qualifies as a moral agent and is thus properly subject to ascriptions of responsibility – namely, one who possesses a capacity for decision. A person of moral character is one who is aware of a given situation and who is responsible for his or her conduct.

According to Aristotle, a decision is a particular kind of desire resulting from free deliberation, one that expresses the agent's conception of what is good. Choice is important – to have desirable ends and relevant means to pursue the ends.[59] Aristotle spelled out the conditions under which it is appropriate to

---

[57] See *Stanford Encyclopedia of Philosophy*, s.v. "moral responsibility," http://plato.stanford.edu .
[58] Aristotle, *Nicomachean Ethics*, ed. and trans. Martin Ostwald (Indianapolis, IN: Bobbs-Merrill, 1962), 1110b15–25.
[59] Ibid., 1111b15–1113b22.

hold a moral agent blameworthy or praiseworthy for some particular action or trait. He proposed that one is an apt candidate for praise or blame if and only if the action or disposition is voluntary. What is forced "has its origin outside the person forced, who contributes nothing."[60] A voluntary action or trait has two distinctive features: (a) the action or trait must have its origin in the agent (that is, the agent must have the ability to decide whether to perform the action or possess the trait; it cannot be compelled externally), and (b) the agent must be aware of what he or she is doing or bringing about. Aristotle emphasizes the will of the agent, that "the origin of the actions is in him," and "it is also up to him to do them or not."[61]

Thus, moral responsibility means that autonomous agents have the understanding of the options before them, have access to evidence required for making judgments about the benefits and hazards of each option, and are able to weigh the relative value of the consequences of their choice. Responsible agents have a sense of history. They understand the connection between past, present, and future. They comprehend causes for action and are able to appreciate likely consequences of a given conduct. In this context, the idea of conscientiousness is relevant. It describes a condition of an active and inwardly driven pursuit of positive goals, duties, and obligations. The concept of conscientiousness applies typically to individuals, but it also can describe the approach of an organization.[62] The goal is to converge between the *ought* and the *is*, that individuals be motivated by ethical standards as well as or instead of profit.

William J. FitzPatrick further explains that all cases of moral responsibility for bad actions must involve a strong form of *akrasia*, or acting against one's better judgment.[63] When individuals do something bad, either they do so in full knowledge that they should not be doing it, which is clear-eyed akrasia, or they are acting in ignorance. In the former case, they will be held responsible; in the latter case, whether they are responsible depends on whether their ignorance is culpable. Their ignorance will be culpable only if they are responsible for some earlier failure that gave rise to that ignorance, and they will be responsible for that earlier failure, again, only if it was a case of clear-eyed akrasia. We do not establish culpability until we arrive at a relevant

---

[60] Ibid., 1110b15.
[61] Ibid., 1110a5–1110a20. For further discussion, see Susan Sauvé Meyer, *Aristotle on Moral Responsibility: Character and Cause* (Oxford: Blackwell, 1993); *Stanford Encyclopedia of Philosophy*, s.v. "moral responsibility," http://plato.stanford.edu/ .
[62] McQuail, *Media Accountability and Freedom of Publication*, 195.
[63] William J. FitzPatrick, "Moral Responsibility and Normative Ignorance: Answering a New Skeptical Challenge," *Ethics* 118, no. 4 (2008): 589–613, 590.

episode of clear-eyed akrasia.[64] Ignorance, whether circumstantial or norma-tive, is culpable if the agents could reasonably have been expected to take measures that would have corrected or avoided it, given their capabilities and the opportunities provided by the social context, but failed to do so either because of akrasia or because of vices such as overconfidence, arrogance, dismissiveness, laziness, dogmatism, incuriosity, self-indulgence, and con-tempt.[65] Failure to recognize the wrongness or imprudence of one's conduct does not relieve one of responsibility.

An understanding of responsibility as protecting individual rights and avoid-ing the infliction of harm on others is the very basis of liberal morality that presupposes the existence of inviolable individual rights. Responsibility in the sense of honoring interpersonal obligations and responding to the needs of others is a matter of personal choice and of social convention.[66] In other words, moral responsibility often is interconnected with social responsibility. Consider the following: Sally is walking outside an army base. She sees that the fence outside an ammunition storage room is broken. Because Sally is a responsible person, she feels a duty to alert someone to what she saw. What would be the most suitable way to address the situation? Would it be by posting the sensitive data on the Internet? Would it be by discreetly calling an army officer and telling the officer what she saw? Social responsibility considerations dictate that Sally not publicize the broken fence because that might lead criminal or hostile elements in society to take advantage of the security blunder and steal weapons from the army base. One must apply common sense and some judgment prior to acting even when the motivation is altruistic.

SOCIAL RESPONSIBILITY

The concept of *social responsibility* assumes, first, that autonomous agents have the understanding of the options before them, have access to evidence required for making judgments about the benefits and hazards of each option, and are able to weigh the relative value of the consequences of their choice.

---

[64] Ibid., 593.

[65] Ibid., 609. Martha Nussbaum clarified that according to Aristotle, akrasia is frequently (not always) caused by an excess of theory and a deficiency of passion. The person who acts akratically against her better judgment is frequently capable of performing correctly in all intellectual ways; "what she lacks is the heart's confrontation with concrete ethical reality." See Martha C. Nussbaum, *Love's Knowledge: Essays on Philosophy and Literature* (New York: Oxford University Press, 1990), 81. For further discussion, see George Sher, *Who Knew? Responsibility without Awareness* (New York: Oxford University Press, 2009).

[66] Lawrence Kohlberg, *The Philosophy of Moral Development: Moral Stages and the Idea of Justice*, vol. 1 of *Essays on Moral Development* (San Francisco: Harper & Row, 1981).

Social responsibility further assumes that people are not islands unto them-selves. We live within a community and have responsibilities to it. The responsibilities are positive and negative – that is, we have a responsibility to better the society in which we live and a responsibility to refrain from acting in a way that might knowingly harm our community. The responsibility is ethical in nature.

The concept's third assumption is that we are rewarded by the social framework in which we live: we care about society, would like to maintain it, and want to contribute to it. The contribution is proactive. We take active steps to do good and to avoid harm.[67] Generally speaking, people care for one another, communicate with respect, and do not stand idly by while seeing others who might be in danger. Both the private and the public sector are morally accountable. As Novak, Trevino, and Nelson have argued, adopting social responsibility norms is the right way to behave.[68] Later I will also argue that social responsibility is vital notwithstanding whether it contributes to an increased sense of trust in the Internet.

Fourth, social responsibility carries burdens and obligations. People should respect their responsibilities, being cognizant of the consequences of their actions. At the same time, people have discretion in the ways in which they carry out their responsibilities in accordance with their capabilities and the circumstances at hand.

Fifth, people are accountable for their conduct. The duty to account for one's conduct is absolute. It must be transparent and comprehensive, with no room for discount or trickery. Accountability exposes the agents to praise or criticism and to rewards or sanctions, in accordance with the issue at hand: how it was conceived, the way it was performed, and the consequences it produced.

Finally, responsible agents avoid, to the best of their abilities, entering into conflicts of interest. Such conflicts might bring about painful compromises and entail harm to others. A chemist who develops a new medication should refrain from accepting the sponsorship of a pharmaceutical company that

---

[67] Burton S. Kaliski, ed., *Encyclopedia of Business and Finance* (New York: Macmillan, 2001); Marvin L. Marshall, "Ensuring Social Responsibility," *Thrust for Educational Leadership* 23, no. 4 (1994): 42–43; Clifford G. Christians and Kaarle Nordenstreng, "Social Responsibility Worldwide," *Journal of Mass Media Ethics* 19, no. 1 (2004): 3–28; Kristie Bunton, "Social Responsibility in Covering Community: A Narrative Case Study," *Journal of Mass Media Ethics* 13, no. 4 (1998): 232–46; William L. Rivers, Wilbur Schramm, and Clifford G. Christians, *Responsibility in Mass Communication* (New York: Harper & Row, 1980).

[68] Michael Novak, *Business as a Calling: Work and the Examined Life* (New York: Free Press, 1996); Linda K. Trevino and Katherine A. Nelson, *Managing Business Ethics: Straight Talk about How to Do It Right* (New York: Wiley, 1999).

might pressure him or her to complete the development trials sooner rather than later and avoid disclosure of all pertinent information regarding the trial's success or failure. Medical development that is solely or primarily driven by profit considerations can be detrimental to medication users. On such matters, responsibility dictates extreme caution and complete openness. Henry Ford rightly said that a business that makes nothing but money is a poor business.[69] Instead of striving only for larger profits for its stakeholders, a responsible enterprise must also consider employees, suppliers, dealers, local communities, and society at large.[70] I will elaborate on Corporate Social Responsibility in chapter 6.

Social responsibility is multifaceted and is expressed in many ways. It is commonly understood to include taking care of one's offspring financially and emotionally, working, caring for society through prosocial actions and beliefs, meeting obligations, being dependable, attending to the needs of others, and taking care of oneself.[71] These avenues of social responsibility contribute to the well-being both of individuals and of society.

The scope of responsibility is of immense importance. Although a person who drives a car is responsible for careful conduct on the road, a bus driver carries a greater responsibility because many more people might be harmed if he or she drives carelessly. Pilots carry even greater responsibility, not only because far more people might be affected by their conduct but also because a plane costs a small fortune. No airline manager would hire a careless person to sit in the cockpit. Responsibility is a *sine qua non* for the job, notwithstanding how good one is at flying airplanes. These considerations of people and costs are substantive in defining the scope of responsibility.

Social responsibility carries a special meaning in the context of media and information and communication technologies (ICTs). In the first half of the 20th century, the American press was expected to adopt social responsibility standards. In 1947, the Commission on Freedom of the Press, headed by Robert Hutchins, met to address growing concerns about the waning credibility of the press. The commission reached two basic conclusions that formed the basis of this theory: (a) the press has a responsibility to society, and (b) the libertarian press that the United States has embraced is not

---

[69] See *Dictionary of Quotes*, http://www.dictionary-quotes.com/a-business-that-makes-nothing-but-money-is-a-poor-kind-of-business-henry-ford/ .

[70] Harold L. Johnson, *Business in Contemporary Society* (Belmont, CA: Wadsworth, 1971), 50.

[71] Hazel Rose Markus, Carol D. Ryff, Alana L. Conner, Eden K. Pudberry, and Katherine L. Barnett, "Themes and Variations in American Understandings of Responsibility," in *Caring and Doing for Others: Social Responsibility in the Domains of Family, Work, and Community*, ed. Alice S. Rossi (Chicago: University of Chicago Press, 2001), 349–99, 397.

meeting that social responsibility.[72] The belief was that the media should be controlled by community opinion, consumer action, and professional ethics, as opposed to being a free marketplace of ideas. The social responsibility theory was also called "glorified libertarianism." Its goal was to impose strict codes of journalistic ethics on the press and simultaneously ensure that the press continued to provide newsworthy content.[73]

The same libertarian theory may be adopted for information and communication technologies. It has been argued that ICTs make humanity increasingly accountable, morally speaking, for the way the world is, will be, and should be.[74] A member of these professions is trained to practice a core skill, requiring autonomous judgment as well as expertise. ICT professionals have an inviolable duty and professional responsibility to serve the best interest of their clients and are expected to maintain certain standards and qualifications. Often, those professionals must respond to wider social and public concerns. The work in ICT is governed by a set of appropriate ethics and is based on knowledge and skill, and workers must follow an accepted code of practice that observes wider responsibilities to clients and society.

A collective of persons, such as a corporation, might be said to be responsible for a state of affairs, but that does not mean that all workers within the corporation have the same responsibilities. In a corporation, each member is responsible primarily for his or her own contributions. Each is personally responsible for the outcome in a partial way because no one individual produces the overall outcome alone. For instance, if a factory produces hazardous materials that spoil the environment, the factory owners will be held responsible for the contamination; we would not expect a production line worker to assume sole responsibility. In every organization, an identified hierarchy determines those who are answerable at each level for the performance of a given task. Responsibility is related to power and authority.

In a small dot-com startup, teamwork is important. All participants know that they are interconnected. The quality of their collective performance will determine the final result. Each member recognizes his or her interdependence on other members for achieving the final result. Therefore, group

---

[72] Scott Lloyd, "A Criticism of Social Responsibility Theory: An Ethical Perspective," *Journal of Mass Media Ethics* 6, no. 4 (1991), 199–209, 199.

[73] Jon L. Mills, "The New Global Press and Privacy Intrusions: The Two Edged Sword," presented at the Second Annual Berkeley–George Washington University Privacy Law Scholars Conference, Berkeley, CA, June 4–6, 2009.

[74] Luciano Floridi and J. W. Sanders, "Artificial Evil and the Foundation of Computer Ethics," *Ethics and Information Technology* 3, no.1 (2001): 55–66. See also Hilmi Demir, ed., *Luciano Floridi's Philosophy of Technology: Critical Reflections* (Dordrecht, Netherlands: Springer, 2012).

members have a vested interest in helping each other, including poorly performing participants, because the weak links may undermine the entire project. In communication projects, this can lead to proactive fault finding in which members recognize their collective responsibility for outcomes.

When something goes wrong, some group members may try to deflect responsibility from themselves, indicating that other members of the group are responsible for the wrongdoing. This practice is, unfortunately, common – far more common than we would like it to be. Responsibility deferral can be directed upward or downward. Upward deferral manifests itself in the statements, "I am not responsible. I only obeyed orders. The boss is responsible." Downward responsibility takes place when a senior official rolls responsibility onto a junior official, blaming him or her for the wrongdoing. The question of where responsibility starts and where it stops is not always easy. Often, when public wrongdoing is at issue, more than one person is eventually found to be responsible.

Consider the example of a little boy who drowned in a swimming pool. To a large extent, we would hold the lifeguard responsible, but we must also scrutinize the judgment of the pool's manager: On which criteria did he or she choose the particular lifeguard for the job? Were there enough lifeguards in the pool, given its size? Were the working conditions adequate? If scrutiny reveals substantial questions regarding the decision-making ability of the pool's manager, we may pursue the matter upward and scrutinize the judgment of the pool's owner in hiring that manager. Was the manager sufficiently qualified for the job? We must probe the manager's background and reasons for conduct and determine whether they were sufficient to provide a safe environment for the swimmers and other pool users.

Sometimes we may find justifiable grounds for responsibility deferral. For instance, sometimes junior officers could not have had the means to know the particular consequences of their conduct because some knowledge was beyond their remit of expertise. For example, junior Web officers cannot take responsibility for their company's policy if they were not involved in the decision-making process and the shaping of the business policy. At other times, however, responsibility deferral is unacceptable. A member of government that opens an unjust war on a neighboring country bears collective responsibility for the decision if he or she was part of the decision-making process. He or she cannot say that the responsibility lies solely with the prime minister or minister of defense. As a participant in the decision, he or she was part of the deliberations leading to the declaration of war. The only way for that member of government to excuse himself or herself from collective responsibility is by resigning. Merely objecting to the decision, while remaining in office, does not suffice. If the individual decides to remain in government, then he or she must accept responsibility for the wrongdoing.[75]

A pertinent statement uttered when people wish to deflect responsibility is "I did not know." We cannot expect people to know all possible outcomes of their decisions, but we can expect them to weigh foreseeable outcomes. If we can reasonably have expected the decision maker to foresee the outcome of the wrongdoing, then such a statement will not stand.

Other statements made to shirk responsibility are "There was nothing I could do" and "I have done all in my power." One must wonder whether this was, indeed, the case. An assessment of a person's conduct requires reviewing what decision-making process was used, which options were weighed, whether the doer was cognizant of the harmful consequences of his or her actions, and whether that person did all in his or her power to avoid the wrongdoing or to minimize it. As stated earlier, ignorance or laziness are not valid excuses for avoiding responsibility.

NETUSER VERSUS NETCITIZEN

In the context of new media, I want to make a distinction between Netusers and Netcitizens.[76] The term *Netuser* refers to people who use the Internet. It is a neutral term that does not convey any clue regarding how a person uses the Internet, and it does not suggest any appraisal of the use.

The term *Netcitizen*, as it is employed here, is not neutral; it implies responsible use of the Internet. Netcitizens are people who use the Internet as an integral part of their real lives – that is to say, their virtual lives are not separate from their real lives. If they invent an identity for themselves on social networks such as Second Life,[77] they do so in a responsible manner. They hold themselves accountable for the consequences of their Internet use. In other

---

[75] In this context, a politician who leads a nation to war to make a personal profit in the stock market lacks moral responsibility. A politician who wages war without prior extensive consultation and deliberation regarding the implications of this grave decision lacks professional responsibility. For further discussion on different types of responsibility, see Ronald Dworkin, *Justice for Hedgehogs* (Cambridge, MA: Harvard University Press, 2011), 102–4. See also Raphael Cohen-Almagor, Ori Arbel-Ganz, and Asa Kasher, eds., *Public Responsibility in Israel* [in Hebrew] (Tel Aviv and Jerusalem: Hakibbutz Hameuchad and Mishkanot Shaananim, 2012).

[76] Netcitizens are also called *Netizens*.

[77] Second Life is a virtual world that was launched on June 23, 2003, by Linden Lab. Its users, called "residents," interact with one another through avatars. Second Life provides residents with opportunities to explore, socialize, participate in individual and group activities, create and trade virtual properties and services, and travel throughout the world. See the Second Life website at http://secondlife.com/whatis/ .

words, Netcitizens are good citizens of the Internet. They contribute to the Internet's use and growth while making an effort to ensure that their communications and Net use are constructive. They foster free speech, open access, and a social culture of respecting and not harming others. Netcitizens are Netusers with a sense of responsibility.

CONCLUSION

The Internet is a vast ocean of knowledge, data, ideologies, and propaganda. It is omnipresent, interactive, and fast. The ease of access to the Internet; its low cost, speed, and chaotic structure (or lack of structure); the anonymity that individuals and groups may enjoy; and the international character of the World Wide Web provide all kinds of individuals and organizations an easy and effective arena for their partisan interests. The Internet contains some of the best-written products of humanity – and some of the worst ones. Although every form of media contains both good and bad content, on the Internet Netusers are able to upload information themselves quickly, without any editorial filter or criticism; thus, the Internet is saturated with content that would unlikely be easily available through conventional media. The Internet serves the positive and negative elements in society.

The Internet's short history provides us a crash course in understanding why a balanced approach is needed to address and resolve conflicting freedoms. Here I would like to invoke Aristotle's Rule of the Golden Mean that for every polarity there is a mean that provides good standards for a life of moderation.[78] Two underpinning principles at the heart of liberal democracy are showing respect for others and not harming others. We must strive to uphold them also on the Internet.

In this book, I stress the concepts of trust and social responsibility. Behaving in a trustworthy way and trusting others as a way to express respect for others as moral beings constitute ethical behavior,[79] as does acting responsibly in all spheres of life and media. Updating and adapting our theoretical frameworks and vocabularies to new circumstances and innovations is important. Doing so supplies us with conceptual instruments with which we are better fitted to

---

[78] Richard McKeon, ed., *The Basic Works of Aristotle* (New York: Random House, 1947). See also A. W. H. Adkins, "The Connection between Aristotle's *Ethics* and *Politics*," *Political Theory* 12, no. 1 (1984): 29–49; Richard Kraut, *Aristotle: Political Philosophy* (New York: Oxford University Press, 2002).

[79] Bjørn K. Myskja, "The Categorical Imperative and the Ethics of Trust," *Ethics and Information Technology* 10, no. 4 (2008): 213–20, 219.

approach contemporary social problems.[80] Luciano Floridi envisages a steady increase in Netusers' responsibility;[81] I hope he is correct in his prediction. We can reasonably expect people to know the difference between good and evil and to act accordingly. Technical solutions can be engineered if all involved parties recognize the challenges and cooperate to overcome them.

The next chapter analyzes each Netuser's responsibility and focuses attention on a growing and quite problematic development: cyberbullying. This phenomenon has evolved largely because of the indifference of Netusers, their lack of concern for others, and their obliviousness to the responsibilities they carry when using the Internet and because of the harmful consequences of their wrongdoing.

---

[80] Anton H. Vedder, "Accountability of Internet Access and Service Providers: Strict Liability Entering Ethics," *Ethics and Information Technology* 3, no. 1 (2001): 67–74, 73.
[81] Luciano Floridi, "A Look into the Future Impact of ICT on Our Lives," *Information Society* 23, no. 1 (2007): 59–64.

4

Agent's Responsibility

He will not enter Paradise whose neighbor is not secure from his wrongful conduct.
                    –Muhammad, Sahih Muslim, Hadith 15

Man is the only animal that blushes. Or needs to.
                    –Mark Twain

The Internet has created new markets and is profoundly changing the way people interact, express themselves, relax, find leisure, explore the world, and think about human phenomena. Made possible by technological advances in computer hardware, software, and telecommunications, the Internet age often allows people to have cyber lives in addition to their offline lives. The two are not necessarily the same.

The object of this chapter is to discuss moral and social responsibilities of Netcitizens. I first discuss the responsibility of people who are using the Internet, Netusers (or Net agents). I continue with a discussion of JuicyCampus and an analysis of the Megan Meier tragedy. The latter illustrates an immoral use of the Internet on social networking websites, abuse of the functions of the Internet without regard to the potential tragic consequences, and exhibition of behavior that is stripped of any sense of moral responsibility. In this context I also address parental responsibility. Finally, I warn against the phenomenon of cyberbullying.

In the physical world, the term *bullying* has tended to describe conduct that occurs when someone takes repeated action to exert control over another person. Traditional bullying is defined as intentional, continued physical, verbal, or psychological abuse or aggression used to reinforce an imbalance of power.[1] It can involve tormenting, threatening, harassing, humiliating,

[1] Dan Olweus, *Bullying at School: What We Know and What We Can Do* (Oxford: Blackwell, 1993); Robin Kowalski, Susan Limber, and Patricia W. Agatston, *Cyberbullying: Bullying in the Digital Age* (Malden, MA: Blackwell, 2008), 17.

embarrassing, or otherwise targeting a victim.[2] The term *cyberbullying* generally refers to online abuses involving juveniles or students. Although in any given instance of cyberbullying, at least one of the parties may not be a youth,[3] discussions about cyberbullying generally revolve around school-age children and often call on schools to address the issue.[4] Studies from different countries have repeatedly demonstrated that cyberbullying is a persistent problem that affects children and youth of both genders.

The discussion considers some troubling episodes of cyberbullying, many of which resulted in suicide, and then analyzes another threat: cyber revenge. Providing details about notable tragic examples of cyberbullying is important so that one can discern common features and then deduce recommendations for a remedy. Because the Internet is a forum for making sexual advances and for sharing sexual experiences, it also serves abusers who exploit intimate vulnerabilities to take revenge. Revenge porn is a growing concern, especially among young adults who change partners. Sometimes, when the termination of relationships is not consensual, the disappointed partner takes revenge by using technology to share with others the past intimate moments.

AGENT'S RESPONSIBILITY

Agents are morally responsible insofar as they have the capacity to choose ends freely and act in accordance with such choices.[5] An agent would be held responsible for speech that directly led to harm. The issue is more complicated when proving a direct link between a Net posting and the real harm is impossible. But please bear in mind that the subject at hand is moral responsibility, not legal responsibility. To prove legal culpability, the prosecution must show that the speech under scrutiny directly led to the harmful action. Preaching is all right. Incitement is illegal. This standard is in accordance with John Stuart Mill's theory, which was the first philosophical analysis of speech to distinguish between advocacy and incitement.[6] However, temporal association is less critical when we speak of moral culpability. Many cases exist in which proving

---

[2] Jacqueline D. Lipton, "Combating Cyber-Victimization," *Berkeley Technology Law Journal* 26, no. 2 (2011): 1103–56.

[3] See, for instance, in relation to the Megan Meier tragedy, Raphael Cohen-Almagor, "Responsibility of Net Users," in *The Handbook of Global Communication and Media Ethics*, vol. 1, ed. Robert S. Fortner and P. Mark Fackler (Oxford: Wiley-Blackwell, 2011), 415–33.

[4] Kowalski, Limber, and Agatston, *Cyberbullying*, 56–88.

[5] John Martin Fischer, "Recent Work on Moral Responsibility," *Ethics* 110, no. 1 (1999): 93–139, 96.

[6] John Stuart Mill, "On Liberty," in *Utilitarianism, Liberty, and Representative Government* (London: J. M. Dent, 1948), 65–170.

legal responsibility would be difficult, but the speech might still be morally wrong, and we will hold the Net speaker responsible. Agents will be held blameworthy for their bad conduct when they clearly aim at doing bad or when they can be held culpable for their ignorance in making bad choices. As Aristotle said, an autonomous agent is aware of what he or she is doing (see chapter 3). In this context, let me reflect on the short-lived existence of JuicyCampus.

JUICYCAMPUS.COM

Words can hurt, and words can move people to action. The anonymity of the Internet is most convenient for spreading malicious, unfounded allegations and for backstabbing. JuicyCampus.com is an example of an online enterprise that was used to impugn the names of young people. The site described itself as "the world's most authentic college website, with content generated by college students for college students. Just remember, keep it Juicy!"[7] The site contained a variety of information, messages, and other materials that its users created and posted, including details about the sexual activities of named individuals, their physical attributes,[8] allegations that they were spreading sexual diseases, attacks on their integrity, accusations of students using others for social climbing, and so on. A Vanderbilt University student who was raped discovered several months after the assault an account of the attack posted on JuicyCampus along with vicious criticism of her character.[9] This irresponsible behavior amounts to dual victimization: first the student was physically raped, and then she was assaulted online. The latter offense to the victim's sensibilities might be no less harmful than the physical harm. The site's managers were well aware of what was done in their forum. Suffering from clear-eyed akrasia, they did not care.

The site's terms and conditions unsurprisingly said the following:

> Please use caution and common sense when viewing the Site. You understand and agree that any Content is the *exclusive responsibility* of the person

---

[7]   The site, http://www.juicycampus.com/ , has been dismantled. If you go to the site now, you are redirected to http://www.kindr.me , a site started by "two passionate entrepreneurs on a mission to make the world a Kindr place." The Kindr people believe "technology should be used to build people up, not tear them down."

[8]   For example, one post stated that D.P. (on the site the full name was explicit) "has the hairiest asshole in the world. You try to go down on her and you get lost in the dirty tangled bush. Seriously that shit makes me want to vom. Like wtf go get your asshole waxed and get those old stinky dingle berries out of there. Like wtf and get a tan holy shit i could use you as a fucking night light. skank."

[9]   "Squeeze the Juice out of Campus," *Daily Cardinal* (University of Wisconsin–Madison), October 14, 2009.

who posted it, and that *you* will be solely responsible for any Content that you post via the Site. You acknowledge that JuicyCampus is *not* responsible for, does not control, does not endorse, and does not verify the Content posted to the Site or available through the Site, and that it makes *no* guarantee regarding the reliability, accuracy, legitimacy, or quality of any such Content. You agree that *you* will bear any and all risk of reliance on the accuracy, validity, or legitimacy of such Content. You agree that JuicyCampus has *no obligation* to monitor the reliability, accuracy, legitimacy, or quality of such Content, nor to enforce any standards ... in connection with such Content. Under *no* circumstances will JuicyCampus be liable in any way to you for any Content, including, but not limited to, any errors or omissions in any Content or any loss or damage of any kind incurred as a result of the use or existence of or exposure to any Content posted or otherwise transmitted via the Site [italics added].[10]

I do not know whether those terms, which were designed to relieve the site managers of any responsibility for content, were composed with thought toward the future or while the site was already operative, when the site managers were cognizant that the site contained falsehoods, allegations, condemnations, factual errors, and offensive language and exhibited anything but trustworthy information and responsible conduct. The comments posted on the site were anonymous, and JuicyCampus did not remove posts on the basis of students' objections. "The second someone's name appears on the site, it's a death sentence," said one student.[11] This, of course, was an exaggeration, but it showed the extent to which people were troubled by this influential website. Behind the shield of anonymity, agents had dusted away all norms of civility.

JuicyCampus closed down on February 5, 2009, apparently because of lack of advertisement revenues.[12] If the story of JuicyCampus has a silver lining, it is that advertisers did not want to be associated with this venture, knowing that parents, administrators, faculty members, and quite a few students did not think the debasing site should exist. JuicyCampus was a nightmare for higher-education officials and some students, who saw the site as a potentially

---

[10] These terms and conditions were posted on the now-disabled site at http://www.juicycampus. com/posts/terms-condition . Interestingly, nearly identical wording appears on the daring site Darelicious at http://www.darelicious.com/pages/terms_and_conditions .

[11] Catherine Holahan, "The Dark Side of Anonymity," *Businessweek*, May 12, 2008.

[12] JuicyCampus published the following on its now-disabled blog: "Unfortunately, even with great traffic and strong user loyalty, a business can't survive and grow without a steady stream of revenue to support it. In these historically difficult economic times, online ad revenue has plummeted and venture capital funding has dissolved. JuicyCampus' exponential growth outpaced our ability to muster the resources needed to survive this economic downturn, and as a result, we are closing down the site as of Feb. 5, 2009." See Jason Kincaid, "JuicyCampus Dries Up," *TechCrunch Daily*, February 4, 2009, http://techcrunch.com/2009/02/04/juicycampus-dries-up/ .

dangerous provocateur that encouraged students to spread hurtful gossip, lies, threats, and racial epithets.[13]

Another gossip site, PeoplesDirt.com, was also shut down for various reasons, including misrepresentation to advertisers as well as racist, homophobic, and abusive posts.[14] In 2009, the site was taken down as a result of an investigation by the Maryland attorney general. A former student at Walt Whitman High School in Bethesda, Maryland, used the site to post death threats, and parents and teachers complained about it repeatedly.[15] In the United States, specific threats against named individuals are not protected under the First Amendment. They fall under the True Threats doctrine. The courts define a *true threat* as a statement made when a "reasonable person would foresee that the statement would be interpreted by those to whom the maker communicates the statement as a serious expression of intent to harm."[16] The words must be explicit, the words must be spoken in a context in which serious harm is imminent, and the speaker must possess the specific intent that the harm occur.

Although JuicyCampus and PeoplesDirt.com are no longer available,[17] plenty of other forums exist where people can say whatever they wish, notwithstanding how damaging, defamatory, or degrading their words might be, without expecting to be accountable for the consequences.[18] Many of the

[13] Jack Stripling, "Juice Runs Dry," *Inside Higher Ed*, February 5, 2009, http://www.insidehighered.com/news/2009/02/05/juicy ; Martha Neil, "Another State AG Probes JuicyCampus Gossip Website," *ABA Journal*, March 25, 2008, http://www.abajournal.com/news/article/another_state_ag_probes_juicycampus_gossip_website . See also Ann Bartow, "Internet Defamation as Profit Center: The Monetization of Online Harassment," *Harvard Journal of Law and Gender* 32, no. 2 (2009): 384–429.

[14] For domain information about the now-defunct site, visit http://whois.domaintools.com/peoplesdirt.com/ .

[15] "Md. AG Goes on Offense against PeoplesDirt.com Web Site," *Southern Maryland News Online*, May 22, 2009, http://somd.com/news/headlines/2009/9994.shtml . See also the complaints about the site on the DC Urban Moms and Dads forum at http://www.dcurbanmom.com/jforum/posts/list/49361.page .

[16] *Planned Parenthood of Columbia/Willamette, Inc. v. Am. Coalition of Life Activists*, 290 F.3d 1058 (9th Cir. 2002); see also *United States v. Cassel*, 408 F.3d 622 (9th Cir. 2005); *United States v. Alkhabaz*, 104 F.3d 1492 (6th Cir. 1997); *United States v. Machado*, 195 F.3d 454 (9th Cir. 1999); *The Secretary, United States Department of Housing and Urban Development, on behalf of Bonnie Jouhari and Pilar Horton v. Ryan Wilson and ALPHA HQ*, HUDALJ 03-0692-8 (decided July 19, 2000), http://portal.hud.gov/hudportal/documents/huddoc?id=Wilson7-19-00.pdf ; Raphael Cohen-Almagor, *The Scope of Tolerance: Studies on the Costs of Free Expression and Freedom of the Press* (London: Routledge, 2006), 257–58.

[17] But see PeoplesDirt.com@ThePeoplesDirt, http://twitter.com/thepeoplesdirt .

[18] See, for example, the LittleGossip website, http://www.littlegossip.com/ , which was closed because of cyberbullying complaints but later reopened. See Stephen Chittenden, "LittleGossip Website Reopens after Bullying Complaints," *BBC News*, December 24, 2010, http://www.bbc.co.uk/news/uk-12074838 . See also the Dirty blog, http://thedirty.com/ , which

Internet service providers that host those forums have neither binding legal obligation nor sufficient incentives to uphold normative standards of decency and privacy. I return to this issue in chapters 6 and 7.

Aristotle described a person of moral character as being someone who is finely aware of others and his or her surroundings and who is richly responsible for his or her conduct.[19] In Jewish literature, we read that one should "cherish another's honor as one's own."[20] Rabbeinu Yonah Gerondi rules, "One must sacrifice his life rather than publicly shame his fellow Jew."[21] Rabbi Elazar HaModai warns us that someone who shames another in public has no share in the world to come.[22] Shaming people is described as "whitening" their faces (causing them to blanch) and is likened to spilling their blood.[23] One who embarrasses another in public and causes that person to blush in shame or deeply offends his or her sensibilities so much so that the person's face becomes white is considered as having murdered that person. The Midrash Shmuel quotes in the name of Rabbi Menachem L'Beit Meir the following description:

> One who is humiliated, his face first turns red, and then turns white, because due to the magnitude of the shame, his 'soul flies away,' as if it wanted to leave the body ... once the blood returns to its source, the face turns white, like someone who has died.[24]

Shame destroys self-esteem and undermines the spirit. By making others feel worthless, we will have made them lose their "Tzelem Elokim" (Godly image); they no longer feel human.

in 2009 was linked with PeoplesDirt.com as an unethical website. See "May 2009 Unethical Websites," *Ethics Scoreboard*, http://www.ethicsscoreboard.com/sites/0905_dirt.html . For more about the fly-by-night nature of such sites, see Camille Dodero, "Revenge-Porn Troll Hunter Moore Wants to Publish Your Nudes with Directions to Your House," *Gawker*, November 29, 2012, http://gawker.com/5961208/revenge+porn-troll-hunter-moore-wants-to-publish-your-nudes-alongside-directions-to-your-house .

19 Aristotle, *Nicomachean Ethics*, ed. and trans. Martin Ostwald (Indianapolis, IN: Bobbs-Merrill, 1962). See also Martha C. Nussbaum, *Love's Knowledge: Essays on Philosophy and Literature* (New York: Oxford University Press, 1990).

20 *Pirkei Avot* 2:15.

21 *Shaarei Teshuvah* 3:139.

22 *Pirkei Avot* 3:15.

23 *Talmud*, Bava Metzia 58b.

24 Daniel Z. Feldman, "Emotional Homicide: The Prohibitions of Embarrassing Others in Public," reprinted from "The Right and the Good" in *Rabbi Jachter's Halacha Files*, vol. 13, no. 17, January 3, 2004, http://www.koltorah.org/ravj/13-17%20Emotional%20Homicide%20-%20Part%201.htm .

To counter JuicyCampus, Connor Diemand-Yauman, 20-year-old president of the Princeton 2010 class, created a new website, OwnWhatYouThink.com, that asked students to pledge not to visit anonymous gossip sites and to stand behind their online statements. Own What You Think sought to unite people and bring personal accountability back into the ways that people communicate and interact with each other. The site was about encouraging individuals to voice their opinions respectfully and constructively while refusing to participate in anonymous and malicious character assassination. This refreshing initiative was also about taking a personal stand for something and encouraging others to do the same. According to the website, "Own What You Think is about collaborating, dissenting, learning, and disagreeing in a constructive manner that allows us to grow as individuals and a society as a whole."[25] "This is about changing the way our generation and our culture look at the way we communicate with one another," Diemand-Yauman explained. "Anonymity = Cowardice."[26]

There are many examples of Netusers' irresponsible conduct assisted by Internet service providers' no-questions-asked, no-content-inspected policy. Often cybersolicitation is at issue. Solicitation online can be for positive causes, such as blood donations for patients in need of bone marrow transplants. However, cybersolicitation can also be for negative and harmful other-regarding purposes as well as harmful self-regarding purposes. Thus, on the Web we find recipes for cannibals and instructions on how to slaughter a human being.[27] In Germany, Armin Meiwes used the Internet to publish an advertisement seeking a well-built male who was prepared to be slaughtered and then consumed. Bernd-Jürgen Brandes answered the ad in March 2001. Meiwes took Brandes to his home, where the latter agreed to have his penis cut off. Meiwes then flambéed the penis and served it up for the two men to eat together. Afterward, Meiwes stabbed his volunteer victim repeatedly in the neck and dissected the corpse. Brandes was the main dish on Meiwes' table for the following days.[28] A responsible business, one that cares about human

25 The original website, http://ownwhatyouthink.com , is no longer available. A new website, which is an outgrowth of the original movement, can be viewed at http://www.owyt.org/history.html .
26 Holahan, "The Dark Side of Anonymity."
27 Bob Arson, "Butchering the Human Carcass for Human Consumption," Church of Euthanasia e-sermon, http://churchofeuthanasia.org/e-sermons/butcher.html .
28 "German Cannibal Tells of Fantasy," BBC News, December 3, 2003, http://news.bbc.co.uk/1/hi/3286721.stm ; "Profile: Cannibal Armin Meiwes," BBC News, May 9, 2006, http://news.bbc.co.uk/1/hi/world/europe/3443803.stm ; "First TV Interview with Human Cannibal: 'Human

dignity and human life and adheres to the basic values that underpin liberal democracies – respect for others and not harming others – does not entertain cannibalistic advertisement.

On the Web, we find extensive discussions of suicide pills[29] and "exit bags" (do-it-yourself suicide kits).[30] In 2005, in Japan alone more than 17,000 Japanese websites offered information on suicide methods.[31] One site calls on people to "save the planet, kill yourself."[32] It advises people to "do a good job" when they commit suicide, saying, "Suicide is hard work. It's easy to do it badly, or make rookie mistakes. As with many things, the best results are achieved by thorough research and careful preparation."[33] The site discusses the pros and cons of death by shooting, hanging, crashing a car, jumping, slitting your wrists, drowning, freezing, overdosing, or gassing yourself with nitrous oxide, exhaust fumes, and oven gas.

Another site describes using guns, overdosing, slashing your wrists, and hanging as the "best methods to commit suicide." Yet another site discusses lethal doses of poison, highlighting their availability, estimating how long death would take, and specifying the amount of pain.[34] Notwithstanding the extent of the agents' liberalism, they should consider the prudence of such postings given the vulnerability of the people such sites might attract. Indeed, some people report that suicide Web forums encouraged them to use suicide as a problem-solving strategy. Paul Kelly, a trustee of the Papyrus charity, which works to prevent suicide in young people, said, "Some of these sites which incite or give advice on suicide are horrifying. They are encouraging vulnerable people to take their own lives."[35] Cases of cybersuicide

Flesh Tastes Like Pork,'" *Spiegel Online International*, October 16, 2007, http://www.spiegel. de/international/zeitgeist/0,1518,511775,00.html .

[29] See, for example, the discussion thread "Suicide by Sleeping Pills," started March 30, 2007, on Kittyradio at http://kittyradio.com/soapbox/mental-health/25048-suicide-sleeping-pills.html .

[30] See "'Exit Bags' Stir Up Death Debate," Beliefnet, http://www.beliefnet.com/News/2002/07/ Exit-Bags-Stir-Up-Death-Debate.aspx ; Wolfgang Grellner, Sven Anders, Michael Tsokos, and Jochen Wilske, "Suicide with Exit Bags: Circumstances and Special Problem Situations in Assisted Suicide" [in German], *Archiv für Kriminologie* 209, no. 3–4 (2002): 65–75, http://www. ncbi.nlm.nih.gov/pubmed/12043438 ; Final Exit Network, http://www.finalexitnetwork.org/ .

[31] Akihito Hagihara, Kimio Tarumi, and Takeru Abe, "Media Suicide-Reports, Internet Use, and the Occurrence of Suicides between 1987 and 2005 in Japan," *BMC Public Health* 7 (2007): 321, http://www.biomedcentral.com/1471-2458/7/321 .

[32] See the website of the Church of Euthanasia at http://www.churchofeuthanasia.org .

[33] Ibid. See also http://www.churchofeuthanasia.org/press/bathchronicle.html .

[34] Maggie Wykes, "Harm, Suicide, and Homicide in Cyberspace: Assessing Causality and Control," in *Handbook of Internet Crime*, ed. Yvonne Jewkes and Majid Yar (Portland, OR: Willan, 2010), 369–90, 371.

[35] Jon Ungoed-Thomas, "Police Hunt Chatroom Users over Web Suicide 'Goading,'" *Sunday Times* (London), March 25, 2007, http://www.thesundaytimes.co.uk/sto/news/ uk_news/article62066.ece .

(i.e., attempted or successful suicides influenced by the Internet) have been documented.[36] In Britain alone, at least 17 deaths since 2001 have involved chat rooms or sites that give advice on suicide methods.[37] The sites facilitate group suicides, providing a forum for like-minded people to meet to arrange their collective deaths.[38] Behavior that encourages suicide constitutes clear-eyed akrasia – behavior that is stripped of any moral and social responsibility that cannot be justified or legitimized.

In the United States in November 2008, a federal statute designed to combat computer crimes was used for the first time to prosecute what were essentially abuses of a user agreement on a social networking site. Previously, this statute was used to address hacking. I discuss this tragic story in some detail, but first let me make two clarifications. First, the issue at hand is not technology but irresponsible behavior that I hope can be changed by raising attention and providing proper education. Technology is merely a means to an end. It can be used and abused. The infrastructure merely facilitates communication. It is people who are blameworthy for misconduct. Second, I do not intend to evoke moral panics and to say that we would be better off without social networking sites. Social networking sites are most important, especially for adolescents. In some respects, they compensate young people for not having secure public spaces to spend time. Whereas people of my generation used to hang out in public parks, teenagers and youth today hang out in virtual spaces and communicate via cyberspace.

Although social networking sites provide easy, often instant communication and exchange to many Netusers, we should also acknowledge the basic characteristics of such interaction and how they affect the way people interact with one another. Social networking blurred traditional boundaries between "private" and "public," between "intimate" and "shared." When Netusers share their private moments with hundreds and even thousands of Net friends, the notion of "intimacy" becomes very different. Some Netusers have fewer inhibitions when exposing themselves and others. Although Netusers may reveal intimate details (including photos) about their lives on social network

---

[36] Lucy Biddle, Jenny Donovan, Keith Hawton, Navneet Kapur, and David Gunnell, "Suicide and the Internet," *British Medical Journal* 336, no. 7648 (2008): 800–802; Ben Cubby, "Lost in a Tragic Web: Internet Death Pacts Increasing Worldwide," *Sydney Morning Herald*, April 24, 2007; S. Beatson, G. S. Hosty, and S. Smith, "Suicide and the Internet," *Psychiatric Bulletin* 24, no. 10 (2000): 434; Susan Thompson, "The Internet and Its Potential Influence on Suicide," *Psychiatric Bulletin* 23 (1999): 449–51.

[37] Mike Harvey, "Horror as Teenager Commits Suicide Live Online," *Times Online*, November 22, 2008, http://www.thetimes.co.uk/tto/news/world/americas/article1998615.ece .

[38] "Man, 34, Seeks Someone to Die With" [in Hebrew], *Walla!*, http://news.walla.co.il/?w=/402/639665 .

sites, they often do not aim to establish close, loving, trusting, intimate relationships with their virtual friends. Indeed, whereas trust is the basis of friendship in the actual world, Netusers default to distrusting others in the virtual world.[39]

Often Netusers do not invest in building relationships with their virtual friends that are based on the love and respect that characterize their offline companion relationships.[40] It is less time consuming to post a general statement on the Facebook "wall" than to personalize individual messages. Trivial information floods the screen on a steady basis ("I am tired"; "I am bored"; "I am sitting on a train southbound from Stockholm"; "Going out 2nite for drinks"). Because Netusers can instantly post messages, they tend to think and reflect less about what they write. The processes of production, editing, and diffusion are now squeezed into brief moments. The ease of pressing a button tempts many to post half-baked thoughts, instinctive reactions, and impulsive rather than thoughtful comments. Thus, we discern a dramatic increase in writing products and a no less dramatic decrease in the quality of content. Messages tend to be impersonal, laconic, shallow, and trivial.

Some Netusers allow themselves to say online things that they would never dare to say offline. The meaning of being *social* is changing. The meaning of *friendship* is in flux. For Aristotle, friendship (*philia*) is part of the essence of living a good life and is key to human happiness. It is intrinsically valuable because the good life is a life with and toward others; it involves being part of the community. And reciprocal sharing of good is the glue of all valuable friendship.[41] In this context, a distinction can be made between *true friends* and *convenience friends*. True friends remain wherever you are, wherever they are; convenience friends disappear when times become less convenient.

True friends will attend to the needs of their friends and care for them. They will be there to share concerns. True friendship requires empathy, commitment, sincerity, unequivocal trust, and reciprocity. Aristotle acknowledged that such friendships are rare because people need time to grow accustomed to each other and they cannot know each other before they have shared experiences, "and they cannot accept each other or be friends until each appears lovable to

[39] Johnny Hartz Søraker, "How Shall I Compare Thee? Comparing the Prudential Value of Actual and Virtual Friendship," *Ethics and Information Technology* 14, no. 3 (2012): 209–19, 216. See also Barbro Fröding and Martin Peterson, "Why Virtual Friendship Is No Genuine Friendship," *Ethics and Information Technology* 14, no. 3 (2012): 201–7; Mark Graham and William H. Dutton, eds., *Society and the Internet: How Networks of Information and Communication Are Changing Our Lives* (Oxford: Oxford University Press, 2014), 348–59.

[40] Dean Cocking, Jeroen van den Hoven, and Job Timmermans, "Introduction: One Thousand Friends," *Ethics and Information Technology* 14, no. 3 (2012): 179–84, 180.

[41] Aristotle, *Nicomachean Ethics*, 1156a5–10, 1156b5–10.

the other and gains the other's confidence."[42] Indeed, in the offline world, a person is lucky to have a few good friends in which he or she can trust. In the online world, a person may have thousands of online friends, many of whom the person does not know or intend to ever meet. The average British 22-year-old claims to have 1,000 or more friends on sites like Facebook.[43] Care, compassion, mutual responsibility, and commitment are difficult to maintain with hundreds of such virtual friends. Quite often online friendship as it is manifested on social networking sites is uncongenial, uncaring, and superficial. With a click of the mouse, you can become a "friend" of another, and by another click you can ignore and delete that person's communication. Sometimes Netusers turn social networking sites into forums of hostility and provocation. The key to a happy life is to surround yourself with people you love and to distance yourself from people who do not deserve your love. Vulnerable people who fail to acquire this key might pay a high price.

THE MEIER TRAGEDY

Lori Drew, 49, her daughter Sarah, who was then 13, and Ashley Grills, 19, a family friend and employee, created on MySpace a fictitious teenage boy, "Josh Evans," to communicate with Sarah's nemesis, Megan Meier, her 13-year-old neighbor. Lori Drew suspected that Meier had spread nasty rumors about her daughter. Meier had a history of depression and suicidal impulses. She received treatment for attention deficit disorder and depression and had been in counseling since third grade.[44] The fictitious Josh's profile and communications were geared to the needs of an insecure and volatile teenage girl and carefully designed to exploit Megan's vulnerabilities and to play on her emotions. During six weeks of online courtship, "Josh" slowly built Megan's trust in him and at one stage "confessed" to her that he loved her "so much."[45] Then, in October 2006, "Josh" suddenly wrote to Megan, "I don't want to be friends with you anymore because you're not nice to your friends."[46] The distraught Megan tried to understand why "Josh" no longer wanted to be her

---

[42] Ibid., 1156b25–30. For further discussion, see the special issue *Friendship in Political Theory*, *Res Publica* 19, no. 1 (2013).

[43] Emma Barnett, "Facebook Generation 'Have More Than 1,000 Friends,'" *Telegraph*, May 23, 2011, http://www.telegraph.co.uk/technology/facebook/8530691/Facebook-generation-have-more-than-1000-friends.html .

[44] Linda Deutsch, "Woman Indicted in Missouri MySpace Suicide Case," *Associated Press Online*, May 16, 2008; Tamara Jones, "A Deadly Web of Deceit," *Washington Post*, January 10, 2008.

[45] Deutsch, "Woman Indicted in Missouri MySpace Suicide Case."

[46] Cynthia McFadden and Mary Fulginiti, "Searching for Justice; Online Harassment," *ABC News* transcript, March 24, 2008.

friend, involving her MySpace friends in a discussion. The query escalated into a barrage of bitter insults and fierce exchanges. Instead of lending support to Megan in that delicate moment, Megan's trusted friends turned against her. Megan received other emails from "Josh" in which he called her "fat" and a "slut,"[47] and said, "You're a shitty person, and the world would be a better place without you in it."[48] Shortly after that last message was sent, Megan wrote back, "You're the kind of boy a girl would kill herself over."[49] Megan hanged herself that same afternoon in her bedroom.

Lori Drew, who masterminded the cyberaffair, was convicted of computer fraud for creating a phony account to trick Megan and inflict emotional distress on the girl. The indictment charged Drew and her co-conspirators with using "the information obtained over the MySpace computer system to torment, harass, humiliate, and embarrass the juvenile MySpace member."[50] Drew filed a motion for judgment of acquittal, which was granted on August 28, 2009. The central question was whether a Netuser's intentional violation of a website's terms of service satisfied the requirement under the US Computer Fraud and Abuse Act of exceeding authorized access. The court ruled that the breach of a website's terms of service alone as the relevant consideration for Computer Fraud and Abuse Act violations placed too much control in the hands of the website operators and gave too little notice to the website users.[51]

The focus of this discussion, however, is on moral and social responsibility. Lori Drew and her co-conspirators are blameworthy and morally culpable for their involvement in this tragedy because they played on Megan's emotions in a crude and cynical way without thinking about where their game might lead. They were fully aware of what they were doing. No one coerced them to take this crude path. They chose it freely, exhibiting a strong form of clear-eyed akrasia, acting (in the case of Lori Drew and Ashley Grills) against their adult better judgment.

Megan's parents discovered the Drew mother and daughter's involvement six weeks after the suicide. Then the Drews mailed the grieving parents a letter, "basically saying that they might feel a little bit of a responsibility but they don't feel no guilt or remorse or anything for what they did."[52] As noted in

---

[47] Tim Jones, "Cyber-Bullying by Classmate's Parents Ends with Teen's Life," *Chicago Tribune*, November 16, 2007.

[48] Lauren Collins, "Friend Game," *New Yorker*, January 21, 2008, 34–41.

[49] Jennifer Steinhauer, "Verdict in MySpace Suicide Case," *New York Times*, November 27, 2008.

[50] Deutsch, "Woman Indicted in Missouri MySpace Suicide Case."

[51] *United States v. Drew*, US District Court, Central District of California, No. CR 08-0582-GW.

[52] Matt Lauer and George Lewis, "Teenager, Megan Meier, Takes Her Own Life after Falling Victim to a Cruel Internet Hoax; Megan's Parents Discuss Importance of Monitoring Children's Online Activities," *NBC News* transcript, November 19, 2007.

chapter 3, failure to recognize the wrongness or imprudence of the agent's conduct does not relieve the agent of responsibility. One month after Megan's death, Lori Drew told sheriff's deputies that the neighborhood had grown hostile because people had "found out her involvement in Megan's suicide."[53] The report recounted Drew's admission that she "instigated and monitored" the fake MySpace profile.[54]

After the suicide, when Lori Drew realized that she might be held responsible for the vicious prank she orchestrated, she tried to minimize the magnitude of her involvement. Ashley Grills said after the tragedy that Drew had suggested talking to Megan via the Internet to find out what Megan was saying about Drew's daughter Sarah. Grills admitted she wrote the message to Megan about the world being a better place without her. The message was supposed to end the online relationship with "Josh" because Grills felt the joke had gone too far. Indeed it did. "I was trying to get her angry so she would leave him alone and I could get rid of the whole MySpace," Grills explained.[55] The result was that Megan left the world. The messages distressed Megan and caused her so much pain that she decided to take her life.

Megan's weaknesses were well known to the Drews. She had accompanied the Drews on several vacations, and Lori Drew knew that Megan was taking medication.[56] Still the deception was carried on for weeks. The Drews must have been aware of the rollercoaster state they had foisted on Megan. The initial idea of knowing what Megan thought about Sarah Drew escalated very quickly into online flirting. Lori Drew pointed her finger at Grills as the mastermind of the hoax. Grills insisted that Lori was deeply involved in the deception. Lori never took responsibility, saying that she did not create or direct anyone to create the fake MySpace account. Grills, in contrast, said that was not true and was willing to take responsibility for her part. Grills said that Kurt Drew, Lori's husband, who also became involved in the tragedy, insisted after the suicide that Grills quickly close the MySpace account and that Lori instructed Grills to keep quiet. Grills maintained that she and the Drews were blameworthy: "I'm partially to blame. They are partially to blame ... I do know what I did, and I take responsibility for it every day."[57]

In the aftermath of Megan Meier's suicide, new websites popped up. Some were meant to commemorate Megan, but a startling site, named

[53] Jones, "Deadly Web of Deceit."
[54] Ibid.
[55] Diane Sawyer and Robin Roberts, "The MySpace Suicide: Ashley Grills Tells Her Story," *ABC News* transcript, April 1, 2008; Deutsch, "Woman Indicted in Missouri MySpace Suicide Case."
[56] Collins, "Friend Game."
[57] Sawyer and Roberts, "The MySpace Suicide."

Meganhaditcoming.com, also appeared. The anonymous blogger claimed to be a former classmate of Megan's. She described Megan in vicious terms as an aggressive, vulgar, and unpopular girl who victimized the Drew girl. More than 5,000 comments were posted within three days – many of them denouncing the blog as "sick" and suggesting it was the work of the Drews.[58]

PARENTAL RESPONSIBILITY

What about parental responsibility? A recent study shows that most parents talk to their children about what they do on the Internet (70 percent) and stay nearby when the child is online (58 percent). But 13 percent of parents seem never to do any of the forms of mediation asked about, according to their children.[59] These parents find it difficult to filter their children's personal material. These parents and their teenage children seem to have enough issues to grapple with, and thus clashing over Internet surfing behavior might not yield the desired benefits. At the same time, parents may be still concerned about their children's online behavior. Without rocking the domestic peace, they would like to achieve peace of mind by knowing with whom their children interact and for what purposes: for example, what do their children do on social networking virtual communities such as MySpace and Facebook? Many parents are concerned about the amount and type of private information youth reveal on their profile pages.[60]

Because of Megan's delicate personality, her parents were proactive in trying to protect her. They authorized Megan's MySpace account, with some restrictions: "1. Your dad and I are the only ones who know the password. 2. It has to be set to 'private.' 3. We have to approve the content. 4. We have to be in the room at all times when you're on MySpace."[61] Only the parents had the password to the account. Megan could not sign on without them. One of the parents tried to be in the room supervising her online conduct.[62] Megan

[58] Jones, "Deadly Web of Deceit."

[59] Sonia Livingstone, Leslie Haddon, Anke Görzig, and Kjartan Ólafsson, *Risks and Safety on the Internet: The Perspective of European Children – Full Findings and Policy Implications from the EU Kids Online Survey* (London: EU Kids Online Network, 2011), 8.

[60] Justin W. Patchin and Sameer Hinduja, "Trends in Online Social Networking: Adolescent Use of MySpace over Time," *New Media and Society* 12, no. 2 (2010): 197–216. See also Sonia Livingstone, Leslie Haddon, Anke Görzig, and Kjartan Ólafsson, *EU Kids Online* (London: EU Kids Online Network, 2011); Sonia Livingstone, Leslie Haddon, and Anke Görzig, eds., *Children, Risk, and Safety Online: Research and Policy Challenges in Comparative Perspective* (Bristol, UK: Policy Press, 2012).

[61] Collins, "Friend Game."

[62] Lauer and Lewis, "Teenager, Megan Meier, Takes Her Own Life after Falling Victim to a Cruel Internet Hoax."

had a timed access to the Internet, which she usually used in the presence of her mother.[63] The vigilant parents could log into the account at any time. They monitored Megan's Internet use. They were aware of her MySpace friends. They were reluctant to authorize the contact with "Josh," because Megan did not know him. Only after she begged to have contact with that "hot" guy did they agree to add him to Megan's list of friends.[64] Still, Tina Meier warned Megan about him, because he could be, for all they knew, a "40-year-old pervert."[65] She even called the police to see whether there was a way to confirm who owned the "Josh" account.[66] She was told nothing could be done unless a crime had been committed.[67]

Megan's parents did not wish to force their daughter to delete "Josh," because they knew such a restriction would upset Megan. They thought that they could monitor the chats and that Megan would not go behind their backs. Megan called her mother when "Josh" suddenly turned against her. Tina returned home and saw her daughter distraught. She switched the computer off, thinking that Megan needed some time to calm down. Yet despite this direct and observant involvement, more vigilant than the involvement of most parents, Megan's parents were unable to prevent the tragedy.

CYBERBULLYING

The story of Megan Meier is tragic, but unfortunately it is not unique. In 2009, 40-year-old Margery Tannenbaum, whose daughter got into some sort of fourth-grade rivalry with a classmate, decided to use craigslist's adult section to harass her daughter's nemesis. The misguided mother posted an ad with an email address that she had created. Once men replied, she forwarded the nine-year-old girl's name and phone number to them. A total of 40 calls came into the girl's house. The astonishing thing was that Tannenbaum was a licensed social worker.[68] The irresponsible mother should have known better. As in the

---

[63] Matt Lauer and John Larson, "Tina Meier Talks about Her Daughter, Megan, Who Committed Suicide over MySpace Relationship That Turned Out to Be Hoax by Adult Neighbor," *NBC News* transcript, November 29, 2007.

[64] Josh Stossel, Elizabeth Vargas, and Deborah Roberts, "The Hoax; MySpace Suicide," *ABC News* transcript, December 7, 2007.

[65] Ibid.

[66] Collins, "Friend Game."

[67] Jones, "Deadly Web of Deceit."

[68] Amy Beth Arkawy, "The Cyber Bully Next Door: Mom Uses Craigslist to Exact Revenge," *News Junky Post*, October 28, 2009, http://newsjunkiepost.com/2009/10/28/the-cyber-bully-next-door mom-uses-craigslist-to-exact-revenge/ ; Brett Singer, "Mom Puts 9 Year Old Girl on Craigs List for Revenge," Strollerderby, May 9, 2009; JoAnne Thomas, "Margery Tannenbaum: Craigslist Ad 9-year-old Girl," RightJuris, July 3, 2009, http://law.rightpundits.com/?p=599 .

Drew case, this was a case of clear-eyed akrasia: the foolish Tannenbaum exploited the Internet with malicious intentions but without weighing carefully the consequences of her conduct. Luckily, the young girl did not have to hear the men's sexual propositions because the calls were intercepted by her parents.

Cyberbullying is defined as using a computer, cell phone, or other electronic device to intimidate, threaten, or humiliate another Netuser.[69] It involves targeted harm inflicted through the use of text or images sent via the Internet or other communication devices. Cyberbullying includes sending embarrassing, offensive, degrading, or threatening text messages or instant messages; stalking a person electronically; stealing passwords; masquerading as another person on social networking sites; spreading malicious rumors; sending threatening or aggressive messages; and sharing private information without permission. Cyberbullying is not limited to texts. It may also include the distribution of embarrassing, violent (footage of fights and assaults), or sexual photographs or videos (including sexting – sharing explicit texts, nude photos, and videos via cell phone); the creation of graphic websites or social networking site pages devoted to harassing a person or ranking the fattest or "sluttiest" student; and online death threats.[70]

Bullies and cyberbullies are often motivated by anger, revenge, or frustration. Sometimes they practice cyberbullying for entertainment or because they are bored or have the opportunity. Some have a wicked sense of humor or wish to receive some sort of recognition from their peers. Some do it by accident – they either sent a message to the wrong recipient or did not think carefully about the consequences of their conduct. The power hungry do it to torment others and for their ego. They get a perverse sense of gratification from tormenting others and causing them distress. Mean girls do it to help bolster or remind people of their social standing, while others see some merit in such behavior.[71] Bullies adopt screen names that do not reveal their identities, and some of them exploit this anonymity to hurt classmates and other Netusers. Such bullies are not likely to say hurtful things in person, but with the Internet as a filter and facilitator, they have no qualms about harassing their victims and pushing their victims to intolerable and most troubling states of mind.[72] Studies estimate that between 13 percent and 46 percent of young victims of

[69] Kowalski, Limber, and Agatston, *Cyberbullying*, 1.

[70] Ruth Gerson and Nancy Rappaport, "Cyber Cruelty: Understanding and Preventing the New Bullying," *Adolescent Psychiatry* 1, no. 1 (2011): 67–71.

[71] "Why Do Kids Cyberbully Each Other?," Stop Cyberbullying, http://www.stopcyberbullying. org/why_do_kids_cyberbully_each_other.html .

[72] Peter Smith, Jess Mahdavi, Manuel Carvalho, and Neil Tippett, *An Investigation into Cyberbullying, Its Forms, Awareness and Impact, and the Relationship between Age and*

cyberbullying did not know their harasser's identity; 22 percent of the bullies did not know the identity of their victim.[73] The public, impersonal structure of social networking sites facilitates making rude, intrusive, and offensive statements that one would be hesitant to make face to face. The offense is exacerbated because often the victims are alone and hesitant to tell others about the aggression they are facing. Not knowing the identity of the electronic bully leaves the victim guessing who the person behind the aggression is: Is it someone known or a complete stranger? This uncertainty creates a suspect and unsafe environment for the bullied.

Bullying is not a new phenomenon. It has been part of life for many generations. In every class, some children always become the target for bully classmates who enjoy ridiculing them and exposing their vulnerabilities to have a "good laugh." Targeting, humiliating, and intimidating other minors typically occurs among teens who know each other from school, a neighborhood, or an after-school activity. Commonly, vulnerable populations attract the attention of bullies because they are perceived as easy targets who have difficulty fighting back. Children with disabilities and special needs are at higher risk of being bullied by their peers.[74] Ethnic minorities are disproportionately targeted. Children and youth with confused sexuality and those who embrace nonconventional (i.e., not heterosexual) sexuality are also targeted.[75] I elaborate later.

Modern technology has amplified the bullying phenomenon tenfold. Cyberbullies can mask their identity and use text messaging, email, instant messaging, message boards, chat rooms, Web pages, webcams, blogs, social networking websites, and audiovisual sharing sites such as Flickr (an online

Gender in Cyberbullying, Research Brief no. RBX03-06 (London: Department for Education and Skills, Department for Education, 2006); Sameer Hinduja and Justin W. Patchin, *Bullying beyond the Schoolyard: Preventing and Responding to Cyberbullying* (Thousand Oaks, CA: Sage, 2009).

[73] Robin M. Kowalski and Susan P. Limber, "Electronic Bullying among Middle School Students," *Journal of Adolescent Health* 41, no. 6 (2007): S22–30; Janis Wolak, Kimberly J. Mitchell, and David Finkelhor, "Does Online Harassment Constitute Bullying? An Exploration of Online Harassment by Known Peers and Online-Only Contacts," *Journal of Adolescent Health* 41, no. 6 (2007): S51–58; Michele L. Ybarra, Marie Diener-West, and Philip J. Leaf, "Examining the Overlap in Internet Harassment and School Bullying: Implications for School Intervention," *Journal of Adolescent Health* 41, no. 6 (2007): S42–50, S48.

[74] Department for Children, Schools, and Families, *Safe to Learn: Embedding Anti-bullying Work in Schools* (Nottingham, UK: DCSF Publications, 2007); Department for Children, Schools, and Families, *Safe from Bullying* (Nottingham, UK: DCSF Publications, 2009).

[75] Lee A. Beaty and Erick B. Alexeyev, "The Problem of School Bullies: What the Research Tells Us," *Adolescence* 43, no. 169 (2008): 1–11; Elise Berlan, Heather Corliss, Alison Field, Elizabeth Goodman, and S. Bryn Austin, "Sexual Orientation and Bullying among Adolescents in the Growing Up Today Study," *Journal of Adolescent Health* 46, no. 4 (2010): 366–71.

photo management and sharing application) and YouTube to cause embarrassment to others. The humiliation can now be posted on many cyberlocations, and the list of technological arenas keeps growing with the invention of new tools and mechanisms. Most cell phones have picture-taking and video-recording capabilities that can easily be uploaded to the Internet. The offensive files could involve pestering, vicious or sexual warnings, or threats.[76] Modern technology facilitates easy and quick dissemination of hurtful and humiliating messages to one or many people. Whereas traditional bullying is a manifestation of an imbalance of power, in which the powerful exploit the advantage they possess to humiliate others, in cyberbullying, the bullies are not necessarily more physically powerful than their victims. The Internet provides a leveling effect where strength is not physical but verbal, where brutality is more about the crudeness of the mind than about the power of the hands, where having social skills to become popular is of little significance. Articulating words through a keyboard can be no less harmful than punching with one's fist. One need not be physically fit nor have social finesse to launch forceful attacks on a victim.

Cyberbullying has a desensitizing effect.[77] Hiding behind a keyboard and a false screen name provides the bully with protection to launch attacks against the victim, whose only salvation might be closing the computer or the cell phone. Because the bully does not see the emotional reaction of the victim, such as crying and shaking, the bully is not fully aware of the consequences of cyberbullying and does not suffer pangs of conscience. Online bullies remain oblivious to what they do and are not moved to stop tormenting their victims.

Indeed, cyberbullying can be relentless. Images of bullying events can be posted on the Internet on multiple sites, thus having lingering painful effect on the victim. Technology can be abused to increase the scale, scope, and duration of bullying. The audience for the bullying can be very large and reached rapidly, and the bullying can follow the victims into their homes, expressed on the screens of their personal electronic devices. Bullying can now take place around the clock, 24 hours a day, seven days a week, without refuge.

[76] David Smahel and Michelle F. Wright, eds., *The Meaning of Problematic Online Situations for Children: Results of Qualitative Cross-Cultural Investigation in Nine European Countries* (London: EU Kids Online, 2014); Sonia Livingstone, Leslie Haddon, Jane Vincent, Giovana Mascheroni, and Kjartan Ólafsson, "Net Children Go Mobile: The U.K. Report," London School of Economics and Political Science, London, 2014; Brian O'Neill, Elisabeth Staksrud, and Sharon McLaughlin, eds., *Towards a Better Internet for Children? Policy Pillars, Players, and Paradoxes* (Göteborg, Germany: Nordicom, 2013); "Cyber Bullying: From Verbal to Virtual," New Age Parents, http://thenewageparents.com/cyber-bullying-from-verbal-to-virtual/ .

[77] Anti-Defamation League, "Cyberbullying: Understanding and Addressing Online Cruelty," ADL Curriculum Connections, New York, 2008, 30, http://www.adl.org/education/curriculum_connections/cyberbullying/cyberbullying_edition_cc_entireunit.pdf .

The Scope of the Phenomenon

Several studies carried out from 2006 onward[78] show that cyberbullying is a concrete and significant phenomenon. Because of measurement differences – time in which the research was conducted as well as the location and age of victims – victimization estimates range greatly, from 9 percent of adolescents in some studies to 34 percent in other studies.[79] Most studies concentrate on youth victimization.

Many such studies were conducted in the United States. In a 2008 study by the American Education Development Center of 22 high schools in west Boston, 16.5 percent of students reported being bullied at school only, 6.4 percent of students reported being bullied online only, and 9.4 percent were bullied both at school and online. The extensive survey of 20,000 students also found that girls were more likely than boys to report being victims of cyberbullying (18.3 percent versus 13.2 percent), and students who did not identify themselves as heterosexual were far more likely to report bullying both online and at school (33.1 percent versus 14.5 percent).[80]

A study among 3,767 middle school students in grades 6, 7, and 8 who attended six elementary and middle schools in the US Southeast and Northwest indicated that 11 percent had been cyberbullied at least once in the preceding two months.[81] Studies show that the most common form of cyberbullying is instant messaging (chat mail and email messages were

---

[78] Older studies exist, such as David Finkelhor, Kimberly J. Mitchell, and Janis Wolak, *Online Victimization: A Report on the Nation's Youth* (Alexandria, VA: National Center for Missing and Exploited Children, 2000), but in the speedy development of the Internet, especially with the growing use of social networking sites, the data presented in them are outdated. Awareness regarding the challenge of cyberbullying grew around 2006, some three years after the introduction of MySpace in July 2003, with research and resulting publications in the following years.

[79] Kowalski and Limber, "Electronic Bullying among Middle School Students"; Kirk R. Williams and Nancy G. Guerra, "Prevalence and Predictors of Internet Bullying," *Journal of Adolescent Health* 41, no. 6 (2007): S14–21; Ybarra, Diener-West, and Leaf, "Examining the Overlap in Internet Harassment and School Bullying"; Wolak, Mitchell, and Finkelhor, "Does Online Harassment Constitute Bullying?"; Michele Ybarra, Dorothy L. Espelage, and Kimberly J. Mitchell, "The Co-occurrence of Internet Harassment and Unwanted Sexual Solicitation Victimization and Perpetration: Association with Psychosocial Indicators," *Journal of Adolescent Health* 41, no. 6 (2007): S31–41; Kowalski, Limber, and Agatston, *Cyberbullying*; Amanda Lenhart, "Cyberbullying: What the Research Is Telling Us," presented at the NAAG: Year of the Child Summit, Philadelphia, May 13, 2009, http://www.pewinternet.org/Presentations/2009/18-Cyberbullying-What-the-research-is-telling-us.aspx .

[80] Martine Powers and David Filipov, "Damage of Online Bullying Severe," *Boston Globe*, November 18, 2011.

[81] Kowalski and Limber, "Electronic Bullying among Middle School Students," S22.

close behind) and that girls are twice as likely as boys to be victims.[82] The Growing Up with Media survey conducted among youth between 10 and 15 years of age attending American private and public schools reported that one in three students (34.5 percent) said that they experienced at least one incident of Internet harassment during the previous year.[83] Other studies showed that girls tend to rely on more indirect forms of bullying, such as spreading rumors and sexually disparaging comments.[84] Another research project showed that almost one in four children between 11 and 19 years of age have been the victims of cyberbullying. The same study showed that approximately 65 percent of kids know of someone who has been cyberbullied.[85]

According to a national phone survey of 935 teenagers conducted by Pew Internet and American Life in November 2006, one in three online teens had experienced online harassment. Teens who shared their identities and thoughts on social networking sites, such as MySpace and Facebook, were more likely to be targets online than were those who did not use such sites. Nearly 4 in 10 social network users (39 percent) had been cyberbullied in some way, compared with 22 percent of online teens who did not use social networks.[86] The American National Crime Prevention Council (http://www.ncpc.org) indicated that more than 43 percent of teens reported being victims of cyberbullying. Nine in 10 teens (92 percent) said they knew the person who was bullying them. Sadly, only 10 percent of these cyberbullying victims told their parents.[87] Most teens do not feel comfortable sharing this delicate concern with their parents. According to the British National Society for the Prevention

---

[82] Miriam D. Martin, "Suicide and the Cyberbully: Two Factor Authentication," *EzineArticles*, May 20, 2010, http://ezinearticles.com/?Suicide-and-the-Cyberbully&id=4328660 . Similarly, a report to the Anti-Bullying Alliance showed that girls were significantly more likely than boys to be cyberbullied, especially by text messages and phone calls. See Smith et al., *An Investigation into Cyberbullying*.

[83] Ybarra, Diener-West, and Leaf, "Examining the Overlap in Internet Harassment and School Bullying," S46.

[84] Kowalski, Limber, and Agatston, *Cyberbullying*; Kaj Björkqvist, Kirsti M. J. Lagerspetz, and Ari Kaukiainen, "Do Girls Manipulate and Boys Fight? Developmental Trends in Regard to Direct and Indirect Aggression," *Aggressive Behavior* 18, no. 2 (1992): 117–27; Laurence Owens, Rosalyn Shute, and Phillip Slee, "'I'm In and You're Out': Explanations for Teenage Girls' Indirect Aggression," *Psychology, Evolution, and Gender* 2, no. 1 (2000): 19–46.

[85] Chris Webster, "What Is Cyberbullying?," 2011, http://www.cyberbullying.info/whatis/whatis.php .

[86] Pauline C. Reich, "The Internet, Suicide, and Legal Responses," in *Cybercrime and Security*, ed. Pauline C. Reich (New York: Oxford University Press, 2008).

[87] "Physical Bullying Down, Cyberbullying Rising," PI Newswire, March 5, 2010, http://www.pr web.com/releases/crime_prevention/bullying_cyberbullying/prweb3689504.htm .

of Cruelty to Children (NSPCC), 28 percent of bullied children did not tell anyone about what they endured.[88]

A study among adolescent girls revealed that one-third of the sample was subject to online harassment, which included name-calling, spreading of damaging gossip, and warnings. Most victims knew their bully and reported that the bully was a friend from school (31.3 percent) or someone else from school (36.4 percent). Some girls were bullied by former boyfriends.[89] Such bullies, who often lack social skills, find solace in the cyberworld. They exploit the Internet to harm others whom they know from school. They exhibit clear-eyed akrasia – that is, acting against one's better judgment[90] – and little sense of social responsibility. I presume they are unaware of the damage they inflict on their victims and do not realize how seriously words can affect others and themselves (because harassment also negatively affects the bullies). Bullied people reported feeling "sad," "angry," "upset," "depressed," "violated," "stressed," "hated," "stupid," "helpless," "exploited," "put down," "frustrated," and "unsafe." Having these feelings makes the bullied more vulnerable to further harassment, thereby creating a sad, vicious circle. The consequences of cyberbullying can be far reaching, permanently damaging the psyche of the victims.[91]

A recent study among European children 9 to 16 years of age who used the Internet indicated that one in five said that someone had acted in a hurtful or nasty way toward them in the past year. One in 20 children reported being bullied online more than once a week. One in 10 reported being bullied a few times during the past year.[92] The bullying took place on the Internet or a mobile phone; 12 percent reported that they had acted in a nasty or hurtful way to others during the past year.[93] The 2013 UK Annual Bullying Survey showed that 21 percent of young people were bullied online.[94] Another study in the United Kingdom that surveyed children six to nine years of age reported that

[88] "Statistics on Bullying," NSPCC, London, March 2013, http://www.nspcc.org.uk/inform/resourcesforprofessionals/bullying/bullying_statistics_wda85732.html .

[89] Amanda Burgess-Proctor, Justin W. Patchin, and Sameer Hinduja, "Cyberbullying and Online Harassment: Reconceptualizing the Victimization of Adolescent Girls," in *Female Victims of Crime: Reality Reconsidered*, ed. Venessa Garcia and Janice Clifford (Upper Saddle River, NJ: Prentice Hall, 2009), 162–76.

[90] William J. FitzPatrick, "Moral Responsibility and Normative Ignorance: Answering a New Skeptical Challenge," *Ethics* 118, no. 4 (2008): 589–613, 590.

[91] Sameer Hinduja and Justin W. Patchin, "Offline Consequences of Online Victimization: School Violence and Delinquency," *Journal of School Violence* 6, no. 3 (2007): 89–112; Kowalski, Limber, and Agatston, *Cyber Bullying*, 85–88.

[92] Livingstone et al., *Risks and Safety on the Internet*, 61.

[93] Ibid., 64.

[94] Ditch the Label, "Annual Bullying Survey 2013," Brighton, UK, 2013, http://www.ditchthelabel.org/downloads/Annual-Bullying-Survey-2013b.pdf .

20 percent of children were the victims of "aggressive or unpleasant" behavior online. This prevalence is partly because children in the United Kingdom use social networks for longer periods than any other countries.[95] In Spain, the figure was 25 percent.[96]

Cyberthreats include direct intimidation or distressing material that raises security concerns. As a result of such communications, Netusers may consider committing a violent act against others or themselves. Bullied victims are at increased risk for committing suicide. Indeed, cyberbullying proved detrimental to victims who could not cope with the malicious attacks and the vile language. According to the US Centers for Disease Control and Prevention, suicide is the third-leading cause of death for 15- to 24-year-olds.[97] I give specific examples later in this chapter.

Lack of social responsibility norms harms the bullies as well as the bullied. Both the bully and the victim may suffer from depression – the number one cause of suicide. Youth who bully others are at increased risk for substance use, academic difficulties, and violence later in life.[98] In 2010, the Associated Press reported at least 12 cases in the United States alone since 2003 in which victims between 11 and 18 years of age killed themselves after enduring cyberbullying.[99] One study shows that a significant number of bullies tend to be bullied; thus a vicious cycle is created. When asked if they had been cyberbullied, 17.3 percent of respondents answered in the affirmative. A similar proportion, 17.6 percent, admitted to cyberbullying others; 12 percent reported being both a victim and a bully.[100] It is important to make potential bullies understand that their involvement in such a practice might hit them back very hard.

---

[95] "Online Bullies Are the Most Dangerous 'Because Abuse Is 24-Hour,'" *Daily Mail*, August 8, 2011, http://www.dailymail.co.uk/sciencetech/article-2023439/Online-bullies-dangerous-abuse-24-hour.html .

[96] Ibid.

[97] Martin, "Suicide and the Cyberbully."

[98] Paul R. Smokowski and Kelly H. Kopasz, "Bullying in School: An Overview of Types, Effects, Family Characteristics, and Intervention Strategies," *Children and Schools* 27, no. 2 (2005): 101–10; Michele L. Ybarra and Kimberly J. Mitchell, "Online Aggressor/Targets, Aggressors, and Targets: A Comparison of Associated Youth Characteristics," *Journal of Child Psychology and Psychiatry* 45, no. 7 (2004): 1308–16; Denise L. Haynie, Tonja Nansel, Patricia Eitel, Aria D. Crump, Keith Saylor, Kai Yu, and Bruce Simons-Morton, "Bullies, Victims, and Bully/Victims: Distinct Groups of At-Risk Youth," *Journal of Early Adolescence* 21, no. 1 (2001): 29–49.

[99] "Facebook App Tackles Cyber Bullying," CBC News, November 1, 2010, http://www.cbc.ca/technology/story/2010/11/01/tech-cyber-bullying-facebook-application.html .

[100] Sameer Hinduja and Justin W. Patchin, "Cyberbulling Fact Sheet: What You Need to Know about Online Aggression," Cyberbullying Research Center, 2009, http://www.cyberbullying.us/cyberbullying_fact_sheet.pdf . See also Smahel and Wright, *Meaning of Problematic Online Situations for Children*; EU Kids Online and Net Children Go Mobile, "New Evidence, New Challenges," presented to the ICT Coalition, Brussels, April 15, 2014.

Cyberbullying and Suicide

A recent study examined the extent to which cyberbullying is related to suicidal ideation among adolescents. Among a random sample of approximately 2,000 middle schoolers, youth who experienced traditional bullying or cyberbullying – as either an offender or a victim – had more suicidal thoughts and were more likely to attempt suicide than those who had not experienced such forms of peer aggression. In addition, victimization was more strongly related to suicidal thoughts and behaviors than was offending.[101] The following tragic stories show that aggression can have grave consequences to the vulnerable and even be fatal.

In the United Kingdom, Sam Leeson, a 13-year-old student from Tredworth, Gloucestershire, hanged himself in his bedroom after apparently suffering months of online bullying on the social networking website Bebo. Leeson was targeted because he was a fan of emo music, which is popular with many children who feel left out of the mainstream. In common with many "emos," Sam wore alternative black or dark clothing and had long hair, which singled him out and attracted the bullies.[102]

After the suicide, a YouTube user who is known only as "imDavidwhoareyou" posted a sick video parodying Leeson's tragic death. Previously that same person had uploaded another video poking fun at Leeson's suicide. A stream of comments has been posted under the videos, the vast majority of them condemning imDavidwhoareyou. In response, he said, "Free speech means I can say something tasteless, and you can say that you find it tasteless. But if you think it was 'sick and really really disgusting,' you haven't been on the internet long enough."[103] The overemphasis on freedom of expression on the free highway

[101] Sameer Hinduja and Justin W. Patchin, "Bullying, Cyberbullying, and Suicide," *Archives of Suicide Research* 14, no. 3 (2010): 206–21; Gianluca Gini and Dorothy L. Espelage, "Peer Victimization, Cyberbullying, and Suicide Risk in Children and Adolescents," *Journal of the American Medical Association* 312, no. 5 (2014): 545–46. For further discussion, see Sameer Hinduja and Justin W. Patchin, *Bullying beyond the Schoolyard*; Shaheen Shariff, *Confronting Cyber-Bullying: What Schools Need to Know to Control Misconduct and Avoid Legal Consequences* (Cambridge: Cambridge University Press, 2009).

[102] Andy Bloxham, "Teenager Sam Leeson Hanged Himself over 'Emo' Taunts," *Telegraph*, June 22, 2008, http://www.telegraph.co.uk/news/uknews/2176009/Teenager-Sam-Leeson-hanged-himself-over-Emo-taunts.html ; "Boy, 13, 'Hanged Himself after He Was Bullied on Bebo for Being a Fan of Emo Music,'" *Mail Online*, June 11, 2008, http://www.dailymail.co.uk/news/article-1025654/Boy-13-hanged-bullied-Bebo-fan-Emo-music.html . See also Mark "Rizzin" Hopkins, "Bebo Suicides and Political Opportunism," Mashable, June 13, 2008, http://mashable.com/2008/06/14/bebo-suicide/ .

[103] "Youtube Nuisance Posts Another Video about Sam Leeson," *Gloucestershire Citizen*, June 18, 2008, http://www.thisisgloucestershire.co.uk/news/Youtube-nuisance-posts-video-Sam-Leeson/article-191577-detail/article.html .

brings about the perception of the Internet as a platform stripped of any shred of social responsibility, on which everything goes.

In Israel, a 16-year-old unnamed teenager hanged himself after relentless bullying both offline and online. The teenager suffered from health problems and was small for his age. His peers exploited his vulnerabilities and referred to him as "dwarf," "grass," and "midget." They called him "faggot" and "dog." They encouraged him to commit suicide. He was beaten at school, and when he went on the Internet, the harassment followed him online into his home. Facebook served as the main scene for the abuse. One post said, "I would have killed you, but because they say 'Let Animals Live,' I'll spare your life."[104]

In Vermont, Ryan Halligan was bullied relentlessly. From preschool through fourth grade, Ryan had problems with speech, language, and motor skills development and received special education services. Ryan's parents first began to notice that their child was being subjected to bullying when he was in fifth grade. Some bullies picked up on Ryan's academic weaknesses and poor physical coordination. The offline and online bullying continued for three years. Ryan said that he hated going to school and asked his parents to move or to home-school him. His parents thought that he needed to learn how to manage the situation as part of growing up. In 2003, when Ryan was in eighth grade, the situation had worsened. The bullies spread rumors that Ryan was gay. To quash the rumor, Ryan approached a pretty, popular girl online and tried to establish a relationship with her. The girl, Ashley, misled him into believing that the relationship might have prospects. But then, in front of her friends, she told him he was a loser, that she did not want anything with him, and that she was only joking online. Ryan said, "It's girls like you who make me want to kill myself."[105] This statement echoes Megan Meier's statement before she committed suicide. Later, Ryan found that Ashley hyperbolized the prank by copying her friends to their private instant message exchanges. Because he believed that she liked him, Ryan revealed personal, embarrassing information about himself that Ashley made public, making him the school laughingstock. Subsequently, Ryan received emails and instant messages from classmates ridiculing and taunting him with even more venom than before. A few weeks

---

[104] Danny Adino et al., "Everyone Knew, All Kept Quiet" [in Hebrew], *Yedioth Ahronoth*, January 6, 2011, 14–15; Alon Hachmon and Limor Nissani-Shoshani, "16-Year-Old Ended Life by Hanging" [in Hebrew], *Post Israel*, January 6, 2011, 4; "Angry at the Monsters" [in Hebrew] *Israel Today*, January 6, 2011, 4–5; Ana Meidan, "My Boy Chose Silence" [in Hebrew], *Yedioth Ahronoth*, February 18, 2011.

[105] Catherine M. Oliverio, "Bullying: The Tragic Death of Ryan Halligan," Denton Publications, April 18, 2009, http://www.denpubs.com/news/2009/apr/18/bullying-the-tragic-death-of-ryan-halligan/ .

later, on October 7, 2003, Ryan Halligan hanged himself in his family's bath-room. He was 13 years old.[106]

Ryan's parents insisted on having some Internet safety rules: no instant messaging or chatting with strangers, no giving any personal information to strangers, no sending pictures to strangers, and no secret passwords. These safety rules were not helpful because Ryan's worst enemies were children from his own school whom he knew. Moreover, Ryan did communicate with a stranger; he had an exchange with another Netuser about death and suicide methods. Together they exchanged ideas about the most suitable way for Ryan to commit suicide. When Ryan announced, "Tonight's the night I think I'm going to do it," the correspondent replied, "It is about blanking [sic] time."[107] Two weeks later, Ryan hanged himself. After the suicide, Ryan's father con-tacted the stranger and asked him whether he discussed the issues of death and suicide with Ryan, and the Netuser denied it. Ryan's father contacted the Netuser's father, alerting the man that his son might also be contemplating suicide. Four years later, that same Netuser continued to post on his personal website many references to death and suicide in a way that seemed to lack any sense of moral or social responsibility. His posts seemed to have played a role in Ryan's death and potentially may contribute to further cases of suicide.

Easy targets for all forms of harassment and bullying are youth who are questioning their sexuality or embracing homosexuality, bisexuality, or trans-gender identity. They are at a greater risk than their peers because they seek acceptance, reassurance, and like-minded people.[108] Thus, they use social networking sites to communicate with people, and by doing so they expose themselves to others who wish to humiliate them. Outing homosexuals against their will is another form of cyberbullying, which can be termed *homophobic*

---

[106] John and Kelly Halligan, "Ryan's Story," http://www.ryanpatrickhalligan.org/ ; Cindy Long, "Silencing Cyberbullies," NEA Today, May 1, 2008; Paul J. Fink, "The Case of a Teenager Who Committed Suicide after Being Bullied Online Shows That the Internet Can Be a Weapon against the Psychiatrically Vulnerable. What Can We Do to Help These Patients?," *Clinical Psychiatry News*, February 1, 2008.

[107] "Interviews: John Halligan," *Frontline*, January 22, 2008, http://www.pbs.org/wgbh/pages/front line/kidsonline/interviews/halligan.html .

[108] Andrew Schrock and danah boyd, "Problematic Youth Interaction Online: Solicitation, Harassment, and Cyberbullying," in *Computer-Mediated Communication in Personal Relationships*, ed. Kevin B. Wright and Lynne M. Webb (New York: Peter Lang, 2011), 368–96. See also Ritch C. Savin-Williams, "Verbal and Physical Abuse as Stressors in the Lives of Lesbian, Gay Male, and Bisexual Youths: Associations with School Problems, Running Away, Substance Abuse, Prostitution, and Suicide," *Journal of Consulting and Clinical Psychology* 62, no. 2 (1994): 261–69; Michele L. Ybarra, Kimberly J. Mitchell, Neal A. Palmer, and Sari L. Reisner, "Online Social Support as a Buffer against Online and Offline Peer and Sexual Victimization among U.S. LGBT and Non-LGBT Youth," *Child Abuse and Neglect*, September 2, 2014, http://dx.doi.org/10.1016/j.chiabu.2014.08.006 .

bullying. It can cause enormous strain that, in turn, can lead to suicidal thoughts and actions. Nonconsensual outing blurs the line between private and public, and it can have tragic consequences.

Tyler Clementi, an 18-year-old student, asked his roommate, Dharun Ravi, to give him some privacy in the room they shared at Rutgers University dorms. Ravi agreed and went down the hall into a friend's room, where he allegedly logged onto his Skype account and connected to a webcam he had set up in their shared room. Ravi and his friend watched Clementi engage in a sexual encounter with another man. Ravi then allegedly streamed the video live and that same night broadcast to the 150 followers of his Twitter feed details of his voyeuristic escapade, outing Clementi in the process and writing with no sense of civility and friendship, "Roommate asked for the room till midnight. I went into molly's room and turned on my webcam. I saw him making out with a dude. Yay." Two evenings later, Ravi allegedly tweeted, "Anyone with iChat, I dare you to video chat me between the hours of 9.30 and 12. Yes it's happening again."[109] The next day, mean-spirited students told Clementi his privacy had been violated via webcam. His world fell apart. Having asked no one for help, Clementi committed suicide. Ravi's alleged clear-eyed akrasia directly led to this most unnecessary death. Civility, decency, privacy, and respect for others are significant. People should think about the likely consequences of their actions.

Around that same month, September 2010, nine other adolescent American boys committed suicide as a result of bullying in various shapes and forms: eye to eye, over the telephone, and over the Internet. They were 13-year-old Asher Brown, 13-year-old Seth Walsh, 14-year-old Caleb Nolt, 15-year-old Justin Aaberg, 15-year-old Billy Lucas, 15-year-old Harrison Chase Brown, 17-year-old Cody J. Barker, 17-year-old Felix Sacco, and 19-year-old Raymond Chase. All were victims of prolonged homophobic harassment. In all cases, school officials did not do enough to stop the mental torture of the vulnerable boys. In all cases, the boys were tormented for long periods because of their sexual identity. They asked for help repeatedly, but were ignored.[110] Many people

---

[109] Ed Pilkington, "Tyler Clementi, Student Outed as Gay on Internet, Jumps to His Death," *Guardian*, September 30, 2010, http://www.guardian.co.uk/world/2010/sep/30/tyler-clementi-gay-student-suicide ; Paul Thompson, "Student Jumps to His Death after Roommate Secretly Films Gay Sex Session and Puts It on the Internet," *Mail Online*, September 30, 2010, http://www.dailymail.co.uk/news/article-1316319/NY-student-Tyler-Clementi-commits-suicide-gay-sex-encounter-online.html# . For further discussion, see Kate Zernike, "Rutgers Webcam-Spying Defendant Is Sentenced to 30-Day Jail Term," *New York Times*, May 21, 2012, http://www.nytimes.com/2012/05/22/nyregion/rutgers-spying-defendant-sentenced-to-30-days-in-jail.html?_r=1&nl=todaysheadlines&emc=edit_th_20120522 .

[110] Bryan Alexander, "The Bullying of Seth Walsh: Requiem for a Small-Town Boy," *Time*, October 2, 2010, http://www.time.com/time/printout/0,8816,2023083,00.html ; Yobie Benjamin, "Bullied Tehachapi Gay Teen Seth Walsh Dies after Suicide Attempt,"

knew what was going on, yet the victims were given neither protection nor security. They became increasingly isolated and desperate until they found solace in death.

I have elaborated on these tragic incidents because the devil is in the details, and we should understand the phenomena by discerning the common denominators and take the necessary measures to ascertain that youth can enjoy a safe environment online, free from harassment and nasty insults. All the people who committed suicide in the preceding cases were between 13 and 19 years of age, when the personality is taking shape and social skills are forming. Teens who are different in one way or another – dress code, sexual orientation, foreignness, health problems – attract the attention of bullies, who prey on their evident vulnerabilities and attack and humiliate them. The findings provide evidence that adolescent peer aggression may have terrible consequences and that it must be taken seriously both at school and at home, and they suggest that teachers and parents should be attentive to alarm signs, should insist on asking questions, and should rush to help even when such help is not requested. Peer harassment contributes to depression, decreased self-worth, anxiety, eating disorders, distress, hopelessness, helplessness, and loneliness, all of which are precursors to self-harm, suicidal thoughts, and suicidal behaviors.[111] Bullied people often have headaches, colds, and other physical illnesses, as well as psychological problems.[112] Other consequences are alcohol use, drug use, and carrying of weapons to school as self-protection.[113] These are alarm bells that should trigger attention and probing. Although the bullied might keep quiet because they are reluctant to expose themselves further, the people around them should not keep quiet. They should raise and confront the issues before it is too late.

SFGate, September 29, 2010, http://www.sfgate.com/cgi-bin/blogs/ybenjamin/detail?entr y_id=73326#ixzz15jQPItgd ; Edecio Martinez, "Seth Walsh: Gay 13-Year-Old Hangs Self after Reported Bullying," CBS News, September 29, 2010, http://www.cbsnews.com/news/ seth-walsh-gay-13-year-old-hangs-self-after-reported-bullying-30-09-2010/ ; "Asher Brown: Gay Texas 13-Year-Old Shoots Himself in the Head after Bullying," Blippitt, September 28, 2010, http://www.blippitt.com/asher-brown-gay-texas-13-year-old-shoots-himself-in-head-after-bullying/ ; Kara Brooks, "Bullied Greensburg Student Takes His Own Life," Fox 59, September 13, 2010, http://www.liveleak.com/view?i=d64_1284735920 .
[111] Kowalski, Limber, and Agatston, Cyberbullying; Elaine M. McMahon, Udo Reulbach, Helen Keeley, Ivan J. Perry, and Ella Arensman, "Bullying Victimisation, Self-Harm, and Associated Factors in Irish Adolescent Boys," Social Science and Medicine 71, no. 7 (2010): 1300–1307.
[112] Stuart Wolpert, "Bullying of Teenagers Online Is Common, UCLA Psychologists Report," UCLA News, October 2, 2008.
[113] Ybarra, Diener-West, and Leaf, "Examining the Overlap in Internet Harassment and School Bullying."

CONCLUSION

Immanuel Kant taught us that individuals respect one another and attribute dignity to each other because doing so is the only practical way for them to pursue their ends in comparative safety and security. John Stuart Mill taught us to ponder the consequences of our action so as to avoid inflicting harm on others. Abusive language might lead to depression and suicide. Stories like those of Megan Meier, Sam Leeson, Ryan Halligan, and Tyler Clementi should be discussed openly and fervently in middle schools, high schools, and colleges. People, especially young people, should be made aware of the power of the word so that the confusion between online and offline responsibility can be settled.

All involved should share responsibility and accountability: parents, schoolteachers and administrators, bullies, and bystanders. Bullying affects us all. Many Netusers have had some experience of it, whether at school or at work. Its impact can be long and lasting. Beat Bullying, a bullying prevention charity, found that up to 44 percent of suicides among those 10 to 14 years of age may be bullying related.[114] Children who are bullied are 2.23 times as likely to think about killing themselves than children who have not been victimized.[115] Not surprisingly, both children and parents say bullying is among their top concerns. Beat Bullying helps children and young people who are so deeply affected by bullying that they can barely face going to school the next morning. Beat Bullying also works with those identified as bullies by addressing their negative behavior and trying to shape their attitudes and change their manners. The aim is for people to take responsibility and develop a sense of ownership over their actions, thereby building foundations for change and improvement in their life opportunities. Schools where Beat Bullying is active report up to an 80 percent reduction in bullying.[116]

Netcitizens can develop a website, blog, or social networking group on Facebook, LiveJournal, or MySpace for friends and community in which they evoke awareness of the problem of cyberbullying and alert readers to potential signs of distress that bullied people show. The warning signs include unexpected or sudden loss of interest in using the computer; nervous, jumpy, anxious, or scared appearance on accepting messages; and

[114] "'Bullying' Link to Child Suicide Rate, Charity Suggests," BBC News, June 13, 2010, http://news.bbc.co.uk/1/hi/uk/10302550.stm .

[115] Karen Kaplan, "Teens Taunted by Bullies Are More Likely to Consider, Attempt Suicide," *Los Angeles Times*, March 10, 2014, http://articles.latimes.com/2014/mar/10/science/la-sci-sn-bullying-cyberbullying-suicide-risk-20140310 .

[116] "Working Together to Prevent Child Suicide and Bullying," Beat Bullying, London, http://www.beatbullying.org/dox/what-you-can-do/what-you-can-do.html .

discontinued interest in going to school or participating in extracurricular or general out-of-school activities. Bullied people might be visibly angry, frustrated, depressed, or gloomy after using the computer.[117] They might become abnormally withdrawn and distant from family, friends, and favorite activities. They might lack appetite or suddenly begin to do poorly in school. They might complain frequently of headaches, stomachaches, or other physical ailments; have difficulty in sleeping or have frequent bad dreams; or appear troubled or suffer from low self-esteem.[118] Netcitizens may point to valuable information on the Internet from Embrace Civility in the Digital Age,[119] the World Association of Newspapers and News Publishers,[120] and the Cyberbullying Research Center.[121]

I should further note the work of Cyberbullying.org, one of the first websites set up in Canada which provides advice for young people on how to prevent and take action against cyberbullying;[122] Chatdanger, a website that informs about potential dangers online, suggests how to stay safe while using the Internet, and provides a wealth of data and links to further sources;[123] the Anti-Bullying Alliance, which brings together more than 60 organizations into one network with the aim of reducing bullying;[124] Kidscape, a helpline for the use of parents, guardians, or concerned relatives and friends of bullied children;[125] and the Olweus Bullying Prevention Program, a comprehensive program designed and evaluated for use in elementary, middle, and junior high schools.[126]

In the United Kingdom, the important charities are the National Society for Prevention of Cruelty to Children, which was established to end cruelty to children;[127] Beat Bullying, mentioned earlier, an international bullying-prevention charity working and campaigning to make bullying unacceptable;[128]

---

[117] "Signs of Bullying," NoBullying, http://nobullying.com/signs-of-bullying/ ; "What Is Cyber Bullying?," Violence Prevention Works!, http://www.violencepreventionworks.org/public/cyber_bullying.page .

[118] Martin, "Suicide and the Cyberbully."

[119] See the organization's website at http://www.embracecivility.org .

[120] Aralynn McMane, "Internet in the Family," *World Association of Newspapers and News Publishers*, January 12, 2011, http://www.wan-ifra.org/articles/2011/01/12/internet-in-the-family .

[121] See the organization's website at http://www.cyberbullying.us/ .

[122] See the organization's website at http://www.cyberbullying.org/ .

[123] See the organization's website at http://www.chatdanger.com/resources/ .

[124] See the organization's website at http://www.anti-bullyingalliance.org.uk/ .

[125] See the organization's website at http://www.kidscape.org.uk/about-kidscape/helpline/ .

[126] See the organization's website at http://www.clemson.edu/olweus/ .

[127] See the organization's website at http://www.nspcc.org.uk/what-we-do/what-we-do-hub_wdh71749.html .

[128] See the organization's website at http://www.beatbullying.org/gb/about-beatbullying/ .

Barnardo's;[129] and ChildLine.[130] These organizations are instrumental in providing information and promoting awareness regarding the possible harms of social networking forums on the Net. Further information is available on websites such as Childnet,[131] CyberAngels,[132] Bully OnLine,[133] i-Safe,[134] NetSmartz Workshop from the National Center for Exploited and Missing Children,[135] GetNetWise,[136] and WiredSafety.[137] These sites provide a wealth of information about cyberbullying, cyberstalking, and cyberabuse, including helplines for victims of any kind of cyberabuse.

Programs such as Tweenangels and Teenangels, operated by WiredSafety,[138] help educate youth about safe and responsible Internet use. Government agencies, such as the US Computer Emergency Readiness Team,[139] and large corporations, such as Microsoft,[140] also provide guidance on how to protect against various Internet threats.

Netcitizens may mentor younger students and step forward to offer help to younger friends and family members, showing the children that they understand the issues at hand, are sensitive to the children's concerns, and are able to keep issues discreet. Netcitizens may share with younger people their experiences and ideas on how to keep safe online. Netcitizens can offer advice about what young people can do if they run into trouble. It is very important to speak to adolescents about the importance of privacy. Adolescents do not need to divulge too much information about themselves. They should be extra careful as to whom they reveal personal information. Netcitizens should warn adolescents against getting into flame wars and advise them never to send messages when angry. Netcitizens should caution young Netusers not to respond to messages and texts from people they do not know personally. Adolescents need to be instructed to listen to their feelings. If postings do

[129] See the organization's website at http://www.barnardos.org.uk/?gclid=CO3-ytiovKgCFYob4QodJoM1Bw .

[130] See the organization's website at http://www.childline.org.uk/Pages/Home.aspx .

[131] See the organization's website at http://www.childnet.org/ .

[132] See the organization's website at http://www.cyberangels.org/ .

[133] See the organization's website at http://www.bullyonline.org/ .

[134] See the organization's website at http://www.isafe.org/ .

[135] See the organization's website at http://www.netsmartz.org .

[136] See the organization's website at http://kids.getnetwise.org/tools/ .

[137] See the organization's website at http://www.wiredsafety.org/ .

[138] See Wired Safety's website at http://www.wiredsafety.org/ . A separate site for Teenangels can be found at http://teenangels.org/ .

[139] Mindy McDowell, "Dealing with Cyberbullies," Security Tip ST06-005, February 6, 2013, US Computer Emergency Readiness Team, Washington, DC, http://www.us-cert.gov/cas/tips/ST06-005.html .

[140] See the bullying information on the Microsoft Safety and Security Center website at http://www.microsoft.com/protect/parents/social/cyberbullying.aspx .

not look right, do not feel right, or make Netusers uncomfortable, they should save the message, not respond, and consult a trusted adult. By spending time with younger friends and members of their family, Netcitizens can show them how to responsibly use technology and how to report cases of cyberbullying, explaining that not reporting cyberbullying only plays into the bully's hands.

Adolescents are common victims, but they are not the only victims. Social networking sites and blogs have increasingly become breeding grounds for anonymous online groups that attack women, people of color, and members of other traditionally disadvantaged groups. We need to teach adults and especially children that silence when others are being hurt is not acceptable. Safety should be maintained both online and offline, and studies should be carried out about the connections between the two. As Stopcyberbullying.org holds, the task is to create a generation of good cybercitizens who control the technology instead of being controlled by it.[141] Agents are morally and socially responsible for all their conduct, whether in a cyber or a real environment.

---

[141] "Take a Stand against Cyberbullying," Stopcyberbullying.org, http://www.stopcyberbully ing.org/take_action/take_a_stand_against_cyberbullying.html . For further discussion, see O'Neill, Staksrud, and McLaughlin, *Towards a Better Internet for Children?*

5

Readers' Responsibility

A person who is only concerned with himself, will wake up one morning and question his worth. A person who gives his time and effort to others will know his worth when he sees the fruits of his labour.

–Yoni Jesner

As use of the Internet has become widespread, it has enabled global daily communication, served as a new central public space, and provided a point of free access to a multiplicity of information sources. Those who express themselves over the Internet enjoy nearly unrestricted freedom, obtained mainly as the result of the Web's architecture. As explained in chapter 2, dissemination of information is far reaching, global, cheap, and fast.

The Internet can be used for positive purposes as well as for negative and wicked purposes. The Web is replete with detailed, accurate manuals designed to teach the average person how to harm others, assemble weapons and explosives, and commit crime. New technologies are facilitating traditional criminal activities and creating avenues for new and unprecedented forms of deviance.[1] On the Internet, people exchange fantasies of how they would like to violently rape and murder young girls.[2] On a webpage called "Place of Dark Desires,"[3] people can watch rape videos, some of which show the raping of preteen girls. People post instructions for producing weapons and bombs and for committing acts of violence, such as how to become a successful

---

[1] Matthew Williams, "Policing and Cybersociety: The Maturation of Regulation within an Online Community," *Policing and Society* 17, no. 1 (2007): 59–82, 60.

[2] Cf. *U.S. v. Baker and Gonda*, 890 F. Supp. 1375, US District Court, E.D. Michigan (June 21, 1995); *U.S. v. Alkhabaz*, 104 F.3d 1492 (6th Cir. 1997). For further discussion, see Jennifer E. Rothman, "Freedom of Speech and True Threats," *Harvard Journal of Law and Public Policy* 25, no. 1 (2001): 283–367; Raphael Cohen-Almagor, *The Democratic Catch: Free Speech and Its Limits* [in Hebrew] (Tel Aviv: Maariv, 2007).

[3] See the webpage at http://rape-videos-movies.com/scenes/8-rape-of-the-preteen-girl/ .

hitman,[4] how to build practical firearm silencers, and how to carry out car bomb attacks.[5] Although some people may celebrate this openness as a democratizing, publicly empowering characteristic of the Internet that promotes intellectual and social progress, the hazards of such websites should not be ignored.

One way to confront the dangers of boundless speech on the Web is by focusing on how people should react on encountering antisocial speech. My aim is to promote reader responsibility. This chapter begins with the story of Kimveer Gill, a 25-year-old man from Laval, Montreal, who carried out his intention to murder young students at Dawson College. It is argued that readers who encounter antisocial and violent content on the Net should share the information with relevant societal agents to forestall dangerous eruptions of violence. The case studies in this chapter show that readers who come across violent statements and suicidal calls for help often fail to take preemptive steps to avert violence and suicide. This lack of responsiveness can result from (a) an underestimation of the seriousness of the statements; (b) personal norms of behavior that will not allow them to alert the authorities; or (c) akrasia (i.e., acting against their better judgment) or other moral weaknesses such as dismissiveness, laziness, dogmatism, incuriosity, self-indulgence, or contempt for human life.

THE MURDEROUS ATTACK

On the morning of September 13, 2006, Kimveer Gill, dressed in black combat boots and a black *Matrix*-style trench coat and armed with three guns, drove his black car to downtown Montreal with murderous intent. Gill walked along a busy street – the Maisonneuve – past the Dawson Daycare Center, which daily oversees 48 toddlers. Gill disliked cigarettes, so when he saw some students smoking outside the main entrance of Dawson College, he shot two of them. Then he went inside to the college's atrium. It was lunchtime, so many students were filling the cafeteria as Gill began shooting at random with his semiautomatic weapon.[6] He killed 18-year-old Anastasia Rebecca De Sousa, a student at the college, and injured at least 20 people (4 of whom were hospitalized in critical condition).[7] The gunman showed no mercy for the

---

[4] For further discussion, see *Rice v. Paladin Enterprises Inc.*, No. 96-2412, 128 F.3d 233 (November 10, 1997).

[5] Yaakov Lappin, "Al-Qaeda's Car Bomb Guide" [in Hebrew], Ynet, January 7, 2007; Alexander Meleagrou-Hitchens, "New English-Language al-Qaeda Explosive Manual Released Online," *ICSR Insight*, International Centre for the Study of Radicalization, London, December 31, 2010.

[6] Jan Wong, "Get under the Desk," *Globe and Mail*, September 16, 2006, A8.

[7] Daniel Renaud, "Gunman Showed No Pity to Girl," *Toronto Sun*, September 15, 2006, 3.

wounded Anastasia and refused to allow a fellow student to help her.[8] Her autopsy revealed that she was shot nine times at close range.[9]

Gill took hostages and used them as human shields while the police were pursuing him. During the gunfire exchange, Gill was hit in the arm. He then committed suicide by shooting himself. This dramatic chain of events took less than 10 minutes from beginning to end.[10]

Gill had no known connection to Dawson College, the largest college in downtown Montreal. Unlike other universities in the city, it is one vast, interconnected building. At noon, the students congregate in two cafeterias.[11] Hence, it is reasonable to assume that Gill deliberately chose Dawson College because he wanted to kill as many people as possible.

KIMVEER GILL'S MENTAL CONDITION AS REFLECTED THROUGH HIS BLOG

Kimveer Gill was a depressed and troubled young man. He was an unemployed loner who lived in his parents' basement in the Montreal suburban neighborhood of Laval.[12] He enjoyed the virtual world of a website named VampireFreaks.com, dedicated to goth culture.

Gill's posts to the VampireFreaks.com website reveal his disturbed nature and provide insight into his predictable end. His screen name was Fatality666.[13] His postings reveal his likes and dislikes:

- **What would you like to achieve this year?** – Stay Alive
- **How do you want to die?** – like Romeo and Juliet, or in a hail of gunfire[14]
- **What is your favorite movie?** – *Natural Born Killers*[15]
- **What is your favorite weapon?** – Tec-9 semi-automatic handgun

Gill noted that the Tec-9 was an illegal weapon in Canada.[16] He wrote the following self-description:

---

[8] Ibid., 2.

[9] Tu Thanh Ha, Ingrid Peritz, and Andre Picard, "Shooter Had Brief Military Service," *Globe and Mail*, September 16, 2006, A9; "Anastasia Rebecca de Sousa," *NationMaster Encyclopedia*, http://www.statemaster.com/encyclopedia/Anastasia-Rebecca-de-Sousa/ .

[10] Andre Picard, "Gunman Shot Student Again and Again," *Globe and Mail*, September 15, 2006, A8.

[11] Wong, "Get under the Desk," A9.

[12] Jain Ajit, "Raging, Alienated, Gill Was a Walking Time Bomb," *India Abroad* (New York), September 22, 2006, A1.

[13] Natalie Pona, "Net Violence Unchecked," *Toronto Sun*, September 15, 2006, 4.

[14] "Profile Posted by Kimveer Gill," *National Post*, September 15, 2006, A4.

[15] Phinjo Gombu, "Web Diary, Photos Reveal Angry Man Who Loved Guns and Hated People," *Toronto Star*, September 14, 2006, A1.

[16] Ibid.

His name is Trench. You will come to know him as the Angel of Death . . . He is not a people person. He has met a handful of people in his life who are decent. But he finds the vast majority to be worthless, no good, conniving, betraying, lying, deceptive.[17]

Gill uploaded more than 50 pictures on his VampireFreaks.com page. These pictures depicted him dressed like his heroes – the gunmen from the Columbine High School shooting – in a long, black trench coat and matching boots, carrying various weapons. In one of the pictures, titled, "You're next," he pointed a handgun at the camera.[18] In another picture, he held a sign to deliver a message: "My Gothic Princess Leaves a Trail of Tears. God Has Forsaken Her. God Will Pay."[19] In his last photo on the VampireFreaks blog, he wore his signature trench coat and held up an automatic weapon. The caption read, "ready for action."[20] On his virtual tombstone he wrote, "Kimveer – Lived fast. Died young. Left a mangled corpse."[21]

Gill posted many messages on VampireFreaks; sometimes he would post entries every 15 minutes. He wrote: "I love VampireFreaks. This is my new home. I shall reside here till the day I die."[22] Excerpts from his blog expose the psychotic personality of a man who was obsessed with hate, death, and guns. For example, on March 15, 2006, Gill wrote:

> I hate this world
> I hate the people in it
> I hate the way people live
> I hate god
> I hate deceivers
> I hate betrayers
> I hate religious zealots
> I hate everything
> I hate so much
> (I could write 1000 more lines like these, but does it really matter, does
>   anyone even care)
> Look what this wretched world has done to me.[23]

[17] Phil Couvrette, "Rampage Shooter an Angry Loner," *Pittsburgh Post-Gazette*, September 15, 2006, A4.
[18] Pona, "Net Violence Unchecked," 4.
[19] Ibid., 4.
[20] "Montreal Shooting: The Blog – Excerpts 'I Hate This World . . . I Hate So Much,'" *National Post*, September 15, 2006, A4.
[21] Phil Couvrette, "College Gunman Liked Columbine Role-Play," *Sun-Sentinel* (Fort Lauderdale, FL), September 15, 2006, 20A.
[22] "Killer Likened Life to a Video Game," *Globe and Mail*, September 15, 2006, A9.
[23] "Montreal Shooting," A4.

His role models were outlaws such as Bonnie and Clyde, as well as Romeo and Juliet – couples who disregarded societal norms and died tragic deaths as a result. He admired the Germans, especially Adolf Hitler, and wrote one entry in German: "I will crush my enemies and eliminate them."[24]

In another post, he wrote, "Give them what they deserve before you go." The word "them" referred to a vast array of people, places, and things. From Gill's postings, we know that among his most hated things were comedies, governments, sunlight, and country music.[25] Gill expressed loathing toward authority figures such as police officers, teachers, and principals; he singled out "jocks" for high school bullying.[26] Furthermore, nine months before his rampage, he wrote specifically that the day in which he planned to seek revenge would be gray: "A light drizzle will be starting up."[27] Indeed, such was the weather on the day of his rampage. About two hours before the rampage, Gill wrote on the website that he had been drinking whisky in the morning ("mmmmmm, mmmmmmmmmm, good!!") and described his mood from the night before as "crazy" and "postal."[28]

Gill did not restrict his violent thoughts to his blog on VampireFreaks. He posted various disturbing and distressing comments on other websites as well. Gill's dark attitude toward the world was confirmed by personality tests he took on the Internet: A test called "Evil-O-Meter" rated him as "pure Evil." Another quiz, "Which dictator are you?," suggested that his personality was consistent with Adolf Hitler's. A personality test based on one of his favorite video games, *Postal*, rated him as having an 84 percent chance of "going postal" (i.e., of being involved in a violent massacre) and an 86 percent chance of killing someone. These outcomes were accompanied by a recommendation to seek professional help immediately.[29] A police source commented in the aftermath of Gill's rampage, "It was very obvious his state of mind was deteriorating greatly over the last three weeks."[30]

All of these materials were visible and easily accessible on the VampireFreaks website. Possibly because of this openness, Gill thought the

[24] Sue Montgomery and Jeff Heinrich, "Acting Out His Fantasy: Dawson College Gunman Posted Visions on His Blog of What He Enacted Wednesday," *Edmonton Journal*, September 15, 2006, A3.

[25] "Killer Likened Life to a Video Game," A9.

[26] Gombu, "Web Diary, Photos Reveal Angry Man Who Loved Guns and Hated People," A1.

[27] Wong, "Get under the Desk," A8.

[28] "A Blog of Violence and Death," *Newsday*, September 15, 2006, A32; "Killer Likened Life to a Video Game," A9.

[29] "Killer Likened Life to a Video Game," A9.

[30] Siri Agrell and Paul Cherry, "Blogs Reveal a Deteriorating Mind, Police Say," *National Post*, September 16, 2006, A9.

police were after him. In February 2006, in his blog, he wrote, "I know you're watching me mother-f-----s. I laugh at thee. There is nothing you can do to stop me. HA HA HA HA HA . . ."[31] Later that month, he claimed that officers were pretending to be "nice little Goth girls" as part of their surveillance.[32] Unfortunately, the police did not monitor Gill's actions. If they had, they would have undoubtedly come across Gill's disturbing virtual footprints and explicit threats:

> Turn this f---ing world into a graveyard
> Crush all those who stand in your way
> Let there be a river of blood in your wake
> Walk through that river with pride.[33]

I will discuss the issue of monitoring websites in chapters 7 (the Internet service provider's responsibility) and 8 (the state's responsibility). Regrettably, none of the readers of the busy VampireFreaks website deemed it necessary to alert law enforcement agencies about Gill's dangerous state of mind as revealed by his postings.

KIMVEER GILL AND VIOLENT VIDEO GAMES

Apart from writing in his blog, Gill enjoyed playing violent video games that exacerbated his fatal thoughts. The vast majority of video games contained violent content and many of them included violence that resulted in serious injury or death.[34] One such game was *Postal 2*. In this game, which was heavily criticized for intentionally displaying high levels of violence and racial stereotyping,[35] the user takes on the persona of "the Postal Dude" – a tall, thin man with a goatee, sunglasses, and a long, black trench coat who believes the world is out to get him. Throughout the game, the player participates in scenarios that involve killing people to advance the plot. Eventually, to complete the game, the player "goes postal," killing everyone in sight. In his blog, Gill described what he expected of a sequel to the game: "there should be more to do, other than just shooting people for no reason, there needs to be a plot and a

---

[31] Siri Agrell, "Troubled Kids 'Gravitating' to Vampire Site," *National Post*, September 15, 2006, A6.

[32] Ibid.

[33] "Killer Likened Life to a Video Game," A9.

[34] For general discussion, see Douglas A. Gentile and Craig A. Anderson, "Violent Video Games: The Newest Media Violence Hazard," in *Media Violence and Children: A Complete Guide for Parents and Professionals*, ed. Douglas A. Gentile (Westport, CT: Praeger, 2003), 131–52, 133.

[35] David George-Cosh, "Gill Attracted to Gun Violence," *National Post*, September 15, 2006, A5. New Zealand has banned the game. See Tony Smith, "New Zealand Censor Pulls Postal 2," *Register*, November 30, 2004, http://www.theregister.co.uk/2004/11/30/nz_postal_2_ban/ .

good story line."[36] Gill reflected on his feelings about the video game character by saying, "Postal dude was sad before he became angry and psychotic, that's the part we've never seen in the game. He was normal, but the world made him the way he became."[37]

Another game Gill liked playing was *Super Columbine Massacre RPG*,[38] released in 2005. The game is based on the Columbine High School shooting orchestrated by Eric Harris and Dylan Klebold on April 20, 1999, in Littleton, Colorado. The two young men killed 13 people and wounded 24 others before committing suicide.[39] In the game, the player becomes Klebold or Harris and embarks on a cartoon slaughter, walking through Columbine High School shooting students and teachers.[40] "Work sucks ... school sucks ... life sucks ... what else can I say?" wrote Gill, maintaining, "Metal and Goth kick ass. Life is a video game; you've got to die sometime."[41]

There is a demonstrable connection between (a) violent video games and movies and (b) violent crimes committed by susceptible adolescents and young people. Exposure to violent media increases hostility, aggression, and suspicions about the motives of others. Such exposure offers violence as a way to address potential conflict situations. Analysis of violent interactive video game research suggests that playing such games increases aggressive thoughts and behavior as well as angry feelings while at the same time decreasing helpful behavior.[42] The interactive component of the video games makes some video games potentially more dangerous than similar books and movies. Playing violent video games can increase a person's aggressive thoughts, feelings, and behavior. Games that simulate the maiming and killing of people harm the psyche of young players and can lead to violence. Young men who are habitually aggressive become more so when

---

[36] Ingrid Peritz, Omar El Akkad, and Tu Thanh Ha, "Seething Misfit Was Obsessed With Guns," *Globe and Mail*, September 15, 2006.

[37] "Killer Likened Life to a Video Game," A9.

[38] See the website for the game at http://www.columbinegame.com/ .

[39] Jennifer Rosenberg, "Columbine Massacre," About.com, http://history1900s.about.com/od/ famouscrimesscandals/a/columbine.htm . Information can also be found at A Columbine Site, http://acolumbinesite.com/ .

[40] Charles Gibson et al., "A Closer Look; A Mind of a Killer," *ABC News* transcript, September 14, 2006.

[41] Tony Harris, Heidi Collins, and Allan Chernoff, "Rounding Up the Enemy; Lone Gunman Opens Fire on Students in Montreal," CNN, September 14, 2006.

[42] American Psychological Association, "Resolution on Violence in Video Games and Interactive Media," August 17, 2005, https://www.apa.org/about/policy/interactive-media.pdf ; Craig A. Anderson, "Violent Video Games Increase Aggression and Violence," testimony before the US Senate Commerce, Science, and Transportation Committee hearing on the impact of interactive violence on children, March 21, 2000, http://www.psychology.iastate.edu/faculty/caa/ abstracts/2000-2004/00senate.pdf .

exposed repeatedly to violent games. Even brief exposure can temporarily increase aggressive behavior.[43]

Another troubling video game–related incident involved stepbrothers William and Joshua Buckner from Tennessee. In June 2002, the brothers decided to go on a shooting spree after playing *Grand Theft Auto III*[44] and killed a woman and a man. In other reported incidents, in July 2004, a British teenager was inspired by the gory video game *Manhunt*[45] – in which players are awarded points for killing people by vicious methods – to bludgeon a 14-year-old boy to death with a hammer, and Devin Moore, age 18, was sentenced to death in Alabama after killing two policemen and a dispatcher, replicating a scene from the video game *Grand Theft Auto: Vice City*.[46] In June 2003, Moore was brought in for questioning to a Fayette police station. He grabbed a pistol from one of the police officers and shot and killed that officer and two others before fleeing in a police car.[47]

Danny Ledonne, game creator of *Super Columbine Massacre RPG*, said he was sorry for Kimveer Gill's attack, but he did not believe the game simulation had anything to do with it. In his opinion, "*Super Columbine Massacre RPG* is a piece of art and social commentary, and it can't be blamed for the actions of a person who has lost touch with reality. . . . I regret people who misconstrue my message or people who use this video game as a means to justify their own immoral behavior."[48]

---

[43] Craig A. Anderson and Karen E. Dill, "Video Games and Aggressive Thoughts, Feelings, and Behavior in the Laboratory and in Life," *Journal of Personality and Social Psychology* 78, no. 4 (2000): 772–90. http://www.apa.org/pubs/journals/releases/psp784772.pdf ; Craig A. Anderson, Nicholas L. Carnagey, Mindy Flanagan, Arlin J. Benjamin, Janie Eubanks, and Jeffery C. Valentine, "Violent Video Games: Specific Effects of Violent Content on Aggressive Thoughts and Behavior," *Advances in Experimental Social Psychology* 36, no. 1 (2004): 199–249, http://www.psychology.iastate.edu/faculty/caa/abstracts/2000-2004/04AESP.pdf ; Douglas A. Gentile, Paul J. Lynch, Jennifer Ruh Linder, and David A. Walsh, "The Effects of Violent Video Game Habits on Adolescent Hostility, Aggressive Behaviors, and School Performance," *Journal of Adolescence* 27, no. 1 (2004): 5–22; Joan Biskupic, "Can States Keep Kids from Violent Video Games?," *USA Today*, October 29, 2010, 1A–2A.

[44] Information about the game is available at http://uk.ps2.ign.com/objects/015/015548.html .

[45] To watch a clip, go to http://www.rockstargames.com/manhunt/main.html .

[46] Information about the game is available at http://www.gamespot.com/grand-theft-auto-vice-city/ . See also "Games and Films Blamed," *National Post*, September 15, 2006, A5. In another upsetting video game, *Night of Capturing Bush*, you can play the role of a warrior who takes down US troops in an Iraqi setting while war songs play in Arabic in the background. The game culminates in a showdown with George W. Bush, the former US president. See "What Are Some Cheat Codes for Xbox on Grand Theft Auto 5?," Answer Party, http://answerparty.com/question/answer/what-are-some-cheat-codes-for-xbox-on-grand-theft-auto-5 .

[47] "Devin Moore," *Murderpedia*, http://www.murderpedia.org/male.M/m/moore-devin.htm .

[48] Brodie Fenlon, "Game Creator Says He's Sorry," *Toronto Sun*, September 15, 2006, 7.

Ledonne's excuses are weak. Although I have no intention to evoke moral panics by blaming technology, I urge the innovative minds behind video games to adhere to some norms of civility and social responsibility beyond the obvious financial incentives. I urge them to think of the social implications, not only the monetary ones. Ledonne's game plot involves entering a school to embark on a shooting spree; students are shot in the library and in the cafeteria, using real photographs, transcripts, and news coverage of the Columbine High School attack. Gill's rampage bore some similarities to the Columbine shootings.[49] Moreover, Gill specifically referred to the video games he played on his blog, saying, "Life is like a video game, you gotta die sometime," a phrase virtually identical to the one originally attributed to 18-year-old Devin Moore during his arrest.[50]

In response to Ledonne's claims, antiviolence activist Val Smith said, "For [Ledonne] to believe that you can put a game out into the public domain and expect that no one will be influenced is naive at best and possibly criminally negligent."[51] Such a passive attitude displays obvious akrasia – in this instance, acting against one's best judgment with keen attention to the financial consequences and little regard for potential antisocial consequences. The result of such negligence is habitual players copycatting the video game violence.

Violent video games were among Gill's favorite hobbies. As noted, he had a lively virtual existence on goth VampireFreaks.com. Let us discuss VampireFreaks.com to help us understand readers' behavior and responsibilities.

VAMPIREFREAKS.COM

The VampireFreaks website was founded in 1999 by a Brooklyn resident, Jethro Berelson, who calls himself "Jet." In 2006, during the Kimveer Gill affair, the site claimed to have 600,000 to 700,000 members and millions of hits.[52] VampireFreaks relates to a blood-lusty subculture of would-be vampires who are distinct from the more pacifist goths.[53] The website featured weblogs and

[49] Ibid.

[50] George-Cosh, "Gill Attracted to Gun Violence," A5.

[51] Fenlon, "Game Creator Says He's Sorry," 7. See also Pierre Boudreau, "Reconstructing Columbine," International Game Developers Association, Montreal, QB, http://legacy.igda.org/og/terms/485/22?page=9 .

[52] Pona, "Net Violence Unchecked," 4; Michele Mandel, "Out for Blood," *Toronto Sun*, September 24, 2006, 5. VampireFreaks.com appears both on Wikipedia's list of major active social networking sites and Traffkd.com's list of social media and social networking sites. See http://en.wikipedia.org/wiki/List_of_social_networking_websites and http://traffikd.com/social-media-websites/ , respectively.

[53] Margaret Philip and Caroline Alphonso, "The Geeks at the Back on Computers," *Globe and Mail*, September 15, 2006, A9.

online journals by people with usernames such as SuicideOfLove, TeenageOddity, RottingNails, RazorBladeChris, DrowningInBlood, WiltedBlood, and LoveInTheBedOfRazors. Apparently, quite a few of its members share feelings of depression, loneliness, and anger mixed with gallows humor.[54] Some of the website members were obsessed with blood, pain, rape, sadomasochism, and necrophilia.[55] One Netuser explained, "It cannot be denied that many people who share our lifestyle are fascinated by blood and death but virtually none of us would ever do anything to hurt another person."[56] As I will discuss, this statement is not altogether accurate.

VampireFreaks serves as a virtual meeting place for goths – a place where they can share a sense of community and belonging. Gill belonged there more than he did to anything in his real life. He pledged his allegiance to goth culture on VampireFreaks.com, proving that allegiance in the things he liked: jittery music, morbid poetry, Mohawk spiky hair, the signature black clothes, and the spirit of social alienation.[57]

Gill was not the first criminal over the past few years who has been connected to VampireFreaks.com. In 2005, Kevin Madden, 19, and Timothy Ferriman, 18, his self-professed vampire friend, were prosecuted for the murder of Johnathon Robert Madden, Kevin's 12-year-old brother, and for the attempted murder of Kevin's stepfather. The killer brother was a vampire fanatic with an obsession with blood sipping as sexual foreplay.[58] Johnathon was stabbed 71 times. During the trial, it was revealed that the 16-year-old former girlfriend of one of the killers blogged on VampireFreaks. The girl's posted profile on VampireFreaks listed among her likes "blood, pain . . . cemeteries and knives."[59] She was the prosecution's key witness and downplayed her interest in vampire fetishism when she testified at the jury trial.[60] When her VampireFreaks postings later became known, she was found to have perjured herself in court, and the judge declared a mistrial. A second trial found the men guilty.[61]

---

54 Robert Remington and Sherri Zickefoose, "12-Year-Old Faces Judge in Triple Murder: Boyfriend, 23, Also Accused in Deaths of Medicine Hat Family," *Edmonton Journal*, April 26, 2006, A2; Mandel, "Out for Blood," 5.

55 Kevin Connor, "T.O.'s Vampires out for Blood," *Toronto Sun*, March 6, 2005, 36.

56 Mandel, "Out for Blood," 5.

57 Jain Ajit, "Raging, Alienated, Gill Was a Walking Time Bomb," A1.

58 Connor, "T.O.'s Vampires out for Blood," 36; "Johnathon Revealed as Brother Sentenced," Canada.com, September 30, 2006, http://www.canada.com/nationalpost/news/story.html?id= e515bb40-019f-4982-9304-92ba7ced56dc&k=27748 .

59 Christie Blatchford, "Johnathon Trial Aborted," *Globe and Mail*, February 16, 2005.

60 Ibid. See also Peter Small, "Teen Found Guilty in Beating Death of Brother," Tribe.ca, February 27, 2006, http://www.tribemagazine.com/board/showthread.php?t=111821 .

61 Agrell, "Troubled Kids 'Gravitating' to Vampire Site ," A6; Michele Mandel, "Match Made in Hell," *Edmonton Sun* (Alberta), September 17, 2006, 7; Siri Agrell, "'Vampire' Blog Derailed

In April 2006, a 12-year-old girl who called herself the "Runaway Devil" and her 23-year-old boyfriend, Jeremy Allan Steinke, were charged with the triple murder of her parents and brother: Marc Richardson, 42, his wife Debra, 48, and their son Jacob, 8.[62] Both the girl and Steinke were part of the VampireFreaks community, where the girl used the online name of Killer-Kitty-X; described herself in her profile as "wiccan" and "insane"; and confessed to liking "hatchets, serial killers, and blood."[63] The seventh grade student adopted a dark goth style, with heavy eyeliner and nail polish. In one picture on her website, she posed holding a gun to the camera as she pledged her love for goth, punk, dark poetry, and death metal music.[64] Her boyfriend matched her likes; he preferred "blood, razor blades, and pain."[65] He presented himself as a 300-year-old werewolf who liked the taste of blood.[66] Steinke and his girlfriend each had personal pages on VampireFreaks and made chilling postings prior to the slayings of the Richardson family.[67] One message, posted from Steinke's souleater52 account days before the Richardson family was killed, made reference to "doing morbid stuff to others! ... which I'm going to do this weekend."[68]

After the Richardson triple killing, many goths were irritated that the media were portraying the subculture as dangerous; they took pains to say that their interest was harmless.[69] Within same month, however, VampireFreaks was once again on the news. Eric Fischer, a 23-year-old goth from New York, was arrested after showing up at a cemetery expecting to have sex with a 13-year-old girl he met on VampireFreaks. It was the second alleged incident in which Fischer used the website to lure young girls. Apparently, no reader raised an alarm. In March 2006, he had been

Murder Trial: Boastful Postings Cast Doubt on Credibility of Star Crown Witness," *Ottawa Citizen*, February 17, 2005, A6.

[62] Lloyd Robertson, "Web Links to Shooting," CTV Television, September 14, 2006.

[63] Petti Fong, "Girl Apologized to Dead Family," *Toronto Star*, July 11, 2007, A4.

[64] Holly Lake, "Linking the Internet and Goth Culture to the Medicine Hat Murders May Be Jumping to Conclusions, Experts Caution," *Ottawa Sun*, April 27, 2006, 5.

[65] Agrell, "Troubled Kids 'Gravitating' to Vampire Site," A6.

[66] James Stevenson, "Slain Boy Found in His Bed Surrounded by Blood Soaked Toys," Canadian Press NewsWire, June 12, 2007.

[67] Ian Austen, "Gunman at Montreal College Left Dark Hints of Rage Online," *New York Times*, September 15, 2006, 10; Ian MacLeod, "Vampire Culture Gets Another Black Mark after Shooting: Website Linked to Medicine Hat Slayings," *Calgary Herald*, September 15, 2006, A3.

[68] Sherri Zickefoose, "Girl Accused in Slayings Back in Court Today," *Calgary Herald*, May 1, 2006, B1; Robert Remington and Sherri Zickefoose, *Runaway Devil: How Forbidden Love Drove a 12-Year-Old to Murder Her Family* (Toronto, ON: McClelland & Stewart, 2009).

[69] MacLeod, "Vampire Culture Gets Another Black Mark after Shooting," A3.

arrested on rape charges after attacking a 16-year-old girl he had met on the website.[70]

In June 2006, three young men were sentenced for a deliberate fire that destroyed the 105-year-old Minnedosa United Church in Minnedosa, Manitoba. One of the men had posted his profile on VampireFreaks. Referring to Jesus Christ, he wrote, "If he comes back, we'll kill him again."[71] In February 2009, 36-year-old Robert Earl Hogan of Hillsboro, Oregon, was sentenced to 10 years in federal prison for luring a 14-year-old girl he had met on VampireFreaks so that he could have sex with her.[72] In August 2009, 28-year-old Derek Campbell of Toronto was charged for meeting a 13-year-old girl he chatted with on VampireFreaks and taking video of her in sexual positions. That same week, the police arrested Arthur Brown of Toronto, a 44-year-old bisexual vampire fanatic, for sexual assault of a 14-year-old girl whom he had met on the website.[73] These incidents did not significantly undermine the trust of VampireFreaks members in their social networking website. It is still a popular website in its niche.

Parry Aftab, head of Wired Safety (a volunteer watchdog organization that monitors websites), said that the goth culture is not at issue in the discussion of Gill's crime but that the role of the VampireFreaks website cannot be easily discounted. She said her organization had contacted VampireFreaks.com repeatedly about countless complaints she had received about the site's content from parents and teens who were harassed and abused. "I think the site is starting to breed a different goth," Aftab said, pointing out that "many of the kids who are highly troubled and those who are making trouble for others, are gravitating to that site." Aftab noted that "the major problem ... is that [the site] seems to normalize aberrant behavior." Aftab maintained, "Some of these kids who are troubled know they'll only get attention on there if they do something different than everyone else. You have to up the ante."[74]

Websites such as VampireFreaks.com create virtual communities and put people in touch. On such sites, when people brag that they are doing

[70] Agrell, "Troubled Kids 'Gravitating' to Vampire Site," A6; "'Vampire' Meets 'Teen' in Graveyard," UPI, April 28, 2006, http://www.upi.com/Odd_News/2006/04/28/Vampire-meets-teen-in-graveyard/UPI-84571146253763/ .

[71] MacLeod, "Vampire Culture Gets Another Black Mark after Shooting," A3.

[72] Trench Reynolds, "Man Sentenced for Luring Teen over VampireFreaks," February 18, 2009, *Trench Reynolds Crime News*, http://trenchreynolds.me/2009/02/18/man-sentenced-for-luring-teen-over-vampirefreaks/ .

[73] Kevin Connor, "'Vampire' Accused of Child Assault," *Toronto Sun*, August 21, 2009, http://cnews.canoe.ca/CNEWS/Crime/2009/08/21/10545726-sun.html . See also Earl Morningstarr, "Vampire Freak Arrested for Child Sex Crime," *Morningstarr*, August 22, 2009, http://www.themorningstarr.co.uk/2009/08/22/vampire-freak-arrested-for-child-sex-crime/ .

[74] Agrell, "Troubled Kids 'Gravitating' to Vampire Site," A6.

something outrageous, others often congratulate them and sometimes encourage such behavior.[75] After the police in Suffolk County, New York, intervened in an incident in which a "vampire" used the VampireFreaks website to arrange to meet a teen in a graveyard, Police Deputy Inspector Mark Griffiths said that VampireFreaks.com attracts people "on the fringe" who are "lonely and depressed."[76]

Thus, it is useful to place criminals within their social milieu, but it must be emphasized that the goth culture as such is not at fault for such crimes; nor should one conclude that law enforcement should specifically target websites associated with the entire goth culture. The places in cyberspace where individuals express their violent thoughts do not belong to one specific cultural group. The individuals who express murderous thoughts through the Internet have a variety of cultural backgrounds.

Certainly, Internet service providers (ISPs) should be attentive when they receive complaints and warnings. After the Gill shooting, a person known as PunchBlades posted praises of Gill as a "saint" and stated that he wished to do the same. "This will happen at Hudson High School Senior," he wrote in his post, "and when it does, I can't wait to die, or help in the process."[77] Yet the one Netuser who tried to do the right thing and report PunchBlades complained online that his efforts were completely ignored by the administrators of VampireFreaks.com. It was only after the story appeared in the newspapers that Quebec police moved in and arrested a 15-year-old teen and charged him with uttering Internet threats. The teen had allegedly written that Gill was not a very good shot and claimed that he could do a much better job and that he could not wait to die at his own school.[78]

READERS' RESPONSIBILITY

On VampireFreaks.com, Gill was not reprimanded for his postings. Quite the contrary, he received moral support from his website friends. On Tuesday, September 12, 2006, just a day before the shooting, Caranya, a 19-year-old member from Indiana, wrote to Gill, "Can I go play with you?? I wanna go hunt down the preppies with you!!"[79] Subsequent postings from visitors to

---

[75] Maurice DuBois, "Parry Aftab of Wiredsafety.org Discusses Monitoring of Web Sites That Could Influence Violent Behaviors among Its Users," *CBS News* transcript, September 16, 2006.

[76] "'Vampire' Meets 'Teen' in Graveyard."

[77] Mandel, "Out for Blood," *Toronto Sun*, 5.

[78] Marci Ien, "Dawson College Students Back in Class Today," CTV Television, September 19, 2006.

[79] Allison Hanes, Sean Silcoff, and Graeme Hamilton, "Gunman Fantasized about Rampage," *National Post*, September 15, 2006, A1.

Caranya's webpage reprimanded her: "Congratulations on inspiring a psycho to go on a murderous rampage killing innocent kids," wrote one. Another posted, "One has to wonder where he was able to get his moral support from."[80] Gill's blog was immediately removed after the killing, but not before a stream of online comments were posted, most of them denouncing Gill. One, however, read, "I've been to Dawson College. The people there are so superficial I actually thought about shooting the school up myself. Thank you, unknown guy with a Mohawk. I salute you."[81] A 16-year-old VampireFreaks member named Melissa from Sherbrooke, Quebec, expressed her surprise that Gill was responsible for the Montreal tragedy. "I found him super cool," she said. "There was nothing strange about his blog."[82]

Despite the violent messages included in Gill's profile and postings, no one reported him to the police. Responsible readers of websites should be alert to problematic postings and speak out when they read warnings of troubled individuals who seem to be on the verge of explosion. Teachers, administrators, parents, and peers are often the firsthand recipients of such expressions. They can help prevent violence by seeking treatment for raging people who show the type of behavior that might erupt into violence.

Responsible Netusers who do not wish to be identified or do not wish to report directly to legal authorities should report to hotlines. Hotlines enable Netusers to report Internet content that they find problematic. I will elaborate on the issue of hotlines in chapter 9.

INTERNET WARNINGS

The catharsis theory holds that venting anger produces a positive improvement in the psychological state of an angry person. The word *catharsis* comes from the Greek word *Katharsis*, which means cleansing or purging. According to the catharsis theory, accumulated aggressive energy needs a release. A person who bottles up rage often seeks ways to let off stream. The catharsis theory holds that releasing aggression is an effective way to purge or reduce angry and aggressive sentiments.[83]

Scientific tests, however, have supported the catharsis theory only in part. Tests confirm the first assumption: people need to vent. The second hypothesis, that releasing aggressive sentiments reduces aggression, has been disputed and negated. Venting involves behaving aggressively, often against safe,

---

[80] Dahlia Lithwick, "Networking Born Killers," *Slate Magazine*, September 23, 2006.
[81] Robertson, "Web Links to Shooting."
[82] Agrell, "Troubled Kids 'Gravitating' to Vampire Site," A6.
[83] David G. Myers, *Social Psychology* (New York: McGraw-Hill, 1993), 447–54.

inanimate objects. It keeps angry feelings alive in memory and actually increases the likelihood of subsequent aggressive responses.[84] Venting, thus, does not reduce anger and aggression. Repeated tests show that subsequent interpersonal aggression remains high after venting, in stark contrast to what the catharsis theory led people to believe.[85] Expressing hostility breeds more hostility.

Usually, killers do not just snap and start shooting. Kimveer Gill was a walking bomb ready to explode, filled with growing rage and hatred. According to Kevin Cameron, a traumatic stress expert, "Serious violence is an evolutionary process."[86] The process begins with bitterness, degenerates into anger and rage, and – if there are no mitigating circumstances – sometimes ends with a violent explosion. People need to vent their hostility, their acrimony, their anger. They provide signs and hints. They find it difficult to contain all the boiling emotions inside them. In the Internet age, it is convenient for them to be able to vent their feelings into the virtual world. Canadian anthropologist Elliott Leyton describes the most desperate and unchecked of these Netusers as people "who had looked upon their own lives and pronounced them unlivable," the result of which is often gore, murder, and suicide – an exaction of revenge for the pain they feel, even at the sacrifice of their lives.[87]

The April 1999 Columbine slaughter set the benchmark for Gill and other killers, not only in terms of violence but also in terms of using of the Internet to publicize their malevolent thoughts and schemes. In profanity-laden postings, killer Eric Harris began making threats months prior to the Columbine shootings. On his personal Web page he wrote, "I hate you people for leaving me out of so many fun things."[88] Relating to natural selection, he wrote that it was "the best thing that ever happened to the Earth. Getting rid of all the stupid and weak organisms."[89] Harris could not have been more explicit: "I will be armed to the f---ing teeth and I WILL shoot to KILL and I WILL f---ing

[84] Brad J. Bushman, "Does Venting Anger Feed or Extinguish the Flame? Catharsis, Rumination, Distraction, Anger, and Aggressive Responding," *Personality and Social Psychology Bulletin* 28, no. 6 (2002): 724–31, 725.

[85] Brad J. Bushman, Roy F. Baumeister, and Angela D. Stack, "Catharsis, Aggression, and Persuasive Influence: Self-Fulfilling or Self-Defeating Prophecies?," *Journal of Personality and Social Psychology* 76, no. 3 (1999): 367–76, 374.

[86] Eric Strachan, "Gill, Games, Goth, and Guns," *Pembroke Observer* (Ontario), September 16, 2006, 18.

[87] Christie Blatchford, "Social Analysis of Violent Acts Could Be Key to Prevention," *Globe and Mail*, September 15, 2006, A10.

[88] Perry Swanson and Kim Nguyen, "Web Rants Raise Red Flags for Violence: But Police Can Do Little to Prevent Attacks," *Gazette* (Colorado Springs), December 16, 2007.

[89] Ibid.

KILL EVERYTHING. It'll be very hard to hold out until April."[90] Warsaw University criminologist Kacper Gradon states that, time after time, the graphic warnings have been in plain view on Internet discussion boards and websites.[91] There is a pattern, and this pattern should not be ignored.

On March 21, 2005, 16-year-old Native American Jeff Weise went on a killing spree. He was armed with three guns and multiple rounds of ammunition and was wearing a bulletproof vest that belonged to his grandfather, who was a police officer. He began the spree by killing his grandfather and his grandfather's female companion. Then, he drove his grandfather's police car to Red Lake High School in the Red Lake Indian Reservation in Minnesota. He murdered the 28-year-old unarmed security guard who stood at the school's front door. Inside the school, he killed a teacher and five students in cold blood and severely injured seven more people before shooting himself in the head.[92]

Weise came from harsh circumstances. During his short life, he lost several close family members: his father committed suicide when he was only eight years old, and two years later, a car accident killed his cousin and paralyzed and brain damaged his mother. Weise himself had tried to kill himself. Red Lake High School expelled him for flunking classes and misbehaving; later, he returned to the school and killed six of his victims there.[93] Weise's online presence reveals his distorted mind and tormented soul. His MSN profile page shows the disturbing nature of his thoughts:

Occupation: Doormat
My MD Category Interests: Military, High Schools, Death & Dying
A Little About Me: 16 years of accumulated rage suppressed by nothing more then brief glimpses of hope, which have all but faded to black. I can feel the urges within slipping through the cracks, the leash I can no longer hold. . . .
Favorite Things: Moments where control becomes completely unattainable. . . .

---

[90] Timothy Apple, "Hiding in Plain Website: Killers from Columbine to Dawson College Have Broadcast Their Intentions Online Long before Going on Their Murderous Rampages," *Globe and Mail*, April 22, 2008.

[91] Ibid. For further discussion on the Columbine shooting, see David L. Altheide, *Terror Post 9/11 and the Media* (New York: Peter Lang, 2009): 117–33.

[92] Rhea R. Borja, "Details of Minn. School Shooting Emerge," *Education Week*, March 22, 2005; "School Killing Rampage: Student Slays Nine, Self on Minnesota Indian Reservation," *San Diego Union-Tribune*, March 22, 2005, http://www.utsandiego.com/uniontrib/20050322/news_7n22shooting.html ; Jodi Wilgoren, "Shooting Rampage by Student Leaves 10 Dead on Reservation," *New York Times*, March 22, 2005, http://www.nytimes.com/2005/03/22/national/22shoot.html .

[93] Ceci Connolly and Dana Hedgpeth, "'The Clues Were All There': School Shooter Depicted as Deeply Disturbed, Ignored Teen," *Washington Post*, March 24, 2005, http://www.sfgate.com/cgi-bin/article.cgi?file=/c/a/2005/03/24/MNG7UBU2GT1.DTL&type=printable .

Times when maddened psycho paths briefly open the gates to hell, and let
chaos flood through. . . .
Those few individuals who care enough to reclaim their place. . . .
Hobbies and Interests: Planning Waiting Hating
Favorite Quote: "We are little flames, poorly sheltered by frail walls against
the storm of dissolution and madness, in which we flicker and sometimes
almost go out." – Wehrmacht Private Paul Baumer, *All Quiet on the Western
Front*.[94]

Weise created a violent, blood-soaked, 30-second animated video titled
"Target Practice."[95] The video showed a man shooting four people and
blowing up a police car before committing suicide. Weise uploaded this
video, under the nickname "Regret," to NewGrounds.com, a multimedia
website, in October 2004.[96] Several weeks later, he posted a second short
video (50 seconds long) titled "Clown," which showed a character who is
eventually strangled by a clown.[97] In a brief bio attached to his "Regret"
NewGrounds.com profile and accompanying his Flash animations, Weise
mentioned some of his favorite movies. Among them were Gus Van Sant's
2003 film, *Elephant*, which is about a Columbine-style school shooting. On
his MSN profile page, he included a still picture from *Elephant* showing
two teenage characters who are dressed in camouflage, carry duffle bags
containing weapons, and are headed for the school door.[98]

Jeff Weise was also active on the neo-Nazi website Nazi.org. Under the
names "NativeNazi" and "TodesEngel" ("Angel of Death" in German), he
posted 34 messages on the website forum.[99] In one message, he stated, "I guess
I've always carried a natural admiration for Hitler and his ideals, and his
courage to take on larger nations."[100] In another message, Weise claimed
that he had been questioned by the police in 2004 regarding an alleged plot
to shoot up the school on Adolf Hitler's birthday, but he denied any

[94] See "School Killer's Animated Terror: Minnesota Teen Posted Bloody Flash Film Late Last
Year," *Smoking Gun*, http://www.thesmokinggun.com/documents/crime/school-killers-anima
ted-terror . The article contains a screen capture made on March 24, 2005, of Jeff Weise's MSN
Profile Page. See also Roddy Kenneth Street Jr., "The Devil's Massacre: How Satan
Devastates Native American Tribes," Worldwide Christian Tracts, 2005, http://www.world
widechristiantracts.net/1/StreetTract_S6T1_v1d_WCTN.html . You can view Jeff Weise's journal, "Thoughts of a Dreamer," at http://weise.livejournal.com .
[95] The animated video can still be seen at http://www.newgrounds.com/portal/view/195194 .
[96] Ibid.
[97] "School Killer's Animated Terror."
[98] Ibid.
[99] Ibid. See also "Teen Rampage Shooter Admired Hitler, Took Prozac," InfoWars.com, March
23, 2005, http://www.infowars.com/articles/us/teen_rampage_leaves_10_dead.htm .
[100] "Minnesota School Shooter on Neo-Nazi Web Sites," Reuters, March 22, 2005, http://www.
infowars.com/articles/us/teen_rampage_leaves_10_dead.htm#websites .

connection.[101] Despite the numerous explicit warning signs, not enough was done to prevent Weise from carrying out his murderous intentions.

In September 2006, 19-year-old Alvaro Castillo killed his father before a going on violent rampage at Orange High School in Hillsborough, North Carolina. Fortunately, there were only a few minor injuries among the students. Castillo's MySpace page lists "handguns, shotguns, and rifles" among his "general interests." One of his pictures depicts him brandishing a pair of scissors as he appears ready to stab an unidentified young male in the head. The caption reads, "Attempted Murder." Like Gill, Castillo was apparently obsessed with the 1999 Columbine massacre. On August 29, just before he took a sawed-off shotgun to his old high school, Castillo sent a videotape and letter to the *Chapel Hill News* claiming to be obsessed with Columbine.[102]

In December 2007, 24-year-old Matthew Murray killed four people at two religious sites in Colorado before taking his own life. Murray had posted numerous online rants, blaming his rage on his mother, Christians, and others. Murray quoted extensively from the Web postings that were published more than eight years earlier by Columbine High School killer Eric Harris. Murray had repeated this stark description of Harris's state of mind: "I'm full of hate and I love it." He also warned, "I'm coming for everyone soon, and I will be armed."[103] Unfortunately, Murray's venting did not prompt any responsible readers to alert the appropriate authorities.

On November 7, 2007, in Jokela, north of Helsinki, Finland, 18-year-old Pekka-Eric Auvinen shot 8 people and wounded 10 others before committing suicide.[104] Auvinen left plenty of visual material on the Internet over the previous hours, days, and weeks, but no reader voiced concerns. Before the shooting, Auvinen placed a video on the YouTube website with the title "Jokela High School Massacre – 11/7/2007." The Jokela High School clip was one of about 89 videos posted on the site under the username "Sturmgeist89" ("Storm Spirit" in German).[105] The video showed a picture of the school, which then receded to reveal an image of a man resembling Auvinen against a red background, pointing a gun at the screen.[106] The video

[101] Ibid.
[102] Lithwick, "Networking Born Killers."
[103] Swanson and Nguyen, "Web Rants Raise Red Flags for Violence."
[104] "Six Pupils, Nurse and Headmistress Killed in a Finnish School Shooting," NewsRoom, This Is Finland, November 8, 2007; "Breaking News: Death Toll in School Shooting Reaches Nine," *Helsingin Sanomat*, November 8, 2007.
[105] The videos were uploaded to the website LiveLeak at http://www.liveleak.com/view?i= bb4_1194559199 (no longer available). See, however, this video posted on Dailymotion: http://www.dailymotion.com/video/x3fc5m_sturmgeist89-practises-blank-shooti_news .
[106] The video was uploaded at http://www.liveleak.com/view?i=369_1194449557 (no longer available).

clip was accompanied by the song "Stray Bullets" by KMFDM, a German industrial rock band.[107] The same song was uploaded to the Internet by Columbine killer Eric Harris.[108]

The YouTube Sturmgeist89 profile indicates that the teenager appeared to be fascinated with killing. Auvinen's postings included video footage of the Columbine High School shootings, the 1993 Waco siege in the United States, the 1995 sarin gas attack in Tokyo, and bombs falling on Baghdad during the 2003 invasion. Many videos showed victims being wheeled away or people running for their lives. Throughout the videos, the word *DIE* constantly flashed on the screen. Other video clips included Nazi war criminals.[109]

In September 2008, 22-year-old Matti Juhani Saari from Kauhajoki, a provincial town in western Finland, opened fire at the town's School of Hospitality. He gunned down 10 people before shooting himself in the head.[110] Once again, the writing was on the wall. Under the username "Wumpscut86,"[111] Saari had uploaded to the YouTube website four videos showing him using his 22-caliber handgun.[112] One of them, titled "Me and My Walther P22 Target,"[113] is believed to have been uploaded just five days before the massacre. In it, the gunman – wearing a black leather jacket and black jeans – is seen shooting a handgun several times at an offscreen target. This video bears a chilling resemblance to the video posted on YouTube by Pekka-Eric Auvinen. Both men had photo-manipulated the images, coloring them in red and black.[114]

On the mornings of the shootings, both men also acted similarly. Auvinen and Saari updated their web profiles just before embarking on their killing sprees. Pekka-Eric Auvinen tried to set the Jokela High School on fire, and

---

[107] For more information about KMFDM, see the band's website at http://www.kmfdm.net/bio/ .

[108] Sakari Suoninen, "Finland School Shooter Admired Hitler, Nietzsche," Reuters, November 8, 2007, http://www.reuters.com/article/worldNews/idUSL0831998120071108?pageNumber=1 .

[109] "Finland in Mourning after Fatal School Shooting," CNN, November 8, 2007, http://edition. cnn.com/2007/WORLD/europe/11/08/school.shooting/ .

[110] Peter Popham and Toby Green, "YouTube Gunman Slaughters 10 Students," *Independent*, September 24, 2008, http://www.independent.co.uk/news/world/europe/youtube-gunman-slaughters-10-students-940306.html .

[111] Wumpscut is the name of a German electro-industrial Goth band. See the band's website at http://www.wumpscut.com/ .

[112] "Profile: Finnish School Suspect," BBC News, September 23, 2008, http://news.bbc.co.uk/2/hi/ europe/7631786.stm . For a screenshot of the killer's full YouTube profile, see http://www. antville.org/static/spacecat90/files/wumpscut86.html .

[113] See Saari's YouTube profile at http://www.antville.org/static/spacecat90/files/wumpscut86. html .

[114] "Updated 19:00: Saari Copied Jokela Killer Pekka-Eric Auvinen in Everything He Did," *Helsingin Sanomat*, September 24, 2008.

there was a fire at the Kauhajoki School of Food Management, apparently set by Saari's Molotov bottle bombs.[115]

Among Matti Saari's favorite videos were clips from the Columbine massacre in 1999.[116] His YouTube page included what seem to be lyrics from a song called "War" by Wumpscut. One part reads, "Whole life is war and whole life is pain, and you will fight alone in your personal war. War, this is war!"[117] On another website, LiveLeak.com, Saari's video depicted him looking down at the camera and pointing a gun toward it. "You will die next," he said in English before shooting four times.[118]

On January 21, 2009, Zhu Haiyang decapitated a fellow student at Virginia Tech. As in the preceding cases, the killer vented his anger and frustration over the Internet. Zhu Haiyang expressed frustration over problems, including stock losses. His post on a Chinese-language blog dated January 7, 2009, read, "Recently I've been so frustrated I think only of killing someone or committing suicide."[119]

On January 8, 2011, Jared Lee Loughner emptied his pistol at a constituent event held by Congresswoman Gabrielle Giffords in Tucson, Arizona. He killed 6 people and injured 13 others.[120] Months before the rampage, Loughner expressed his anger and frustration on the Internet, announcing his murderous intentions. On his MySpace page, Loughner ranted about the government spending illegal money and his lack of trust in the police; he referred several times to suicide and to killing authorities. On December 20, 2010, he wrote, "I HAVE THIS HUGE GOAL AT THE END OF MY LIFE: 165 rounds fired in a minute!"[121]

On July 22, 2011, Anders Behring Breivik perpetrated Norway's largest massacre since World War II. He used a car bombing to hit a government building in central Oslo, killing 8 people and injuring 15 others. Later that

---

[115] Ibid.

[116] "Profile: Finnish School Suspect."

[117] See Saari's YouTube profile at http://www.antville.org/static/spacecat90/files/wumpscut86. html .

[118] "'You Will Die Next': Finland School Killer Matti Saari," OddCulture.com, September 23, 2008.

[119] Sue Lindsey, "Police Search Homes of Va. Tech Victim, Suspect," Associated Press, January 23, 2009; "Virginia Tech Student Decapitated," Sky News, January 23, 2009, http://news.sky. com/story/663943/virginia-tech-student-decapitated .

[120] Sarah Garrecht Gassen and Timothy Williams, "Before Attack, Parents of Gunman Tried to Address Son's Strange Behavior," *New York Times*, March 27, 2013, http://www.nytimes.com/ 2013/03/28/us/documents-2011-tucson-shooting-case-gabrielle-giffords.html .

[121] "Tucson Gunman before Rampage: 'I'll See You on National T.V.'" CBS News, April 11, 2014, http://www.cbsnews.com/news/jared-loughner-who-shot-gabrielle-giffords-in-tucson-ranted-online/ .

same day, Breivik opened fire at a Norwegian Labour Party youth camp on the island of Utøya, killing 69 people – mainly children and adolescents. After the massacre, it was found that Breivik had expressed his hatred of Muslims on a Norwegian anti-Muslim website named Document.no.[122] Shortly before the attack, Breivik posted on the Internet a manifesto that included his lengthy operational diary. The manifesto contained detailed information about Breivik's life, upbringing, and political ideology; advice on weapons training, explosives testing, body armor, and tactical and logistical planning; ideas on how to build a paramilitary organization; lists of the categories of people in European society who needed to be killed; a system of medals and ribbons to be awarded to fighters in the coming civil war; and tombstone designs for those who were killed.[123]

On May 23, 2014, Elliot O. Rodger killed 6 people and wounded 13 others in a shooting spree in Santa Barbara, California. Rodger was a lonely and frustrated young man. He vented his frustrations on the Internet in a series of 21 videos, which he uploaded onto YouTube. Shortly before the killing spree, Rodger had posted a video in which he recounted the isolation and sexual frustrations of his life and voiced his intent to go on "a mission of retribution." He spoke of the women who rejected him, the happiness he saw around him, and his life as a virgin at the age of 22. He called his message "Elliot Rodger's Retribution" and said it was the last video he would post. "It all has to come to this," Rodger said. "Tomorrow is the day of retribution. The day I will have my retribution against humanity. Against all of you. For the last eight years of my life, ever since I hit puberty, I've been forced to endure an existence of loneliness, rejection and unfulfilled desires. All because girls have never been attracted to me. In those years I've had to rot in loneliness."[124] Apparently, law enforcement officers knew about the videos. Police officers visited Rodger three weeks before his rampage and questioned him about the

---

[122] James Ridgeway, "Oslo Shooting: Read Anders Behring Breivik's Internet Comments Here," *Mother Jones*, July 23, 2011, http://www.motherjones.com/politics/2011/07/anders-behring-breiviks-online-comments .

[123] Jerome Taylor, "Faces of Hatred: Norway Mass Killer's Life Laid Bare," *Independent*, July 25, 2011, http://www.independent.co.uk/news/world/europe/faces-of-hatred-norway-mass-killers-life-laid-bare-2319892.html ; Scott Stewart, "Norway: Lessons from a Successful Lone Wolf Attacker," *Stratfor Security Weekly*, July 28, 2011, http://www.stratfor.com/weekly/20110727-norway-lessons-successful-lone-wolf-attacker#axzz3FagZkeZ5 .

[124] Ian Lovett and Adam Nagourney, "Video Rant, Then Deadly Rampage in California Town," *New York Times*, May 24, 2014, http://www.nytimes.com/2014/05/25/us/california-drive-by-shooting.html?emc=edit_th_20140525&nl=todaysheadlines&nlid=33802468&_r=0 . Elliot Rodger's manifesto, "My Twisted World: The Story of Elliot Rodger," has been posted at http://www.scribd.com/doc/225960813/Elliot-Rodger-Santa-Barbara-mass-shooting-suspect-My-Twisted-World-manifesto .

disturbing videos. They did not watch the videos, and they were not aware of his "Day of Retribution" video until after the deadly rampage.[125]

The common denominator connecting these episodes of violence is that all the killers had vented their murderous intentions on Internet social networking sites. Social networks create virtual friends. Many members have hundreds and even thousands of friends. Friendship involves showing care and compassion and being available when support is needed. As Aristotle said, friendship is about possessing moral character, having awareness of others and responsibility for one's conduct.[126] Aristotle distinguished between (a) friendship between base people based on pleasure or utility and (b) true friendship between good people. Only the friendship of good people is immune to slander.[127] He said that "if people are friends, they have no need of justice, but if they are just they need friendship in addition; and the justice that is most just seems to belong to friendship."[128]

The staggering thing about the episode involving Elliot Rodger is that his therapist and mother were concerned about the videos he posted online and alerted the police, but the investigating officers did not pursue the alert as seriously as they should have.[129] Likewise, in similar violent episodes, such as Gill's, or in the many cases of cyberbullying that have ended in suicide (discussed in chapter 4), virtual friends paid little or no attention to the pain of the distressed person and failed to make any attempt to provide help. In other cases, discussed next, Netusers even contributed to the misery that drove their "friends" to commit suicide and urged them to finish their lives.

SUICIDE

In *Groundwork for the Metaphysic of Morals*, Immanuel Kant asks whether suicide can be reconciled with the idea of humanity as an end in itself and answers in the negative. If one escapes from one's burdensome situation by destroying oneself, one is using a person merely as a means to keeping oneself in a tolerable condition up to the end of one's life. However, Kant argues, "a man is not a *thing* [*Sache*], so he isn't something to be used *merely* as a means, and must always be regarded in all his actions as an end in himself. So I can't

---

[125] Michael R. Blood and Martha Mendoza, "Police Knew but Didn't View Elliot Rodger's Videos," Associated Press, May 30, 2014, http://www.utsandiego.com/news/2014/May/29/police-knew-but-didnt-view-elliot-rodgers-videos/2/ .

[126] Aristotle, *Nicomachean Ethics*, ed. and trans. Martin Ostwald (Indianapolis, IN: Bobbs-Merrill, 1962), books VIII and IX.

[127] Ibid., book VIII, chapter 4.

[128] Ibid., book VIII, chapter 1.

[129] Blood and Mendoza, "Police Knew but Didn't View Elliot Rodger's Videos."

dispose of a man by maiming, damaging, or killing him – and that includes the case where the man is myself."[130]

In a number of troubling episodes, people have used the Internet to announce their intention to commit suicide. Sometimes readers behave responsibly and save lives. For instance, in one case, a British 16-year-old boy sent a private suicide threat through Facebook to a girl 3,600 miles away, in Maryland. The girl told her parents, who in turn called the local state police. The only details the police were given were scant facts from the boy's Facebook profile. The police in Britain managed to narrow down the boy's home to eight possible addresses and dispatched officers to each one. Three hours after his message was sent, the teenager was found at the fourth address, alive but suffering from an overdose of prescription pills. The teen was rushed to the nearest hospital, where he made a full recovery. Chief Superintendent Brendan O'Dowda commented, "Without the girl in Maryland, this wouldn't have happened. It is a credit to her to have been brave enough to have instigated this."[131] This case exemplifies the positive use of technology and international cooperation that saved a life. In addition, online forums exist that provide support or succor for suicidal teens.[132]

However, a significant number of readers and watchers on the Internet have not only neglected trying to stop attempted suicides but also encouraged people to go through with their attempts. Such behavior exemplifies akrasia at its worst. Netusers are becoming detached from the impact of what they say online. They seem to believe that they are merely playing in a movie in which there are no consequences for their misconduct. The anonymity and privacy that the Internet offers desensitizes people, making some of them devoid of compassion, care, and any sense of responsibility or accountability.

In 2008, 19-year-old Abraham Biggs of Florida committed live public suicide by overdosing on pills. Biggs suffered from manic depression and had threatened to commit suicide in the past. He exchanged messages with readers on Bodybuilding.com and broadcast his last actions on Justin.tv, detailing the amount of drugs he intended to take. Apparently, he wanted to share his last moments with others and make death, usually a very private issue, public. The moderators of the forum did not take him seriously because of his unfulfilled past threats. Hundreds of people were watching Biggs online

[130] Immanuel Kant, *Groundwork for the Metaphysic of Morals*, trans. Jonathan Bennett ([1865] 2008), 29, http://www.redfuzzyjesus.com/files/kant-groundwork-for-the-metaphysics-of-morals.pdf .

[131] Robin Henry, "Facebook Friend Saves a Life," *Times Online*, April 5, 2009, http://www.thetimes.co.uk/tto/technology/article1859442.ece .

[132] For instance, Aasra is a crisis intervention center for the lonely, distressed, and suicidal. See the organization's website at http://www.aasra.info/aboutus.html .

as he swallowed pills before collapsing on his bed and appearing to fall unconscious. While the video was still streaming, viewers finally called the police, who broke the door, found the body, and switched off the camera. Up to 1,500 people were viewing.[133] In a message posted on Bodybuilding.com, Biggs had described how he felt and explained why he intended to end his life. In response, some other users of the site egged him on. Not only did people neglect trying to stop Biggs or calling the police, but they also aired abuses and encouraged him to commit suicide.[134] "You want to kill yourself?" said one. "Do it, do the world a favour and stop wasting our time with your mindless self-pity."[135] Similar words pushed another unstable young person, Megan Meier, to commit suicide (see chapter 4).

As Biggs was lying in bed after taking the lethal pills, many forum members continued to insult him.[136] Other viewers debated whether the lethal dose was sufficient to kill Biggs.[137] As in the Meier tragedy, here was an asymmetric situation, where one person was communicating with many, subjecting himself to a potentially vulnerable position that could engender humiliation. The entire situation could have been prevented if those who had read Biggs's earlier blog entries had behaved responsibly and alerted the police. Instead, the viewers of his final video watched passively as he killed himself. It was 3:00 a.m. when Biggs first said that he would attempt suicide. He died 12 hours later. The akratic Netusers must have been aware of the gravity of the situation, but they were not moved by it. It is striking that so many lingering hours elapsed and so many people watched before someone decided to do the responsible thing and notify the police.[138] As for Justin.tv, its moderators

[133] Bobbie Johnson, "Police Investigate as Teenager Appears to Kill Himself on Video Website," *Guardian*, November 21, 2008, http://www.guardian.co.uk/technology/2008/nov/21/internet-video-overdose-teenager ; Mike Harvey, "Horror as Teenager Commits Suicide Live Online," *Times Online*, November 22, 2008, http://www.thetimes.co.uk/tto/news/world/amer icas/article1998615.ece .

[134] Tudor Vieru, "Florida Teen Commits Suicide on Web Cam," *Softpedia*, November 23, 2008, http://news.softpedia.com/news/Florida-Teen-Commits-Suicide-on-Web-Cam-98467. shtml ; Emily Friedman, "Florida Teen Live-Streams His Suicide Online," ABC News, November 21, 2008, http://abcnews.go.com/Technology/MindMoodNews/story?id= 6306126&page=1 .

[135] Johnson, "Police Investigate as Teenager Appears to Kill Himself on Video Website."

[136] "Florida Teen Commits Suicide Live on Web Cam," Fox News, November 21, 2008), http://www.foxnews.com/story/0,2933,455784,00.html .

[137] Rasha Madkour, "College Student, 19, Kills Himself on Live Web Cam," *San Diego Union-Tribune*, November 22, 2008, http://www.signonsandiego.com/uniontrib/20081122/news_1 n22suicide.html .

[138] Vieru, "Florida Teen Commits Suicide on Web Cam." For a discussion on suicide and ethics, see Raphael Cohen-Almagor, *Speech, Media, and Ethics: The Limits of Free Expression* (Houndmills, UK: Palgrave, 2005), 105–23.

were either unaware or did not care. The site got many hundreds of hits during the airing of the video.

Justin.tv removed the video of Biggs and the transcript of the chat screen that had been running along with it. Those who egged on Biggs to commit suicide rapidly deleted their posts.[139] In a statement on the company's website, chief executive Michael Siebel commented that "a tragedy ... occurred within our community today" and that "we respect the privacy of the broadcaster and his family during this challenging time.... We have policies in place to discourage the distribution of distressing content, and our community monitors the site accordingly. This content was flagged by our community, reviewed and removed according to our terms of service."[140] The message board on Bodybuilding.com, where Biggs left his suicide note, appears to have been deleted, but many of the comments left in response to it were unsympathetic. Later reports confirmed that reckless website viewers encouraged Biggs to end his life.[141]

In 2003, Brandon Vedas, a 21-year-old from Phoenix, Arizona, collapsed from a fatal overdose after he was urged to take the lethal drugs by other irresponsible Netusers. One akratic Netuser told Vedas, "That's not much.... Eat more. I wanna see if you survive or if you just black out."[142] Vedas's brother Rich said after Brandon's death, "These people treat it like somehow it's not the real world. They forget it's not just words on a screen."[143] Rich Vedas noted that while some viewers expressed remorse, many of them did not grasp the reality of the situation, "that my brother has died as a result of what happened that night."[144]

The common denominator of these and other similar cases[145] is that the Net readers allowed preventable suicides to occur, either by akratic passivity or by verbal prompting. Instead of attempting to hinder the suicides, they became

---

[139] Saskia E. Polder-Verkiel, "Online Responsibility: Bad Samaritanism and the Influence of Internet Mediation," *Science and Engineering Ethics* 18, no. 1 (2012): 117–41; Therese Lisieux, "Abraham Biggs Suicide: Biggs Bodybuilding.Com Candyjunkie Suicide Photos," November 20, 2008, http://celebgalz.com/abraham-biggs-suicide-biggs-bodybuilding-com-candyjunkie-suicide-photos/ .

[140] Harvey, "Horror as Teenager Commits Suicide Live Online."

[141] Johnson, "Police Investigate as Teenager Appears to Kill Himself on Video Website."

[142] Kelley Beatty, "Tragedy in a Chatroom," WiredSafety Group, February 18, 2003, https://www.wiredsafety.org/resources/editorial/0001.html .

[143] Ibid.

[144] "Online 'Suicide' Brother's Pain," BBC News, February 19, 2003, http://news.bbc.co.uk/1/hi/technology/2773547.stm .

[145] See, for example, Boaz Wollinitz, "The Boys Egged on Facebook – and the Boy Committed Suicide" [in Hebrew], *Walla!*, January 5, 2011, http://news.walla.co.il/?w=/10/1776345 . In another case, in January 2009, a young Polish man streamed his suicide live over an Internet chat site. One Netuser witnessing the act immediately alerted police, but officers who arrived on the scene found him dead. See "Polish Man Streams His Suicide on the Internet," CNN, January 27, 2009,

encouraging participants. They failed to show even a scant amount of either human compassion or respect for human life.

CYBERSTALKING

Stalking is a pattern of "repeated, unwanted attention, harassment, and contact."[146] According to the National Center for Victims of Crime, it is "a course of conduct directed at a specific person that would cause a reasonable person to feel fear."[147] Stalking may include following or waiting secretly for the victim; continuously sending the victim unwanted, intrusive, and frightening communications; damaging the victim's property; making direct or indirect threats to harass, intimidate, or harm the victim or the victim's family, friends, and pets; repeatedly sending the victim unwanted items; and obtaining the victim's personal information with the intent to follow and harm the victim.[148]

Cyberstalking, also called *online stalking* and *Internet stalking,* is a crime-facilitating speech. It is a form of stalking that involves computer communication. Using the obvious advantages of the Internet, stalkers can operate anonymously from the privacy of their homes, retrieving information about the victim by using Web crawlers without having to venture out into the physical world. Both offline and online stalkers are driven by a desire to exert some control and power over their target. According to the British Crime Survey, more than 1 million women and 900,000 men are stalked in the United Kingdom every year.[149] In the United States, in 1999, 1 out of every 12 women (8.2 million) and 1 out of every 45 men (2 million) reported being stalked at some time in their lives.[150] In 2010, 1 in 6 women in the United States (19.3 million women, or

http://edition.cnn.com/2009/WORLD/europe/01/27/poland.website.hanging/index.html?iref=24 hours .

[146] "Get Informed," Avalon: A Center for Women and Children, Williamsburg, VA, http://www. avaloncenter.org/get-informed .

[147] "Stalking," National Center for Victims of Crime, Washington, DC, http://www.victimsof crime.org/our-programs/stalking-resource-center/stalking-information#what .

[148] Ibid. See also Michael L. Pittaro, "Cyber Stalking: An Analysis of Online Harassment and Intimidation," *International Journal of Cyber Criminology* 1, no. 2 (2007): 180–97.

[149] "Researchers Seek to Find True Level of Cyberstalking," BBC News, September 24, 2010, http:// www.bbc.co.uk/news/uk-11404284 . See also Samantha Fenwick, "Cyber-stalking Laws: Police Review Urged," BBC News, April 30, 2011, http://www.bbc.com/news/13200185 .

[150] US Department of Justice, 1999 *Report on Cyberstalking: A New Challenge for Law Enforcement and Industry* (Washington, DC: US Department of Justice), http://www.clintonlibrary.gov/assets/ storage/Research%20-%20Digital%20Library/ClintonAdminHistoryProject/11-20/Box%2015/1225 098-justice-appendix-b-vol-2-3-4.pdf.pdf ; Patricia Tjaden and Nancy Thoennes, "Stalking in America: Findings from the National Violence against Women Survey, *Research in Brief,* April 1998, https://www.ncjrs.gov/pdffiles/169592.pdf ; "Cyberstalking Is a Growing Problem," Berkman

16.2 percent) and 1 in 19 men (5.9 million men, or 5.2 percent) reported being stalked during their lifetime. For both female and male victims, stalking was often committed by people they knew or with whom they had had a relationship. Two-thirds of the female victims of stalking (66.2 percent) reported stalking by a current or former intimate partner and nearly one-quarter (24.0 percent) reported stalking by an acquaintance. About 1 in 8 female victims (13.2 percent) reported stalking by a stranger. Repeatedly receiving unwanted phone calls or unwanted voice or text messages was the most commonly experienced stalking tactic for both female and male victims. Cyberstalking included unwanted emails, instant messages, and messages through social media.[151]

The Manhattan District Attorney's Office estimates that about 20 percent of its sex crimes involve cyberstalking. Women are twice as likely as men to be victims of stalking by strangers and eight times as likely to be victims of stalking by intimates.[152] Los Angeles law enforcement officials believe that email and the Internet are involved in about 20 percent of the stalking crimes that they handle.[153] Citizens' initiatives to tackle cyberstalking are important because many victims will not go to the police, preferring to share their concerns with other sources. One such initiative is Working to Halt Online Abuse (WHOA), which was established by Jayne Hitchcock in 1997 to fight online harassment by educating the public and law enforcement personnel and by empowering victims. Hitchcock says that WHOA receives hundreds of cyberstalking reports each year. WHOA helps victims halt online abuse by contacting the victims' ISPs and convincing them to cease their harassment, threats, and malicious rumors.[154] WHOA receives an estimated 50 to 75 cases of online harassment or cyberbullying per week.[155]

The Murder of Amy Boyer

The tragic story of Amy Boyer serves as a poignant example of the dangers attached to cyberstalking. Boyer was 20 years old when she was murdered by

---

Center for Internet and Society, Cambridge, MA, http://cyber.law.harvard.edu/vawoo/cyberstalking_problem.html .

[151] Michele C. Black, Kathleen C. Basile, Matthew J. Breiding, Sharon G. Smith, Mikel L. Walters, Melissa T. Merrick, Jieru Chen, and Mark R. Stevens, *The National Intimate Partner and Sexual Violence Survey: 2010 Summary Report* (Atlanta: National Center for Injury Prevention and Control, Centers for Disease Control and Prevention, 2011), 29–34.

[152] US Department of Justice, *1999 Report on Cyberstalking*; Tjaden and Thoennes, "Stalking in America"; "Cyberstalking Is a Growing Problem."

[153] US Department of Justice, *1999 Report on Cyberstalking*.

[154] See WHOA's website at http://www.haltabuse.org/about/about.shtml .

[155] "Online Harassment/Cyberstalking Statistics," WHOA, http://www.haltabuse.org/resources/stats/ .

21-year-old Liam Youens. Boyer had no idea that Youens had been stalking her for more than eight years. The young man had been infatuated with her since they had met in the eighth grade. Although Youens had made very little impression on Boyer, Youens became obsessed with the girl who had, in his mind, turned him down, even though they had not exchanged more than a few words. Like Kimveer Gill, Youens was unemployed and lived with his parents. He surfed the Internet for hours at a time and learned to use it for his base purposes. Whatever information Youens could not acquire on the Internet, he found by paying a private information broker through Docusearch.com.[156] Youens purchased Boyer's date of birth, Social Security number, and home and job addresses. He became Boyer's shadow, but she was absolutely oblivious to his obsession. Like Gill, Youens announced his murderous intentions on the Internet and glorified the Columbine killers. Besides Boyer, Youens also wanted to kill a man named Owen and all the students from Nashua High School who had ever mocked him.[157] On October 15, 1999, Youens murdered Boyer and then killed himself.

The Amy Boyer episode raises questions about the social responsibility of ISPs, of readers, and of information broker companies. The next two chapters discuss ISPs' liability. The current topic is readers' responsibility. Youens set up two websites dedicated to Boyer. One was descriptive in nature, containing her personal information and photos. The second website served the purpose of venting. Youens wrote excessively about his feelings toward and plans for Amy Boyer. The obsessive Youens wrote, "in the last month I've dreamt about her every single night. The last dream I had Amy was pregnant, so I stabed [*sic*] the fetus through her, then cut her throat down to the bone, and broke her neck with my hand."[158]

It is not known how many Netusers read Youens's website and became aware of his murderous intentions during the two and half years of the site's operation. On his website, Youens mentioned that he consulted "Pieter" about his plans. At one point, Docusearch, the data company he paid to find Boyer's

[156] Herman T. Tavani, *Ethics and Technology: Controversies, Questions, and Strategies for Ethical Computing* (Hoboken, NJ: Wiley, 2011), 2.

[157] Robert Douglas, "The Murder of Amy Boyer," *News for Public Officials*, April 13, 2005, http://www.davickservices.com/murder_of_amy_boyer.htm ; "Web Link To Murder," CBS News, October 19, 1999, http://www.cbsnews.com/news/web-link-to-murder/ ; Amy Lynn Boyer website, http://www.netcrimes.net/Amy%20Lynn%20Boyer_files/Amy%20Lynn%20Boyer.htm ; "Understanding Amy Boyer's Law: Social Security Numbers, Crime Control, and Privacy," Privacilla.org, December 2000, http://www.privacilla.org/releases/AmyBoyer.html ; "The Amy Boyer Case," Electronic Privacy Information Center, Washington, DC, http://epic.org/privacy/boyer/ .

[158] See the reproduction of Youens's website at http://www.netcrimes.net/Amy%20Lynn%20Boyer_files/liamsite.htm .

details, was unable to trace her work address, and Youens wrote: "hmmm . . . aparently those idiots cant find her employer if any. i could have set fire to her house and shot her earlier, but now i fear she isnt there anymore. and plan Owen is nothing because he dropped out (thankgod). Pieter recomends i go on a rampage, but i dont know. wish i had bought a vest and 100rnd drum."[159] Youens was able to execute his wicked plan without interference. No one alerted law enforcement.

The Association of Independent Information Professionals (AIIP) provides an example of a responsible data company.[160] To become a member of AIIP, an individual must agree, in writing, to accept AIIP's Code of Ethical Business Practices, which states, in part, that members must "uphold the profession's reputation for honesty, competence, and confidentiality" and "accept only those projects which are legal and are not detrimental to our profession."[161] Docusearch, in contrast, does not seem to show any concern for ethics. It prides itself on being "America's Premier Resource for Private Investigator Searches & Lookups."[162] Its website states, "We provide articles relevant to specific searches and present feature columns written by leaders in the investigative community. If you require the personal touch, you may always contact us. Happy Hunting!"[163] The casual statement, "Happy Hunting!," is anything but ethical or professional. It typifies Docusearch's behavior, and it is disconcerting and irresponsible.

With such companies, a person can pay for retrieving all the information he or she needs for identity theft or, in this case, stalking for the purpose of murder. No one questioned Youens about his intentions. Youens himself was struck by the ease with which he was able to purchase Boyer's personal information while concealing his evil intentions. He wrote, "I found an internet site to do that, and to my surprize everything else under the Sun. . . . It's accually obsene what you can find out about a person on the internet. I'm waiting for the results."[164] Indeed, the ease of access to personal information is troubling. Financial gains dismiss any shred of social responsibility. The private investigator–information broker did not take any constructive steps to find out who Youens was or why he needed the information about Boyer. Had Docusearch merely conducted a Google search on "Amy Boyer Liam

---

[159] Ibid.

[160] For more information about AIIP, see the organization's website at http://www.aiip.org/ .

[161] The full Code of Ethical Business Practices is available at http://aiip.org/content/code-ethical-business-practices .

[162] "America's Premier Resource for Private Investigator Searches and Lookups," Docusearch.com, http://www.docusearch.com/about.html .

[163] "How We Can Help: The Basics," Docusearch.com, http://www.docusearch.com/basics.html .

[164] Douglas, "The Murder of Amy Boyer."

Youens," it would likely have found Youens's website documenting his intent to murder Boyer.

Moral and social responsibilities were not the only factors of concern in the Boyer case. Legal responsibility was pertinent, too. After the murder, Amy's mother and stepfather filed suit for wrongful death against Docusearch and the investigators with whom the company had subcontracted to obtain Boyer's personal information. In February 2003, the New Hampshire Supreme Court found Docusearch civilly liable for the criminal acts of its stalker-customer; it deemed stalking and identity theft to be foreseeable risks of selling personal information.[165] If a private investigator's disclosure of information creates a foreseeable risk of criminal misconduct against the person whose information is disclosed, the private investigator should be held liable. In obtaining Boyer's Social Security details from a credit reporting agency without Boyer's knowledge or consent, Docusearch invaded Boyer's privacy. The court found that Docusearch had "a duty to exercise reasonable care in disclosing a third person's personal information to a client."[166] When Docusearch sold Boyer's Social Security number and employment address, stalking and identity theft were "sufficiently foreseeable" risks.[167] This is especially true since the investigator did not know Youens and his purpose in seeking the information. Thus, Docusearch was negligent in selling Boyer's personal information, and Docusearch was liable for the harm that came to her.

In 2000, as part of the Violence Against Women Act, Congress extended the federal interstate stalking statute to include cyberstalking and adopted what came to be known as "Amy Boyer's Law." The law prohibits the sale or display of an individual's Social Security number to the public, including through the Internet, without prior consent. The law allows a person harmed by wrongful release of a Social Security number to sue the wrongdoer for equitable relief and monetary damages in a US District Court. In addition, the Social Security commissioner can impose a civil penalty of $5,000 for each violation, with increased penalties (maximum of $50,000) if the privacy violations constitute a general business practice.[168]

Obsessed Net abusers use cyberstalking to harass people they dislike; to take revenge; to settle accounts; and to victimize and to mock peers, teachers, colleagues, neighbors, and others. In 2007, Tyler Yannone and Lauren

---

[165] *Remsburg v. Docusearch, Inc.*, No. 2002-255, 2003 N.H. LEXIS 17 (N.H. Feb. 18, 2003).

[166] Ibid.

[167] Ibid. For further discussion, see Richard A. Spinello, "Informational Privacy," in *The Oxford Handbook of Business Ethics*, ed. George G. Brenkert and Tom L. Beauchamp (New York: Oxford University Press, 2010), 366–87.

[168] See 42 U.S.C. section 1320 B-23 P.L 106-553.

Strazzabosco, two students attending Mooresville High School in North Carolina, were charged with cyberstalking after creating a MySpace profile that depicted a school administrator as a pedophile.[169] A similar situation occurred a year later when a Providence High School freshman was charged with cyberstalking after he set up a website that suggested that a male teacher was a pedophile. Four other students were disciplined pursuant to the Charlotte-Mecklenburg Schools code of student conduct, which bans students from distributing "any inappropriate information, relating in any way to school issues or school personnel, distributed from home or school computers."[170] In 2010, 17-year-old Shannon Marie Mitchell was arrested and charged with aggravated cyberstalking for posting messages about her boyfriend's former girlfriend on an adult website. Mitchell was apparently jealous of the 15-year-old ex-girlfriend, so she posted photos of the young girl, along with her phone number, describing the various sexual acts that the girl was ready to perform. The result was a landslide of calls to the girl's phone from patrons of the porn site, as well as police involvement.[171]

CONCLUSION

The Internet is open to all. It has served killers. It should also serve the positive elements in society to prevent murders.

The fundamental principle of social responsibility rests on the duty to make humanity itself our end. The way to do this is to promote the ends that autonomous human beings freely choose as long as they do not harm others. The important lesson learned from the tragic incidents described is the urgent need to promote responsibility among readers. Netusers who entertain murderous thoughts use the Net to vent their hostilities. Those murderers

---

[169] "2 North Carolina High School Students Charged with Cyber Stalking for Fake MySpace Page," Fox News, March 29, 2007, http://www.foxnews.com/story/0,2933,262524,00.html ; "Mooresville Teens Accused of Creating Fake Web Sites Depicting School Administrators," WSOCTV.com, March 23, 2007, http://www.wsoctv.com/news/11349703/detail.html ; Mark Turner, "Mooresville Students Charged with Cyberstalking," http://www.markturner.net/2007/03/30/mooresville-students-charged-with-cyberstalking/ .

[170] "South Charlotte Teen Charged With Cyberstalking of Teacher," WSOCTV.com, January 31, 2008, http://www.wsoctv.com/news/15186700/detail.html ; Chris Riedel, "The Fight against Cyberbullying," *Journal*, January 5, 2008, http://thejournal.com/articles/2008/05/01/the-fight-against-cyberbullying.aspx .

[171] Pete Kotz, "Shannon Marie Mitchell, 17, Refused to Stop Cyber-Stalking Boyfriend's Ex, Gets Jail," *True Crime Report*, March 4, 2010, http://www.truecrimereport.com/2010/03/shannon_marie_mitchell_17_refu.php ; Kristin Giannas, "Ocala Girl Charged with Aggravated Cyberstalking," WCJB-TV, March 5, 2010, http://www.wcjb.com/news/6258/ocala-girl-charged-with-aggravated-cyberstalking .

did not see others as ends in themselves. They gave their own lives a meaning by killing others and then killed themselves. They socialized with others through the Internet, even when the others were obscure and unknown to them. The Internet's seemingly anonymous nature, its wide dissemination, and its ease of use make it a perfect platform for hostile venting. Such signs should not go unnoticed or ignored. We should devise appropriate mechanisms for readers to report others' antisocial and murderous intentions. If they so desire, readers should be able to maintain their privacy. But it is the responsibility and duty of Netusers to report violent postings. After all, human lives are at stake.

Obviously, not all shooters display online anger before committing their crimes. It is equally true that violent and gory speech or images are produced online by vast numbers of individuals who do not go on to attack people. But the ascending frequencies in which killing sprees happen and the fact that sometimes those who entertain murderous intentions vent their frustrations and aggression online require action on national and international levels. Preemptive measures could prevent the translation of murderous thoughts into murderous actions. Such cooperation, through voluntary and organized operations, must include all sectors: civil society organizations and businesses, and especially ISPs and website administrators and owners.

In the next two chapters, I discuss the most intricate issue of ISP responsibility. I examine the arguments for and against a proactive stance in overseeing content and applying discretionary judgment as to what is published on the Net. In this context, I also discuss relevant business aspects of ISP and Web-hosting conduct and then examine the implications of applying social responsibility principles to fight social problems that have become rampant on the Internet: terror, child pornography, cyberbullying, and hate.

6

Responsibility of Internet Service Providers and Web-Hosting Services, Part I: Rationale and Principles

Commerce is as a heaven, whose sun is trustworthiness and whose moon is truthfulness.

–Bahá'u'lláh

For many years, the red-light district in Amsterdam has been one of the most notorious pornographic centers in the world. Dozens of sex shops offer merchandise for all human preferences, tastes, and perversions. I went to a few of them and asked for child pornography. The shop-keepers all said that they did not carry such material. They had thou-sands of magazines and videos; how did they know, with such unshaken confidence, that no such material had found its way into the store? When I stubbornly asked, "Are you sure?," the response was a suspicious stare and angry conviction that no child pornography existed in their stores. They had taken measures to ascertain that such illegal material would not be available to the public.

Likewise, rowdy Internet service providers (ISPs) and Web-hosting services (WHSs) support rowdy websites. Would expecting them to scrutinize their sites to avoid certain illegal material be unrealistic? Does responsibility dictate for them to do this?

The issue of responsibility of ISPs and host companies is interesting and complex. Their actions and inactions directly affect the information environment. This chapter and the one that follows are concerned with social responsibility of ISPs and Web-hosting services WHSs. I distinguish between *access* providers that have no control over the messages and information they carry and *service* providers that offer content. Should the latter make the effort to monitor their sites for problematic informa-tion, or are they relieved of any responsibility? This is arguably the most intriguing and complicated issue of Net responsibility. I argue that

ISPs and WHSs should aspire to take responsibility for content. In the next chapter, I assert that ISPs and WHSs should adopt a proactive stance in countering antisocial and violent content. This policy would benefit not only their customers and society at large but also their business. In chapter 8, I further contend that they should respect and abide by the laws of the countries in which they operate.

In *The Philosophy of Information*, Luciano Floridi addressed the issue of the truthfulness of data, which he termed *alethic neutrality*.[1] I, in turn, wish to speak of different meanings of neutrality: (a) net neutrality as a nonexclusionary business practice, highlighting the economic principle that the Internet should be open to all business transactions; (b) net neutrality as an engineering principle, enabling the Internet to carry the traffic uploaded to the platform; and (c) net neutrality as content nondiscrimination, accentuating the free speech principle. I call the last *content net neutrality*. While endorsing the first two meanings of net neutrality, I argue that ISPs and WHSs should scrutinize content and discriminate against not only illegal content (for instance, child pornography and terrorist activity on the Net) but also content that is morally repugnant. Here the concept of responsibility comes into play. Many morally repugnant types of net speech, such as child pornography, terrorism, and criminal activities, are covered by law and are widely considered illegal, whereas others, such as hate speech and bigotry, are not necessarily so. *Hate speech* is defined as a bias-motivated, hostile, malicious speech aimed at a person or a group of people because of some of their actual or perceived innate characteristics. It expresses discriminatory, intimidating, disapproving, antagonistic, or prejudicial attitudes toward the disliked target group.

Article 1(1) of the International Convention on the Elimination of All Forms of Racial Discrimination defines *racial discrimination* as

> any distinction, exclusion, restriction or preference based on race, colour, descent, or national or ethnic origin which has the purpose or effect of nullifying or impairing the recognition, enjoyment or exercise, on an equal footing, of human rights and fundamental freedoms in the political, economic, social, cultural or any other field of public life.[2]

Under the convention, the dissemination of hate speech is prohibited because it is aimed to injure, dehumanize, harass, intimidate, debase, degrade, and victimize the targeted groups and to foment insensitivity and brutality against

[1] Luciano Floridi, *The Philosophy of Information* (Oxford: Oxford University Press, 2010), 80–107.
[2] United Nations, International Convention on the Elimination of All Forms of Racial Discrimination, adopted and opened for signature and ratification December 21, 1965, entry into force January 4, 1969, http://www.ohchr.org/EN/ProfessionalInterest/Pages/CERD.aspx .

them. Liberal democracies have the responsibility to protect weak third parties. They should be concerned about the predicament of vulnerable people who are targeted because of their race, religion, ethnicity, culture, or sexual preference. Media organizations and Internet facilitators should share this responsibility and concern.

CORPORATE SOCIAL RESPONSIBILITY IN INFORMATION AND COMMUNICATION TECHNOLOGY

The satiric journalist Ambrose Bierce defined the *corporation* as an ingenious device for obtaining profit without individual responsibility.[3] Conversely, the Corporate Social Responsibility (CSR) movement was established to evoke responsibility in corporations. Its main principles dictate (a) integrated, sustainable decision making that takes into consideration the positive and negative potential consequences of decisions; (b) obligation on the part of corporations not only to consider different stakeholders (individuals or groups that have a vested interest in the firm) and interests but also to incorporate them into the decision-making processes; (c) transparency, which is vital for ensuring accountability; (d) consistent respect for societal and environmental ground rules, set by the company in light of public concerns (e.g., giving equal treatment to women and minorities, welcoming people with disabilities, avoiding use of certain harmful chemicals); (e) precautionary steps to be taken before implementing agreed-upon decisions; (f) liability for decisions and enactment of remedial measures to redress harm inflicted as a result of conduct; and (g) investment in the community to benefit the public good.[4]

Archie Carroll articulated in his seminal work that beyond a firm's obvious economic and legal obligations, the social responsibility of businesses encompasses ethical and discretionary responsibilities. Business is expected, by definition, to make a profit. Society expects business to obey the law. In addition, ethical responsibilities include adherence to ethical norms. By *ethical norms*, Carroll means adherence to fairness, justice, and due process.

[3] Ambrose Bierce, *The Devil's Dictionary* (New York: Neale, 1911), http://www.alcyone.com/max/lit/devils/ .

[4] Michael Kerr, Richard Janda, and Chip Pitts, *Corporate Social Responsibility: A Legal Analysis* (Markham, ON: LexisNexis, 2009); Kenneth E. Goodpaster, "Corporate Responsibility and Its Constituents," in *The Oxford Handbook of Business Ethics*, ed. George G. Brenkert and Tom L. Beauchamp (New York: Oxford University Press, 2010), 126–57; Gabriel Abend, *The Moral Background: An Inquiry into the History of Business Ethics* (Princeton, NJ: Princeton University Press, 2014). See also David Weissbrodt's review of Kerr, Janda, and Pitts, *Corporate Social Responsibility*, in *Human Rights Quarterly* 32 (2010): 207–15; William B. Werther and David B. Chandler, *Strategic Corporate Social Responsibility: Stakeholders in a Global Environment* (Thousand Oaks, CA: Sage, 2010).

By *discretionary responsibilities*, Carroll refers to philanthropic contributions and nonprofit social welfare activities.[5] Carroll's pyramid of CSR depicted the economic category at the base and then built upward through legal, ethical, and philanthropic categories. In his view, a company with good CSR practices should strive to make a profit while obeying the law, and it should behave ethically as a good corporate citizen.[6]

Information and communication technology (ICT) and other CSR can be practiced in four forms. In its purest form, CSR is practiced for its own sake. The firm expects nothing back from its ethical and responsible activities, and it becomes socially responsible because that is the noble way for corporations to behave. Following a Kantian perspective, a corporation is viewed as a moral community in which all stakeholders both create the rules that govern them and are bound to one another by those same rules. A second, less pure form of CSR is manifested when it is undertaken for enlightened self-interest: firms undertake ethical conduct with the belief that such conduct pays. The benefit could be tangible or intangible, but in either case, payback is expected. This second form of CSR is related to the third form, in which ethical and responsible conduct is seen as a sound investment. According to the "sound investment theory," the stock market reacts to firms' actions and will reward socially responsible behavior. Michael Porter argues that companies ought to invest in CSR as part of their business strategy to become more competitive.[7] Of course, no business can solve all of society's problems or bear the cost of doing so. Instead, each company should identify the issues that intersect with its particular business and be attentive to consumers' concerns.[8] The fourth form of CSR, also related to enlightened self-interest, is ethical conduct practiced to avoid interference from external political influences. In this

---

[5] Archie B. Carroll, "A Three-Dimensional Conceptual Model of Corporate Social Performance," *Academy of Management Review* 4, no. 4 (1979): 497–505; Archie B. Carroll, *Business and Society: Managing Corporate Social Performance* (Boston: Little, Brown, 1981). See also Andrew Crane, Dirk Matten, Abagail McWilliams, Jeremy Moon, and Donald S. Siegel, eds., *The Oxford Handbook of Corporate Social Responsibility* (Oxford: Oxford University Press, 2009).

[6] Archie B. Carroll, "Corporate Social Responsibility," *Business and Society* 38, no. 3 (1999): 268–95; Archie B. Carroll and Ann K. Buchholtz, *Business and Society: Ethics, Sustainability, and Stakeholder Management*, 8th ed. (New York: South-Western College, 2011), especially chapters 2 and 6.

[7] Michael Porter, "CSR: A Religion with Too Many Priests?," *European Business Forum*, September 22, 2003.

[8] Michael E. Porter and Mark R. Kramer, "Strategy and Society: The Link between Competitive Advantage and Corporate Social Responsibility," *Harvard Business Review*, December 2006, 78–92. For further discussion, see Michael C. Jensen, "Value Maximization, Stakeholder Theory, and the Corporate Objective Function," *Business Ethics Quarterly* 12, no. 2 (2002): 235–56.

case, firms become socially responsible to prevent the authorities from forcing them to be so through legislation.[9] Yahoo!, Google, and other Internet giants would certainly prefer self-regulation to regulatory legislation.

Many businesses embrace the idea of CSR because of the advantages and benefits that it brings. Companies are becoming more socially responsible because such conduct enhances their public image and reputation, increases customer loyalty, results in a more satisfied and productive workforce, diminishes legal problems, and contributes to a stronger and healthier community.[10]

CSR in ICT makes humanity increasingly accountable, morally speaking, for the way information is transferred.[11] Members of these professions are trained to practice a core skill, requiring autonomous judgment as well as expertise. ICT professionals have an inviolable duty to abide by the terms of service and see that their clients are satisfied. Their work is based on knowledge and skill. Certain standards and qualifications are expected to be maintained, following an accepted code of practice that observes wider responsibilities to clients and society.[12] But code words and intentions are not enough. A business cannot build a reputation on mere declaration of intentions. Reputation is built by acting on the code of conduct and applying it.

Online intermediaries encompass conduits such as ISPs, platforms such as video-sharing sites and social networking sites that allow Netusers to access online content and interact with each other, and Web-hosting companies that provide Internet connectivity and space on a server they own for use by their clients. Their responsibility is defined in terms of obligations accepted by employers in relation to their employees and by suppliers in relation to customers and clients. Many responsibilities are customary, subject to negotiation according to the interests and balance of power of the parties involved, and dictated by competition necessities. The acceptance and fulfillment of responsibilities by business actors is mainly determined by

---

[9] Wan Saiful Wan-Jan, "Defining Corporate Social Responsibility," *Journal of Public Affairs* 6, no. 3 (2006): 176–84, 178. See also Stefan Tengblad and Claes Ohlsson, "The Framing of Corporate Social Responsibility and the Globalization of National Business Systems," *Journal of Business Ethics* 93, no. 4 (2010): 653–69; John R. Boatright, *Ethics and the Conduct of Business* (Upper Saddle River, NJ: Pearson Prentice Hall, 2009).

[10] Archie B. Carroll and Ann K. Buchholtz, *Business and Society: Ethics and Stakeholder Management*, 6th ed. (Mason, OH: South-Western, 2006), 42–45.

[11] Luciano Floridi and J. W. Sanders, "Artificial Evil and the Foundation of Computer Ethics," *Ethics and Information Technology* 3, no. 1 (2001): 55–66.

[12] Compare these responsibilities to the responsibilities of the press; see Denis McQuail, *Media Accountability and Freedom of Publication* (New York: Oxford University Press, 2003), 191; Raphael Cohen-Almagor, *Speech, Media and Ethics: The Limits of Free Expression* (Houndmills, UK: Palgrave, 2005), 87–123.

considerations of long-term self-interest and maintenance of good customer relations, although ethical principles may also play a part.[13] Let me illustrate the tension between freedom of information, on the one hand, and moral and social responsibility, on the other, which can have significant business implications not only for the company at hand but also for other information and communication companies in the future. Striking a balance and finding the Aristotelian Golden Mean is not an easy task. Indeed, some companies pay only lip service to this need (in the form of terms of service) or forfeit it altogether.

RESPONSIBILITY OF ISPS AND WHSS

An *Internet access provider* is any organization that arranges for a Netuser or a company to have access to the Internet. An *Internet service provider* is a company or other organization that provides services, content, or technical services for the use or operation of content and services on the Internet, usually for a fee, thereby enabling users to establish contact with the public network. Many ISPs also provide email service; storage capacity; proprietary chat rooms; and information regarding news, weather, banking, or travel. Some offer games to their subscribers. A *Web-hosting service* is a service that runs various Internet servers. The host manages the communications protocols and houses the pages and the related software required to create a website on the Internet. The host machine often uses the Unix, Windows, Linux, or Macintosh operating systems, which have the Transmission Control Protocol (TCP) and Internet Protocol (IP) built in.[14]

General agreement exists in both the United States and Europe that the access provider should not be held responsible for the content of messages. Once the connection to the network has been made, the access provider has no influence either on what material moves through the wires or where it goes. Access providers are much like the postal system in that they do not know the contents of the messages they deliver. In Europe, this position has been codified in the E-Commerce Directive of the European Union.[15] In the

---

[13] McQuail, *Media Accountability and Freedom of Publication*, 191. I expand this discussion in chapter 8. For further deliberation, see Bryan Horrigan, *Corporate Social Responsibility in the 21st Century: Debates, Models, and Practices across Government, Law, and Business* (Northampton, MA: Edward Elgar, 2010).

[14] Preston Gralla, *How the Internet Works*, 8th ed. (Indianapolis, IN: Que, 2007), 173.

[15] Directive 2000/31/EC of the European Parliament and of the Council of June 8, 2000, on certain legal aspects of information society services, in particular electronic commerce, in the Internal Market, http://eur-lex.europa.eu/LexUriServ/LexUriServ.do?uri=CELEX:32000 L0031:EN:NOT . For further discussion, see Mark F. Kightlinger, "A Solution to the *Yahoo!*

United States, so-called common carrier provisions allow certain carriers of communications to carry all manner of traffic without liability (e.g., telephone service providers), and more recently Congress granted limited immunity to access providers for violations of copyright law in the Digital Millennium Copyright Act.[16]

WHSs, however, are a different story. A host provider may be a portal or a proprietary service that gathers a large amount of information for user access. Closer to a virtual forum site or bazaar than to a postal system, WHSs provide Web space, help subscribers find material more easily, and establish "bulletin boards" and email services. Generally, the host provider does not have anything to do with the content placed on the server, but it has a good deal to do with its organization in the "marketplace."[17]

Because the host provider offers more than a connection service, the question of liability is more complicated. Legal systems have to determine when the value added by the host provider's services begins to make it look less like an access provider and more like a content provider. The task is made all the more difficult as new technologies create new business opportunities for inventive entrepreneurs and the services offered by host providers change.[18]

I intend to probe the issue of ISP and host company responsibility from the ethical and social perspectives; others have provided in-depth legal analysis.[19]

Problem? The E-Commerce Directive as a Model for International Cooperation on Internet Choice of Law," *Michigan Journal of International Law* 24, no. 3 (2003): 719–66; Lee A. Bygrave, "Germany's Teleservices Data Protection Act," *Privacy Law and Policy Reporter* 5 (1998): 53–54, http://folk.uio.no/lee/oldpage/articles/Germany_TDPA.pdf .

[16] The Digital Millennium Copyright Act of 1998, Pub. L. No. 105-304, 112 Stat. 2860 (October 28, 1998); National Research Council, *Global Networks and Local Values: A Comparative Look at Germany and the United States* (Washington, DC: National Academies Press, 2001), 119–20.

[17] National Research Council, *Global Networks and Local Values*, 120.

[18] Ibid. For further discussion, see Jan Van Dijk, *The Network Society* (London: Sage, 2012), 151–57; Milton L. Mueller, *Networks and States: The Global Politics of Internet Governance* (Cambridge, MA: MIT Press, 2010).

[19] Website hosting can give rise to liability for trademark infringement and copyright infringement but generally will not give rise to liability for defamation. The leading American legal precedents on Internet offensive speech are *Cubby, Inc. v. CompuServe*, 776 F. Supp. 135 (S.D. N.Y. 1991); *Stratton Oakmont, Inc. v. Prodigy Services Co.*, 1995 WL 323710 (N.Y. Sup. Ct. May 24, 1995); *Blumenthal v. Drudge*, 992 F. Supp. 44 (D.D.C. 1998); *Zeran v. AOL*, 129 F.3d 327 (4th Cir. 1997), cert. denied, 524 U.S. 937 (1998); *Doe v. America Online*, 25 Media Law Rep. (BNA) 2112, 1997 WL 374223 (Fla. Circ. Ct. June 26, 1997), 718 So. 2d 385 (Fla. Dist. Ct. App. 1998). At present, American courts tend to hold that ISPs are not liable for content posted on their servers, which, under section 230(1) of the Communications Decency Act (1996), is discussed in chapter 7 of this volume. For analysis, see Matthew C. Siderits, "Defamation in Cyberspace: Reconciling *Cubby, Inc. v. Compuserve, Inc.* and *Stratton Oakmont v. Prodigy Services Co.*," *Marquette Law Review* 79, no. 4 (1996): 1065–82; Sarah B. Boehm, "A Brave New World of Free Speech: Should Interactive Computer Service Providers Be Held Liable for the Material They Disseminate?," *Richmond Journal of Law and Technology* 5, no. 2 (1998): 7;

Now, it is one thing to argue that a *moderator* of a specific site should be held liable for content on the sites he or she manages. This would seem a fair demand for small sites and arguably also for large sites. It is another thing to argue that *ISPs* should be held liable for content. Those who object to the idea of holding ISPs responsible for content on their servers argue that the Internet is like a telephone carrier. Both provide communication service. You cannot hold a phone carrier, say Verizon, liable for using the phone line to plan a crime. Verizon provides a service, which can be abused. The same – so it can be argued – is true for Internet providers. On both the phone and the Internet, we find the basic components of communication: a sender, a message, a channel, and a receiver. Generally speaking, both the phone and the Internet also provide opportunity for feedback.

All this is true. But the Internet is different from the phone in some critical technological, organizational, and geographic ways that make the comparison unconvincing. Let me uncover some of the major differences. First, the Internet is based on a technology that uses connectionless protocols for host-to-host resource sharing called packet switching. Packet switching represents a significantly different communications model. As explained in chapter 2, packet switching is based on the ability to place content in digital form.[20] Diffused routers, rather than main switches, are the key to delivering the packets to the intended destination.[21]

Second, the history of the Internet is based on innovation and flexibility. Access is not controlled at a central switchboard. Netusers do not require permission to develop new ideas and to add new functionalities. The Domain Name System allows the diffusion of information as long as the common protocols are used, thereby enabling individuals and entities to connect to the World Wide Web.

Third, the volume, scope, and variety of data that the sender is able to transmit over the Internet are much larger than what can be sent over the phone. Over the phone, two or more people can speak. On the Internet,

---

Michelle J. Kane, "Internet Service Provider Liability: Blumenthal v. Drudge," *Berkeley Technology Law Journal* 14 (1999): 483–501; Mitchell P. Goldstein, "Service Provider Liability for Acts Committed by Users: What You Don't Know Can Hurt You," *John Marshall Journal of Computer and Information Law* 18, no. 3 (2000): 591–641; Jonathan A. Friedman and Francis M. Buono, "Limiting Tort Liability for Online Third-Party Content under Section 230 of the Communications Act," *Federal Communication Law Journal* 52, no. 3 (2000): article 12; Joel Voelzke, "Don't Shoot: I'm Just the Web Site Host!," *Computer and Internet Lawyer* 20, no. 5 (2003): 4–15.

[20] Luciano Floridi explains the noteworthy advantages of the binary system of data encoding in *Information: A Very Short Introduction* (Oxford: Oxford University Press, 2010), 27–28.

[21] John Mathiason, *Internet Governance: The New Frontier of Global Institutions* (Abingdon, UK: Routledge, 2009), 7.

people can speak, chat, send email, buy merchandise, gamble, share and transfer files, and post various forms of data such as music, photos, and video clips. The Internet presents information in textual, audible, graphical, and video formats, evoking the appearance and function of print mass media, radio, and television combined. The Internet is a global system of interconnected private and public computer networks that interchange data. It is a "network of networks" that consists of millions of local and global connections that are linked by copper wires, fiber-optic cables, wireless connections, and other technologies.

Fourth, the lack of centralized control means that preventing any agency that is determined to abuse the Internet for its own purposes from doing so is difficult. There is no phone line to cut.

Fifth, unlike telephone service, the Internet is borderless. The technology encompasses continents. Data do not start and stop at national boundaries, whereas phone companies certainly recognize borders and charge far more money for international calls.

Last, the flexible, multipurpose nature of the Internet, with its potential to operate as a set of interpersonal, group, or mass media channels, makes it a unique communication system. The Internet provides many more avenues than any other form of communication to preserve anonymity and privacy with little or no cost involved.

The Internet is a vast source of information. In this respect it might be likened to a large first- and secondhand bookstore or a large library. An owner of a bookstore cannot be held responsible for the content of every book in the store. The bookstore owner does not read and inspect all the books. Similarly, it can be argued, an Internet provider should not be held accountable for content on its server. But if a bookstore owner is informed that a specific book contains child pornography, some other illegal material, or material that violates copyright law and then does not take the book off the shelves, the owner may be held legally responsible for violating the law. And the owner is also morally responsible. It can be argued that the case of the Internet is similar.

ISPs and WHSs have discretion regarding whether their services are open to all or limited in one way or another. Similarly, bookstore owners have discretion about the books they offer to buyers. Many would like to maintain a quiet and tranquil atmosphere in the store. The books, accordingly, will be for the general readership. Other bookstore owners might opt for a more rowdy atmosphere. They will entertain books containing socially problematic material. The likely result would be that the general readership would refrain from visiting those bookstores. Those stores would become niche stores, for particular readers.

In both kinds of store, their owners would like to keep the business going. They would listen to warnings about the illegality of certain books. The same, it can be argued, is true of ISPs and WHSs: provide a notice first, allowing the provider or host to make a decision about the consequences to which he or she might be held liable. If the provider or host does not act on the warning, it will have to face the consequences. Indeed, on copyright issues, ISPs are expected to assume responsibility. They should also assume moral and social responsibility when violent, antisocial activities are taking place on their servers. Most ISPs and Web-hosting companies would not like their servers to be transformed into forums where people concoct criminal activities.

Exceptions exist, of course. In June 2009, the US Federal Trade Commission (FTC) ordered the shutdown of Pricewert, described as a "rogue" or "black hat" ISP that acted as a hosting center for many high-tech criminals. The FTC alleged that Pricewert was paid to host "child pornography, botnet command-and-control servers,[22] spyware, viruses, Trojans, phishing-related sites, illegal online pharmacies, investment and other web-based scams."[23] Pricewert evoked suspicion because its hosting history had 8 changes on 6 unique name servers over 10 years; its IP history showed 16 changes on 7 unique IP addresses over 10 years.[24]

Web-hosting companies such as 1st Amendment[25] and Go Daddy,[26] as well as blog-hosting companies like Xanga,[27] are friendly to racial propaganda, acting in an irresponsible, akrasian way. A Texas company called the Planet hosted the notorious hate site Stormfront. The Planet had very loose terms of service, such that it wittingly would allow the hosting of such an infamous website.[28] The First Amendment to the US Constitution (discussed in chapter 8) and

---

[22] Any such computer is referred to as a *zombie* – in effect, a computer "robot" or "bot" that serves the wishes of some master spam or virus originator. Most computers compromised in this way are home based. See Margaret Rouse, "Botnet (Zombie Army)," *Search Security*, February 2012, http://searchsecurity.techtarget.com/sDefinition/0,,sid14_gci1030284,00.html .

[23] *Federal Trade Commission v. Pricewert LLC*, US District Court Northern District of California San Jose Division (June 2, 2009); Federal Trade Commission, "FTC Shuts Down Notorious Rogue Internet Service Provider, 3FN Service Specializes in Hosting Spam-Spewing Botnets, Phishing Web Sites, Child Pornography, and Other Illegal, Malicious Web Content," press release, June 4, 2009, http://www.ftc.gov/opa/2009/06/3fn.shtm ; see also Ellen Messmer, "ISP Pricewert Protests Shutdown," *PCWorld*, June 6, 2009, http://www.pcworld.com/businesscenter/article/166258/isp_pricewert_protests_shutdown.html .

[24] Domain information for the website is available at http://whois.domaintools.com/pricewert.com .

[25] See the company's website at http://www.1stamendment-hosting.com .

[26] "Site Overview: Godaddy.com," Alexa, http://www.alexa.com/siteinfo/godaddy.com .

[27] See the company's website at http://www.xanga.com .

[28] In 2010, the Planet merged with SoftLayer. See SoftLayer, "SoftLayer and the Planet Begin Merged Operations," press release, November 6, 2010, http://www.softlayer.com/press/release/

profit conveniently go hand in hand. Social responsibility and respect for people are secondary.

Since its creation in January 1995, Stormfront has served as a supermarket of Net hate, giving a voice to many forms of antisemitism and racism. In its first two years, Stormfront featured the writings of the former Ku Klux Klan Grand Wizard David Duke,[29] William Pierce of the neo-Nazi National Alliance,[30] representatives of the Holocaust-denying Institute for Historical Review,[31] and other extremists. In addition to many articles, Stormfront housed a library of neo-Nazi graphics available for downloading, a list of phone numbers for racist computer bulletin boards, and links to other hateful websites.[32]

Many of the hate sites are fanatically religious in nature. Religion is perceived as the rock around which life should be organized. Religion provides radicals with the answer – indeed the only answer. The argument is that we have little choice in making decisions because everything has already been decided for us by God. After all, it is far better to trust the consistent and enlightened almighty who knows all than to trust the reason of fallible humans.

The elaborate sites hate African Americans and nonwhite immigrants, Muslims, Jews, and gays. They are quite eclectic, offering a wide range of racist publications.[33] Some of them publish in a number of languages. Stormfront contains discussions in many European languages.[34] The website of the National Alliance is being published in 16 languages. Extensive websites contain documents, journals, newspapers, videos, radio programs, television shows,

---

501/softlayer-and-the-planet-begin-merged-operations ; GI Partners, "GI Partners Announces Completion of Merger between SoftLayer and the Planet," press release, November 10, 2010, http://www.gipartners.com/news/gi-partners-announces-completion-of-merger-between-soft layer%C2%AE-and-the-planet%C2%AE . Senior Anti-Defamation League directors spoke with the owner of the Planet to no avail. They said it was a waste of time. Discussion with senior Anti-Defamation League directors, New York, March 22, 2010.

[29] For more information about David Duke, see his official website at http://www.davidduke.com/ .

[30] For more information about William Pierce, a speech clip is available on YouTube at http://www.youtube.com/watch?v=Mu4-RRhs9aM . See also "Extremism in America: William Pierce," Anti-Defamation League, http://www.adl.org/learn/ext_us/Pierce.asp ; Christopher Reed, "William Pierce," *Guardian*, July 25, 2002, http://www.guardian.co.uk/news/ 2002/jul/25/guardianobituaries.booksobituaries1 .

[31] For more information, see the organization's website at http://www.ihr.org/ .

[32] Jordan Kessler, *Poisoning the Web: Hatred Online* (New York: Anti-Defamation League, 1999), 4; Seth Stephens-Davidowitz, "The Data of Hate," *New York Times*, July 12, 2014. For more information on Don Black and Stormfront, see the prepared statement of Howard Berkowitz, "Hate Crime on the Internet," Hearing before the Committee on the Judiciary, US Senate, Washington, DC, September 14, 1999.

[33] See, for instance, the works listed in the Racial Nationalist Library at http://www.racerealist. com/1b.htm .

[34] See the organization's discussion forum at http://www.stormfront.org/forum/ .

books, games, survival information, home-schooling information, cartoons, artwork, jokes, quotes, poems, free stickers, and merchandise. In addition, there are also antireligious,[35] antiabortion,[36] antiliberal, anticommunist, and antifeminist sites.[37]

Some Nazi sites explain and propagate their business on the Internet by advising interested parties to open their business in the United States because there they will not be prosecuted. For example, Zensurfrei.com advertises (in German and English):

> 1GB web-site + 50GB data transfer per month + ten genuine POP3 email addresses. Domain Name registration is FREE for first year! – No setup charge! No registration charge! Only 20,00 €/month (US$20.00/month), ie 60,00 €/quarter (US$60.00/quarter)!
>
> ANONYMOUS WEB-SITES ARE POSSIBLE! The domain name is registered in the name of a US firm. Even our firm does not need to know *your* identity. (Payment can be sent with an anonymous letter with reference to your web-site.)
>
> Political repression is increasing in Europe! European webmasters can reduce their risk by moving their web-sites to the USA! ZENSURFREI establishes your web-site with one of the largest and most reliable servers in the USA. Pay by the quarter or by the year. We accept Euro banknotes or US Dollar banknotes, no coins.

[35] For sites attacking all religions, see Exposing Satanism at http://exposingsatanism.org/ ; Truth and Grace at http://truthandgrace.com/ ; Peace of Mind at http://peace-of-mind.net/ ; and Odinist (a pagan site promoting "Faith, Folk, and Family") at http://www.odinist.com/ . For anti-Christianity sites, see Altar of Unholy Blasphemy at http://www.anus.com/altar/ and Set Free at http://www.jcnot4me.com/ . For anti-Islam sites, see Faith Freedom at http://www.faithfreedom.org/ and Truth and Grace at http://truthandgrace.com/ISLAM.htm . For anti-Hinduism sites, see Jesus Is Lord at http://jesus-is-lord.com/ ; Most Holy Family Monastery at http://www.mostholyfamilymonastery.com/H.O.W._of_JP2_and_V2sect_regarding_pagans_and_infidels.html ; Peace of Mind at http://peace-of-mind.net/ ; and Truth and Grace at http://truthandgrace.com/Hindu.htm ; see also the Hindu American Foundation's Hinduphobia Archive at http://hafsite.org/blog/category/hinduphobia/ .

[36] For examples of antiabortion sites, see Creator's Rights Party at http://www.tcrp.us/ ; the Army of God at http://www.armyofgod.com/ ; and Final Conflict at http://dspace.dial.pipex.com/finalconflict/a14-6.html . Until it was shut down, the Nuremberg Files website instigated violence against abortionists. See *Planned Parenthood of the Columbia/Willamette, Inc. v. American Coalition of Life Activists*, US Court of Appeals for the Ninth Circuit (May 21, 2002); *Planned Parenthood of the Columbia/Willamette Inc. et al. v. American Coalition of Life Activists*, 41 F. Supp. 2d 1130 (D. Or. 1999); *Planned Parenthood of the Columbia/Willamette, Inc. v. American Coalition of Life Activists*, F. 3d, 2001 WL 293260 (9th Cir. 2001). However, the same information can be found at present at another site, Alleged Abortionists and Their Accomplices, http://www.christiangallery.com/atrocity/aborts.html .

[37] Richard Delgado and Jean Stefancic, *Understanding Words That Wound* (Boulder, CO: Westview, 2004), 125.

It's fast, easy and convenient! On request we will – without cost or obligation – check to see if your desired domain name (ending with .com, .net or
.org) is still available. (It is best to tell us a few alternative domain names to
check out at the same time.)

If you order a web-site from us, we register the domain name and establish
your web-site account. This is almost always completed within 24 hours! You
can start downloading your files right away. (Although the domain name does
not start to function for a few days, your web-site is nonetheless already
accessible via the IP address.)[38]

ROWDINESS

I opened this chapter by posing the question whether expecting Internet
intermediaries to scrutinize their sites to avoid certain illegal material is
unrealistic. At present, the widespread answer in the United States is yes.
The issue of trust seems secondary; the thinking is that people who wish to visit
such sites should police them. The concern that others might stumble acci-
dently onto those sites is deemed insignificant. The further question is
whether this objection is principled or practical. Does it stem from consideration for the constitutional First Amendment right to free speech or from
considerations of cost and traffic volumes? Solutions to practical issues can
be found. It is far more difficult to convince ardent, principled opposition to
change its view.

I have said that the Internet may be likened to a large library. In many
research libraries, books known to be problematic for their content are kept in
designated areas, under the eyes of an experienced librarian. If someone
wishes to read a book from that section, the individual has to sign for it and
read it in the same room where the book is kept. This way, the library is
balancing the right to free expression and flow of information with societal
interests in maintaining peace and order. People who have library cards can
still have access to the information, but they may be asked questions about
their purpose for reading the book, and a record is kept that they have read the
book in question. Similar arrangements can be made on the Net. Some
problematic material would have restricted access and people would have to
sign up to read it, providing some details about their identity and why they
wanted to read that particular piece of information.

Thus, certain forms of speech that are at present shielded under the First
Amendment should be in restricted Net areas for which people can register to
obtain access. If you wish to read a manual about how to kill your wife without

---

[38] See the company's website at http://www.zensurfrei.com .

leaving any trace, you will need to leave verifiable details.[39] Similarly, people interested in recipes for rape drugs, recipes for bombs, and manuals on how to kill people[40] should accept some privacy interference.[41] I fully realize that the principle of Net neutrality would oppose registering problematic sites that contain crime-facilitating speech. However, the issue is not the neutrality of the Net but whether we should be neutral regarding this kind of content. From a moral standpoint and from a position of CSR, we cannot be neutral regarding such alarming speech. At the very least, this speech requires some precaution. This precaution would certainly promote the level of trust in technology. In the conclusion to this volume, I propose a new browser for the Internet called CleaNet.

VAMPIREFREAKS.COM

In chapter 5, I discussed Kimveer Gill and his murderous attack on Dawson College in Quebec. Prior to the attack, he announced his intentions on VampireFreaks.com. I have also mentioned other disturbing incidents that brought VampireFreaks.com to the media's attention for the wrong reasons. One would assume that website owners would be proactive in providing a safe and secure environment for their members. After all, ICT professionals have an inviolable duty and professional responsibility to serve the best interest of their clients and of society (see chapter 3). "Jet," the VampireFreaks.com owner and operator, responded to Gill's murderous rampage almost immediately. He relieved himself of any responsibility, saying, "Just because someone goes around shooting people and happens to be a member of VampireFreaks, doesn't mean that this website has influenced him to do such a horrible thing."[42] Jet also said that the website frowns on illegal behavior and bans nudity, hatred, and Nazi paraphernalia.[43]

In this context, it is important to note that after the murder, Kimveer Gill's profile was taken off VampireFreaks.com. Jayson Gauthier, the provincial police force's spokesman, said that an American police force had demanded

---

[39] Shaun Pierce, *PowerBlog*, http://powerballplace.blogspot.com/2005/03/how-to-kill-your-wife-and-get-away.html .

[40] See the case of *Rice v. Paladin Enterprises, Inc.*, 128 F.3d 233 (4th Cir. 1997). In that case, a US appeals court ruled that a publisher, Paladin Enterprises, was not protected by the First Amendment when it published *Hit Man: A Technical Manual for Independent Contractors*. The book is no longer available in print, but the text is available on the Internet.

[41] I speak only of detailed manuals geared to educate people about how to commit the perfect violent crimes. Novels, thrillers, and satiric works that contain artistic merit are not at issue.

[42] Siri Agrell, "Troubled Kids Gravitating to Vampire Site," *National Post*, September 15, 2006, A6.

[43] Joe Mahoney, "Killer's Grim Net Warning," *Daily News* (New York), September 15, 2006, 33.

the shutdown after a request from the Canadian authorities.[44] This measure should have arguably been taken *before* the murder. Gauthier also said that no police department had been aware of Gill until the shootings.[45]

As in the Gill case, YouTube removed the Pekka-Eric Auvinen video and suspended the user's account within hours after the massacre in Finland (see chapter 5).[46] The authorities apparently were unaware of the violent warnings prior to the rampage.

This was not the case in Matti Juhani Saari's massacre. LiveLeak.com decided that the threatening video should not be removed after the rampage.[47] LiveLeak.com's ISP transfers any responsibility it might have for the contents of its website to its surfers.

Jet of VampireFreaks.com claimed he was doing whatever he could to prevent the posting of offensive or dangerous material: "We do monitor user messages and profiles for violent, hateful and offensive material. However on a site with over 600,000 users, it is impossible to monitor everything on the site."[48] In a statement on the VampireFreaks homepage, Jet offered condolences to the victims and called the shootings a "tragic event," but wrote:

> Just because someone goes around shooting people and happens to be a member of vampirefreaks, doesn't mean that this website has influenced him to do such a horrible thing.... The goth scene is a very friendly, nurturing, non-violent community and we are very supportive of our users and do not condone any illegal activities.... In fact I believe we are more mature and responsible than other scenes, in that we value intelligence, part of goth culture is thinking for yourself and being more aware of the world, rather than just following the mainstream trends.... Don't let a few bad seeds ruin our reputation, we are a great community.[49]

The VampireFreaks.com's terms of use state:

1. *Administration and Monitoring of Site*. Due to the nature of social networking sites and the very large amount of content posted frequently, we

[44] Ian Austen, "Gunman at Montreal College Left Dark Hints of Rage Online," *New York Times*, September 15, 2006, 10.

[45] Ibid.

[46] "Student Kills 8, and Himself, at Finnish High School," Reuters, November 7, 2007, http://www.nytimes.com/2007/11/08/world/europe/08finland.html?_r=1&ex=1195189200an den=bfc f75c691f75a30andei=5070andemc=eta1&oref=slogin .

[47] "Finnish School Shooter Video 2," LiveLeak.com, http://www.liveleak.com/view?i=33 d_1222168770 .

[48] Austen, "Gunman at Montreal College Left Dark Hints of Rage Online."

[49] Jano Gibson, "Angel of Death's Web Warnings," *Sydney Morning Herald*, September 15, 2006, http://www.smh.com.au/news/world/angel-of-deaths-web-warnings/2006/09/15/11578 27122312.html .

are unable to monitor everything on the Site, and thus vampirefreaks.com acts as a passive conduit for user interaction. We do, however, respond to reports, notices, and complaints in a timely manner. Though **Synth-Tec takes no responsibility for monitoring or editing user content**, we reserve the right, at our sole discretion, to change, edit, or remove any user uploaded content, including but not limited to text, images, musical works, sound recordings, and video.[50]

The website does not allow uploading "excessive gory material"; exploiting anyone in a sexual or violent manner; engaging in, promoting, or condoning any type of harmful or illegal activity; or spamming and sending sexual messages to minors.[51] Jet said that in the past he had alerted the police a few times of users whose profiles were suspicious.[52] But despite incidents that clearly violated the site's terms of service, Jet does not take any measures to monitor his site. He is aware of what is posted on the site, he has capacity for decision, and yet he chooses to allow complete freedom. Jet is not acting from ignorance. He fails to take moral and social responsibility for whatever reason: overconfidence, unshaken belief in the First Amendment, laziness, dogmatism, or simply because he wishes to save money. Trust is not an issue. His assumption is that VampireFreaks subscribers are unconcerned about the dangers. They do not care. Neither does Jet. But social responsibility is important, notwithstanding whether it contributes to an increased sense of trust in the medium.

ISPs AND WEB-HOSTING RULES AND REGULATIONS

Most ISPs do not wish to be rowdy. They opt for some form of regulation by adopting certain abiding rules. A number of ISPs offer guidelines regarding prohibited Internet content and usage, terms for service cancellation, and Netuser responsibilities. Obviously, ISPs and WHSs have the right and the duty to report potentially criminal activities to the appropriate law enforcement agency. In addition, ISPs and WHSs may prohibit posting legally seditious or offensive content. It is in their sole discretion to prescreen, refuse, or move any content that is available via their service. ISPs and WHSs reserve the right to terminate service if in their discretion a client has violated the terms and conditions of service. The terms and conditions specify what they

[50] See the full text of the terms of use and agreement at http://vampirefreaks.com/site/termsof service.php . See also Siri Agrell and Paul Cherry, "Blogs Reveal a Deteriorating Mind, Police Say," *National Post* (Canada), September 16, 2006, A9.
[51] See the website's terms of service at http://vampirefreaks.com/site/termsofservice.php .
[52] Agrell and Cherry, "Blogs Reveal a Deteriorating Mind, Police Say."

perceive to be antisocial activity, such as child pornography, terrorism, and crime. These issues are patently illegal. What this realm of antisocial yet legal activity includes may vary from one ISP to another. Some ISPs and WHSs abhor having racist organizations on board. Others are tolerant and even supportive of this activity. See the following examples:

- Rules for CompuServe, a division of AOL, prohibit the "distribution of content that is harmful, abusive, racially or ethnically offensive, vulgar, sexually explicit, or in a reasonable person's view, objectionable. Community standards may vary, but there is no place on the service where hate speech is tolerated."[53]
- Lycos's terms of service state that users may not "upload, post, email, otherwise transmit, or post links to any Content, or select any member or user name or e-mail address, that is unlawful, harmful, threatening, abusive, harassing, tortuous, defamatory, vulgar, obscene, pornographic, libelous, invasive of privacy or publicity rights, hateful, or racially, sexually, ethnically or otherwise objectionable."[54]
- Yahoo!'s rules do not allow users to "upload, post, transmit or otherwise make available any Content that is unlawful, harmful, threatening, abusive, harassing, tortuous, defamatory, vulgar, obscene, libelous, invasive of another's privacy, hateful, adult-oriented, or racially, ethnically, or otherwise objectionable."[55]
- BasicISP requires its users to agree not to "post or transmit any unlawful, threatening, abusive, libelous, defamatory, obscene, pornographic, profane, or otherwise objectionable information of any kind, including without limitation any transmissions constituting or encouraging conduct that would constitute a criminal offense, give rise to civil liability, or otherwise violate any local, state, national or international law."[56]
- Tumblr's Community Guidelines declare the website's deep commitment to supporting and protecting freedom of speech but at the same time draw lines around "narrowly defined but deeply important categories of content and behavior that jeopardize our users, threaten our infrastructure, and damage our community."[57] The guidelines state that Tumblr is not for malicious speech; harm to minors; promotion or glorification of self-harm; gore, mutilation, bestiality, or necrophilia; unflagged NSFW (not suitable

---

[53] See the company's terms of use at http://info.aol.co.uk/compuserve/terms.adp .
[54] See the company's terms of service at http://info.lycos.com/resources/terms-of-service/ .
[55] See the company's terms of service at https://info.yahoo.com/legal/us/yahoo/localfreewebsite/details.html .
[56] See the company's terms of service at http://www.basicisp.net/TOS/DSLTOS.aspx .
[57] See the Community Guidelines at https://www.tumblr.com/policy/en/community .

for work) blogs (usually involving sex); uploading of sexually explicit videos; nongenuine social gesture schemes; deceptive or fraudulent links; misattribution and nonattribution; username or URL abuse or squatting; spam; mass registration or automation; unauthorized contests, sweep-stakes, or giveaways; themes distributed by third parties; copyright or trade-mark infringements; confusion or impersonation; harassment; privacy violations; disruptions, exploits, or resource abuse; and unlawful uses or content.[58]

- DataPipe prohibits "tortuous conduct such as posting of defamatory, libelous, scandalous, or private information about a person without their consent, intentionally inflicting emotional distress, or violating trademarks, copyright, or other intellectual property rights."[59]

DataPipe has a page in which it names its favorite charities and its commit-ment "to the communities in which we live and the planet we share."[60] Several ISP associations have developed various codes concerning, among other things, the protection of minors.[61] Codes of conduct provide guidance for Netusers as to what is legitimate and what is not, and they promote the accountability of ISPs.

ISPs and WHSs may have the right in their sole discretion to prescreen and refuse any content that is available via their service. Yahoo! declares that it may or may not prescreen and that it has "the right (but not the obligation) in their sole discretion to pre-screen, refuse, move, or remove any Content that is transmitted or made available via the Service. Without limiting the foregoing, Yahoo and its designees shall have the right to remove any Content (whether or not provided by you) that we believe in good faith may be unlawful or violates the [terms of service] or that has been alleged to infringe any intellec-tual property or that is otherwise objectionable in our sole opinion, without being liable to you in any way for any loss or damage arising from such

---

[58] Ibid.

[59] See the company's terms of service at http://www.datapipe.com/legal/terms_of_service/ .

[60] "Datapipe Gives Back," http://www.datapipe.com/about_us/datapipe_gives_back/ .

[61] For further discussion, see Directorate for Science, Technology, and Industry Committee for Information, Computer and Communications Policy, "Digital Broadband Content: Digital Content Strategies and Policies," report presented to the Working Party on the Information Economy, Organisation for Economic Co-operation and Development, Paris, 2006, http://www.biac.org/members/iccp/mtg/2008-06-seoul-min/DSTI-ICCP-IE(2005)3-FINAL.pdf . See also Monroe E. Price and Stefaan G. Verhulst, "The Concept of Self-Regulation and the Internet," in *Protecting Our Children on the Internet: Towards a New Culture of Responsibility*, ed. Jens Waltermann and Marcel Machill (Gütersloh, Germany: Bertelsmann Foundation, 2000), 133–98.

removal."[62] Some ISPs assert the right to terminate service under any circumstances and without prior notice, especially if content violates the terms of service agreement or if law enforcement or other government agencies request removal. Some ISPs reserve the right to remove information that does not meet the standards they set.[63] However, if such content is not removed by the ISP, neither it nor its partners accept any liability.

The development of ISP and WHS codes of conduct is notably encouraged in the European Union (EU). Its recommendations aim to provide guidelines for national legislation and urge European ISPs to develop codes of good conduct. The recommendations also offer guidelines for the development of national self-regulatory policies regarding the protection of minors and human dignity.[64] In this context, I should mention EuroISPA, the pan-European association of European Internet Service Providers Associations. It is the world's largest association of ISPs, representing more than 2,300 ISPs across the EU and European Free Trade Association countries – including ISPs from Austria, Belgium, the Czech Republic, Finland, France, Germany, Ireland, Italy, Norway, and the United Kingdom. It was established in 1997 to represent the European ISP industry on EU policy and legislative issues and to facilitate the exchange of best practices between national ISP associations. Its secretariat is located in Brussels. The association is recognized as the voice of the EU ISP industry and is the largest umbrella association of ISPs in the world.[65]

In the United States, the prevailing view among media professionals and scholars is that the Free Speech Principle shields even the vilest forms of hate speech. They essentially rest the argument on three broad principles: universalism, net neutrality, and the unique Internet environment. I wish to take issue with these arguments.

As for universalism, I said at the outset that the scope of this book is limited to liberal democracies emerging during the past century or so. The issue at hand is not about *any* offensive speech or *any* speech deemed to be hateful in the eyes of the beholder. Instead, I refer only to speech perceived to be hateful in accordance with liberal principles of showing respect for others and not harming others. My argument will not help those who are offended by the

---

[62] See the company's terms of service at https://info.yahoo.com/legal/vne/yahoo/utos/ .

[63] AOL's Web Services Agreement says: "We may use any legal and technical remedies available to us to prevent any breach or enforce this Agreement and reserve the right to remove or not publish any Content without prior notice." See the agreement at http://help.aol.co.uk/help_uk/documentLink.do?cmd=displayKC&docType=kc&externalId=26301 .

[64] Yaman Akdeniz, *Internet Child Pornography and the Law: National and International Responses* (Aldershot, UK: Ashgate, 2008), 249.

[65] See the organization's website at http://www.euroispa.org/about/who-we-are/ .

speeches of human rights activists; those who are highly disturbed by women shouting commands, appearing in bikini swimsuits, or singing; and those who perceive according equal citizenship rights to homosexuals as abhorrent. I do not doubt that those people are greatly upset by such ideas, speeches, and scenes. But my reasoning is limited in scope to the working of the Internet in liberal democracies.

Another guiding principle inspired by the First Amendment aimed to bolster the special status that freedom of expression enjoys is net neutrality. The underlying belief is that the Internet should remain an open platform for innovation, competition, and social discourse, free from unreasonable discriminatory practices by network operators. All content, sites, and platforms should be treated equally, free from any value judgment. In justifying this philosophy, American new media experts explain that the Internet was built on and has thrived as an open platform where individuals and entrepreneurs are able to connect and interact, choose marketplace options, and create new services and content on a level playing field. Richard Whitt, Google's Washington telecommunications and media counsel, writes, "No one seems to disagree with that fundamental proposition," arguing for the need to "protect that unique environment" and supporting the adoption of "rules of the road" to ensure that the broadband on-ramps to the net remain open and robust.[66] Jack Balkin, from Yale Law School, claims that the open Internet is crucial to freedom of speech and democracy because it allows people to actively participate in decentralized innovation, forming new digital networks free from prior government constraints. People can reach all audiences and find a way around gatekeepers with great new tools and applications.[67]

I take issue with these arguments – that of net neutrality and of the unique nature of the Internet environment as a public domain – positing that any speech should be freely available on the Internet. I argue that screening of content may be valuable and that the implications affording the Internet the widest possible scope can be very harmful. Unlike Balkin, I think that limitless freedom of speech might undermine democracy (see my discussion of the democratic catch in chapter 3). Indeed, one of the dangers we face is that the principles that underlie and characterize the Internet might undermine freedom. Because the Internet is a relatively young phenomenon, people who use and regulate it lack experience in dealing with the pitfalls involved in its working. Freedom of expression should be respected as long as it does not

---

[66] Richard Whitt, "Time to Let the Process Unfold," *Google Public Policy Blog*, October 22, 2009, http://googlepublicpolicy.blogspot.com/2009/10/time-to-let-process-unfold.html .

[67] Chris Naoum, "Web Content Producers Favor Net Neutrality, Reject Regulation of Search Engines," *BroadbandBreakfast*, December 16, 2009.

imperil the basic values that underlie our democracy. Freedom of expression is a fundamental right, but it should not be used in an uncontrolled manner. Like every young phenomenon, the Internet needs to develop with prudence and care. Because we lack long experience, we are unsure of the means suitable to be used to fight unequivocal antidemocratic and dangerous practices, such as the various manifestations of religious extremism, radical ideologies, and terrorism. But we should not stand idly by in the face of threat of violence.

Thus, although I accept the concept of net neutrality, I reject the concept of *content net neutrality* for reasons that I explain next. First, we should ensure that no confusion arises between the two.

CONTENT NET NEUTRALITY

Net neutrality is one of the core principles of the Internet. In October 2009, a group of the world's largest Internet companies wrote a letter of support to the US Federal Communications Commission. The letter is part of an ongoing debate about "network neutrality" – or how data are distributed on the Web. The letter, signed by the chief executives of Google, eBay, Skype, Facebook, Amazon, Sony Electronics, Digg, Flickr, LinkedIn, and craigslist, among others, says that maintaining data neutrality helps businesses compete on the basis of content alone:

> An open internet fuels a competitive and efficient marketplace, where consumers make the ultimate choices about which products succeed and which fail. . . . This allows businesses of all sizes, from the smallest start-up to larger corporations, to compete, yielding maximum economic growth and opportunity.[68]

This document is yet another step in a sustained and until now quite successful effort to grant Internet companies the widest possible freedom and independence to conduct their affairs in a way that best serves their commercial interests. Their responsibility, as these large companies see it, is to provide their customers with efficient service.

Net neutrality is also about the organization of the Internet. No one application (World Wide Web, email, messenger) is preferred to another. All applications should be treated by Internet intermediaries equally and without discrimination. Information providers (such as websites and online services, which may be affiliated with traditional commercial enterprises but

---

[68] "Big Names Support Net Neutrality," BBC News, October 20, 2009, http://news.bbc.co.uk/1/hi/8315918.stm .

may also be individual citizens, libraries, schools, or nonprofit entities) should have essentially the same quality of access to distribute their offerings. "Pipe" owners (carriers) should not be allowed to charge some information providers more money for the same pipes or to establish exclusive deals that relegate everyone else (including small noncommercial or startup entities) to an Internet "slow lane." This principle should hold true even when a broadband provider is providing Internet carriage to a competitor.[69] To this, I agree. The public is interested in having a neutral platform that supports innovations and the emergence of the best technological applications.

However, the American Library Association, one of the advocates of net neutrality, also holds that the principle of net neutrality maintains that consumers and citizens should be free to get access to – or to provide – the Internet content and services they wish and that consumer access should not be regulated on the basis of the nature or source of that content or service.[70] Similarly, the Norwegian Post and Telecommunications Authority holds that Netusers are entitled to an Internet connection that enables them to send and receive content of their choice as well as to an Internet connection that is free of discrimination with regard not only to type of application and service but also to content.[71] I find this part of net neutrality that concerns content much more complicated and problematic. It should be separated from the principle of net neutrality. I call it *content net neutrality*.

Content net neutrality holds that we should treat all content that is posted on the Internet equally. ISPs and WHSs should not privilege or in one way or another discriminate between different types of content. Now, it is unclear what the implications of such a view are. One possible implication against which content net neutrality warns is that a specific search engine might pay ISPs fees to ensure that responses from its website would be delivered to the user faster than the results from a competing search engine that had not paid special fees. Another possible wrong implication against which we all protest is that an ISP might accord a lower priority to packets transmitting, say, video feeds, unless the customer paid a special fee for higher-speed access. The most alarming scenarios involve outright blockage of content by source or by type. An example of blockage by source often cited in news stories is that of the Canadian ISP Telus, which blocked subscribers' access to a website of

---

[69] American Library Association, "Network Neutrality," American Library Association, Chicago, http://www.ala.org/advocacy/telecom/netneutrality . See also Christopher T. Marsden, *Net Neutrality: Towards a Co-regulatory Solution* (London: Bloomsbury, 2010).

[70] American Library Association, "Network Neutrality."

[71] Post-og teletilsynet, "Network Neutrality: Guidelines for Internet Neutrality," Version 1.0, February 24, 2009.

the Telecommunications Workers Union, with which it was in conflict.[72] Labor disputes should never constitute grounds for content discrimination. The example of type-based blocks much mentioned in the debate is that of the Madison River (North Carolina) phone company, which blocked voice over IP (VoIP) traffic from Vonage as an anticompetitive move to protect its own long-distance conventional telephony service.[73] This kind of brute discrimination, motivated by narrow economic interests, is also illegitimate. Such incidents demonstrate the skewed incentives that ISPs might have in controlling content and applications. The present debate is about the extent to which ISPs should be allowed to control the size of the pipes: Can ISPs actively control the bandwidth available to certain websites on the basis of the type of content they provide, thus influencing the Internet speed available to Netusers?

Is the Internet like the electric grid? Columbia law professor Tim Wu explains the logic behind net neutrality by arguing that a useful way to understand this principle is to look at other networks, such as the electric grid, which are implicitly built on a neutrality theory. The general purpose and neutral nature of the electric grid is one of the things that make it extremely useful. The electric grid does not care if you plug in a toaster, an iron, or a computer. Consequently, it has survived and supported giant waves of innovation in the appliance market. The electric grid worked for the radios of the 1930s, and it works for the flat-screen televisions of the 2000s. For that reason, the electric grid is a model of a neutral, innovation-driving network.[74]

However, does this mean that, as you do not expect to control the content of the electric grid, so also you should not aim to control the Internet's content? If this is a plausible deduction, then this comparison is misleading. The electric grid transmits power that enables the functioning of electric equipment. It does not have content, messages, propaganda, instructions, or means to abuse you or to harm you. The Internet, in contrast, has all this. As Luciano Floridi rightly writes, a digital interface is a gate through which a user can be present

---

[72] M. E. Kabay, "The Net Neutrality Debate," *Ubiquity* 7, no. 20 (2006): article 4.

[73] Federal Communications Commission, "FCC Chairman Michael K. Powell Commends Swift Action to Protect Internet Voice Services," news release, March 3, 2005, http://tinyurl.com/hscav .

[74] Tim Wu, "The Broadband Debate: A User's Guide," *Journal on Telecommunications and High Technology Law* 3, no. 1 (2004): 69–95, http://www.jthtl.org/content/articles/V3I1/JTHTLv3 ii_Wu.PDF ; Tim Wu, professor, Columbia Law School, testimony before the House Committee on the Judiciary Task Force on Telecommunications and Antitrust in the hearing on "Network Neutrality: Competition, Innovation, and Nondiscriminatory Access," 109th Cong., 2nd sess., April 25, 2006, http://www.gpo.gov/fdsys/pkg/CHRG-109hhrg27225/html/CH RG-109hhrg27225.htm .

in cyberspace.[75] Regarding the electric grid, you cannot develop subjective notions. The Internet, which contains the best and worst products of its customers, may lead you to develop subjectivity. The Internet contains the power to influence your life in constructive and destructive ways. As thinking people, we are able to differentiate between right and wrong, good and evil; as morally responsible beings, we must discriminate between contents. We cannot be neutral about contents if we wish to continue leading free, autonomous lives. The only meaningful aspect in the comparison between the Internet and the electric grid is that in both cases we insist on some measures to ensure our security. Those measures do not need to include subjectivity when we consider the electric grid. They do require subjectivity when we consider the Internet.

In a hearing before the Task Force on Telecom and Antitrust of the House Committee on the Judiciary, Wu said that the "instinct" of protecting consumers' rights on networks is very simple: let customers use the network as they please.[76] With due appreciation to instincts, which often serve as good guides for conduct, by nature they are not thoughtful. Sometimes, after reflecting and pondering, we act against our instincts – and for good reasons. I think there are ample reasons to doubt whether allowing customers to use the network as they please is a good policy to follow. Although many Netusers appreciate this policy and would not abuse it, some people might opt for abuse. We should respect the users and protect ourselves against the abusers.

However, in an earlier article, Wu explains that the basic principle behind a network antidiscrimination regime is to give Netusers the ability to use nonharmful network attachments or applications and to provide innovators with the corresponding freedom to supply them. ISPs should have the freedom to reasonably control their network (i.e., "police what they own"), and at the same time, the Internet community should view with suspicion restrictions premised on internetwork criteria.[77]

What does "reasonably control their network" mean? First, ISPs prohibit Netusers from using applications or conduct that could hurt the network or other Netusers. For instance, Akamai's Acceptable Use Policy states:

[75] Floridi, *Information*, 13. In his comments on a draft of this chapter, Sam Lehman-Wilzig noted that the electric grid charges the consumer by the amount of electricity used, which is precisely what the opponents of content net neutrality are proposing.

[76] Wu, House Committee on the Judiciary Task Force on Telecommunications and Antitrust, Hearing on "Network Neutrality: Competition, Innovation, and Nondiscriminatory Access." See also Timothy Wu, *The Master Switch: The Rise and Fall of Information Empires* (New York: Knopf, 2010).

[77] Tim Wu, "Network Neutrality, Broadband Discrimination," *Journal of Telecommunications and High Technology Law* 2 (2003): 141–78, 142, 145.

"Customer shall not use the Akamai Network and Services to transmit, distribute, or store material that contains a virus, worm, Trojan horse, or other component harmful to the Akamai Network and Services, any other network or equipment, or other Users."[78] Blocking denial-of-service attacks and spam are also within what we perceive as legitimate network management.

Second, some companies market equipment aimed at facilitating application-based screening and control for broadband networks. Companies such as Check Point[79] and Symantec Gateway Security[80] provide traffic management features with highly developed security management tools. Allot Communications provides facilities to manage traffic and produces a fully integrated, carrier-class platform capable of identifying the traffic flows of individual subscribers.[81] The Blue Coat Advanced Threat Protection solution integrates technologies to deliver a comprehensive defense that fortifies the network. The solution blocks known advanced persistent threats, proactively detects unknown and already-present malware, and automates postintrusion incident containment and resolution. It coordinates the use of malware scanning to block known threats and to identify for deeper analysis content that is deemed suspicious.[82]

Third, ISPs may prohibit Netusers from inflicting harm on others by upholding and promoting crime-facilitating speech designed to encourage harmful conduct. The effort, observes Wu, quite rightly, is to strike a balance between prohibiting ISPs, absent a showing of harm, from restricting what Netusers do with their Internet connection while giving them general freedom to manage bandwidth consumption. This nondiscrimination principle works by recognizing a distinction between local network restrictions, which are generally allowable, and internetwork restrictions, which are suspect. The effort is to develop forbidden and permissible grounds for discrimination in broadband usage restrictions.[83] Wu has in mind illegal activities. I argue that ISPs and WHSs should also consider prohibiting hate speech, which is legal in the United States. Although racist Nazi speech is protected under the US First Amendment, the same speech is not protected in most European countries. Morally speaking, libertarians also agree that such speech is repugnant.

---

[78] See Akamai's policy at http://www.akamai.com/html/policies/acceptable_use.html .
[79] See the company's website at http://www.checkpoint.com/products/enterprise/ .
[80] See the company's website at http://www.symantec.com/avcenter/security/Content/Product/Product_SGS.html and http://www.symantec.com/products-solutions/products/ .
[81] See the company's website at http://www.allot.com/index.php?option=com_content&task=view&id=2&Itemid=4 .
[82] See the company's website at https://www.bluecoat.com/advanced-threat-protection and https://www.bluecoat.com/products/content-analysis-system .
[83] Wu, "Network Neutrality, Broadband Discrimination," 168.

Fourth, ISPs can block unlawful transfer of copyrighted works. Open Internet principles prima facie apply to lawful content, services, and applications – not to activities such as unlawful distribution of copyrighted works, which has serious economic consequences. The enforcement of copyright and other laws and the obligations of network openness can and must coexist. For network openness obligations and appropriate enforcement of copyright laws to coexist, an ISP should reasonably be able to refuse transmission of copyrighted material if the transfer of that material would violate applicable laws.[84] Thus, legitimate network management includes maintaining the technical quality of the network, preventing abuse, and complying with legal dictates. To ensure legitimate network management, ISPs must disclose network management practices to Netusers and regulators, who should assess these practices against net neutrality principles.[85]

In 2010, the Council of Europe Committee of Ministers declared its commitment to network neutrality but at the same time stressed that exceptions can and should be made in accordance with the human rights protection framework.[86] There is a need to take into account the protection of the right to life and the specific protection of children. The former includes the prohibition of advocacy of national, racial, or religious hatred amounting to incitement to discrimination, hostility, or violence; direct and public incitement to commit genocide; and incitement to terrorism. The latter includes the prohibition of images of sexual exploitation of children.[87]

ANTIPERFECTIONISM

Conceptually, both net neutrality and content net neutrality emphasize diversity and plurality. Diversity entails openness and more opportunities for living a valuable and richer life. Pluralism is perceived to be indispensable for having the potential for a good life. Methodologically, the idea of neutrality is placed within the broader concept of antiperfectionism. The implementation and

[84] "Comments of the Motion Picture Association of America, Inc., in Response to the Workshop on the Role of Content in the Broadband Ecosystem, Before the Federal Communications Commission, Washington, DC, In the Matter of A National-Broadband Plan for Our Future," GN Docket No, 009-51, October 30, 2009; Gigi B. Sohn, "Content and Its Discontents: What Net Neutrality Does and Doesn't Mean for Copyright," speech to the Yale Information Society Project, Yale Law School, New Haven, CT, October 27, 2009.

[85] Trans Atlantic Consumer Dialogue, "Resolution on Net Neutrality," DOC No. INFOSOC 36-08, March 2008.

[86] Council of Europe, *Declaration of the Committee of Ministers on Network Neutrality*, adopted September 29, 2010, http://wcd.coe.int/ViewDoc.jsp?id=1678287 .

[87] Wolfgang Benedek and Matthias C. Kettemann, *Freedom of Expression on the Internet* (Strasbourg, France: Council of Europe Publishing, 2014), 81–82.

promotion of conceptions of what people may perceive as good ways of life, though worthy in themselves, are not regarded as a legitimate matter for governmental action. The fear of exploitation, of some form of discrimination, leads to the advocacy of variety and pluralism. Consequently, ISPs and WHSs are not to act in a way that might favor some ideas over others. ISPs and WHSs ought to acknowledge that every person has his or her own interest in acting according to his or her own beliefs; that everyone should enjoy the possibility of having alternative considerations; and that no single belief about moral issues and values should guide us all and, therefore, each has to enjoy autonomy and to hold his or her ideals freely.

The concept of antiperfectionism comprises the political neutrality principle and the exclusion of ideals doctrine.[88] The *political neutrality principle* holds that ISPs' and WHSs' policies should seek to be neutral regarding ideals of the good. It requires them to make sure that their actions do not help acceptable ideals more than unacceptable ones and to see that their actions will not hinder the cause of ideals deemed false more than they do that of ideals deemed true. The *exclusion of ideals doctrine* does not tell ISPs and WHSs what to do. Rather it forbids them to act for certain reasons. The doctrine holds that the fact that some conceptions of the good are true or valid should never serve as justification for any action. Neither should the fact that a conception of the good is false, invalid, unreasonable, or unsound be accepted as a reason for a political or other action. The doctrine prescribes that ISPs and WHSs refrain from using one's conception of the good as a reason for state action. They are not to hold partisan (or nonpartisan) considerations about human perfection to foster social conditions.

Advocates of content net neutrality, in their striving to convince us of the necessity of the doctrine, are conveying the assumption that the decision regarding the proper policy is crucial because of its grave consequences. Content net neutrality entails pluralism, diversity, freedom, public consensus, noninterference, vitality, and so on. If we do not adhere to content neutrality, then we might be left with none of these virtues. This picture leads to the rejection of subjectivity (or perfectionism), whereas I suggest a rival view that observes conduct of policies on a continuous scale between strict perfectionism, on the one hand, and complete neutrality, on the other. The policy to be adopted does not have to be either the one or the other. It could well take the Aristotelian Golden Mean between freedom of expression and social responsibility. It could allow plurality and diversity without

---

[88] Joseph Raz, *The Morality of Freedom* (Oxford: Clarendon Press, 1986), 110–11; see also Raphael Cohen-Almagor, "Between Neutrality and Perfectionism," *Canadian Journal of Law and Jurisprudence* 7, no. 2 (1994), 217–36.

resorting to complete neutrality, involving some form of perfectionism without resorting to coercion. For perfectionism does not necessarily imply exercise of force, nor does it impose the values and ideals of one or more segments of society on others or strive to ensure uniformity, as neutralists fear. On this issue my view seeks to find a fine balance between the views of my Oxford teachers Ronald Dworkin and Joseph Raz.[89] I call this view the *promotional approach*.

THE PROMOTIONAL APPROACH

My midground position is influenced – even dictated – by two principles. In chapter 3, I suggested that any liberal society is based on the idea of respect for others in the sense of treating citizens as equals and on the idea of not harming others. Accordingly, restrictions on liberty may be prescribed when threats of immediate violence are voiced against some individuals or groups. Thus, I submit that ISPs and WHSs should adhere to the promotional approach rather than to neutrality. I reject content net neutrality on important social issues that concern the safeguarding of democracy within which all media (including new) operate and flourish. Ethics requires us to care about the consequences of our actions and to take responsibility for them. As Floridi and Sanders rightly note, ethics is about constructing the world, improving its nature, and shaping its development in the right way.[90]

Liberal thinkers see the aim of a just governmental system as furthering liberty and egalitarian values.[91] They differ over the permissible ways by which the common good may be promoted. If we take the theories of Rawls and Dworkin as examples, both hold that people should be free to choose any conception of the good that does not violate the principles of justice, no matter how different it may be from the prevailing conceptions widely held by their community. Rawls endorses procedural neutrality in order to warrant rights to basic liberties. In "The Priority of Right and Ideas of the Good," Rawls writes that even if political liberalism can be seen as neutral in procedure and in aim,

[89] Ronald M. Dworkin, *Taking Rights Seriously* (London: Duckworth, 1977); Joseph Raz, *Value, Respect, and Attachment* (Cambridge: Cambridge University Press, 2001); Raz, *Morality of Freedom*.

[90] Luciano Floridi and J. W. Sanders, "Internet Ethics: The Constructionist Values of *Homo Poieticus*," in *The Impact of the Internet on Our Moral Lives*, ed. Robert J. Cavalier (Albany: State University of New York Press, 2005), 195–214, 195–96.

[91] John Rawls, *A Theory of Justice* (Boston: Belknap Press, 2005); Ronald M. Dworkin, *A Matter of Principle* (Oxford: Clarendon Press, 1985); Bruce A. Ackerman, *Social Justice in the Liberal State* (New Haven, CT: Yale University Press, 1980); Charles Larmore, *Patterns of Moral Complexity* (Cambridge: Cambridge University Press, 1987).

it may still affirm the superiority of some forms of moral character and encourage some moral virtues.[92]

Dworkin's aim in advocating concrete neutrality is identical, but his theory provides a wider range of individual liberties than that of Rawls. Dworkin sees neutrality as derived from every person's right to equal concern and respect and insists on moral neutrality to the degree that equality requires it.[93] Dworkin was a staunch First Amendment scholar who did not consider social responsibility seriously. My promotional approach holds that ISPs and WHSs should not be neutral regarding different conceptions of the good. They should safeguard the basic tenets of democracy that enable and facilitate their operations. It is within ISPs' and WHSs' interest to adhere to the basic ideas of showing respect for others and not harming others and to apply judgment in promoting those ideas in daily operations. True, the promotional approach may involve some form of "soft" paternalism, but it is certainly not the invasive, gross paternalism that libertarians might fear.[94] In the long run, it would prove advantageous and profitable to Internet business.

Indeed, adopting norms of social responsibility could contribute to ISPs' and WHSs' reputation and marketing. A significant positive relationship exists between CSR activities and consumers' purchasing decisions.[95] Stewart Lewis argues that CSR, referring to practices that improve the workplace and benefit society beyond what companies are legally mandated to do, is established as a fundamental addition to stakeholders' criteria for judging companies and calls for a reappraisal of companies' brand and reputation management.[96]

---

[92] John Rawls, "The Priority of Right and Ideas of the Good," *Philosophy and Public Affairs* 17, no. 4 (1988): 251–76, 263.

[93] As a result, Dworkin also argues that governments must provide a form of material equality for everyone. They should ensure citizens an initially equal distribution and should assist them in increasing their welfare. See Ronald Dworkin, *Sovereign Virtue: The Theory and Practice of Equality* (Boston: Harvard University Press, 2002); Dworkin, *Taking Rights Seriously*.

[94] On paternalism, see Joel Feinberg, "Legal Paternalism," *Canadian Journal of Philosophy* 1, no. 1 (1971): 105–24; Gerald Dworkin, "Paternalism," *Monist* 56, no. 1 (1972): 64–84; Gerald Dworkin, "Moral Paternalism," *Law and Philosophy* 24, no. 3 (2005): 305–19; Cass Sunstein and Richard Thaler, "Libertarian Paternalism Is Not an Oxymoron," *University of Chicago Law Review* 70, no. 4 (2003): 1159–1201, 1166–87; Raphael Cohen-Almagor, "Between Autonomy and State Regulation: J. S. Mill's Elastic Paternalism," *Philosophy* 87, no. 4 (2012): 557–82. For critique, see generally "Soft Paternalism: The State Is Looking After You," *Economist*, April 6, 2006 , http://www.economist.com/node/6772346?story_id=6772346 , and specifically on ISPs, see Ugo Pagallo, "ISPs and Rowdy Web Sites before the Law: Should We Change Today's Safe Harbour Clauses?," *Philosophy and Technology* 24 (2011): 419–36.

[95] Ki-Hoon Lee and Dongyoung Shin, "Consumers' Responses to CSR Activities: The Linkage between Increased Awareness and Purchase Intention," *Public Relations Review* 36, no. 2 (2010): 193–95.

[96] Stewart Lewis, "Reputation and Corporate Responsibility," *Journal of Communication Management* 7, no. 4 (2003): 356–94.

Upholding norms of CSR benefits both the firm and the societies in which it operates.

In the next chapter, I illustrate my reasoning with a few examples. Terror and child pornography constitute illegal speech that necessitates proactive steps to redeem its most harmful consequences. Cyberbullying and hate, in contrast, are forms of speech that are protected in the United States under the First Amendment yet, morally speaking, should be discriminated against in one form or another.

7

Responsibility of Internet Service Providers and Web-Hosting Services, Part II: Applications

The mind of the superior man is conversant with righteousness; the mind of the mean man is conversant with gain.

–Confucius

Suppose you are managing a large American Internet service provider (ISP). One day, FBI agents knock on your door and tell you that your company has been hosting a jihadi site that urged attacks against American targets. How would you feel? What would you say?

This chapter examines the implications of social responsibility for business providers and host companies that entertain problematic speech. The main question is whether ISPs and Web-hosting services (WHSs) should be proactive: that is, should they merely cooperate on receipt of information from various sources, or should they also, to promote trust among their subscribers, scrutinize their sphere for problematic, antisocial, and potentially harmful material? Many ISPs and WHSs are reluctant to do so for a very practical reason: the costs involved in employing professional staff for the task. Cynically, they might argue that morality is one thing, but here we are talking about money.

I argue that ISPs and WHSs should adopt a proactive stance in combating antisocial and violent content. They cannot be neutral toward such a phenomenon. Absolute content net neutrality constitutes clear-eyed akrasia and shedding of moral and social responsibility. The prime troubling examples that have significant presence on the Internet are crime, terrorism, child pornography, cyberbullying, and hate speech.

CRIME

One of the gravest challenges we are facing today is cybercrime – crime that is facilitated by computers or computer networks. Criminal uses of Internet communications include cybertheft, online fraud, online money laundering, cyberextortion, cyberstalking, and cybertrespass. Cybercriminals send spam, host and distribute malware, sell fake and harmful software, install backdoor Trojans (programs that perform like a real program that a user may wish to run but also perform unauthorized actions behind the scene), set up malicious websites, and design schemes to steal valuable information and money.

A California-based Web-hosting company named Atrivo chose to host some of the worst cybercriminals who were operating online. Following complaints, Atrivo's upstream provider cut its traffic, which immediately reduced the volume of spam.[1] Soon enough, however, the volume of spam increased as the cybercriminals found a new place to roost. They shifted their business to another Californian company, McColo, which in addition to spewing spam hosted child pornography, fake security products, and counterfeit pharmaceutical sites and was involved in stealing credit and debit card details.[2] Surprisingly, although McColo was operating for a long time, and many people in the security community had raised awareness about the problem, US law enforcement did not intervene.[3]

McColo came to the attention of a *Washington Post* reporter, Brian Krebs. Krebs's research revealed that McColo played a key role in managing some of the world's major malware warehousing and botnets (software robots that run automatically and autonomously, infecting personal computers so they forward spam and viruses to other computers). Krebs approached the large commercial ISPs that provided McColo with their bandwidth to reach the Internet. Confronted with the damning data and the risk of bad publicity to their business, these ISPs terminated all their associations with McColo. In November 2008, McColo was shut down. Benny Ng, director of infrastructure at Hurricane Electric, one of the companies that served McColo, explained, "Having a company like McColo on your network doesn't look good. . . . As an operator of an international Internet backbone service, you just can't have that."[4] When McColo was unable to continue its operation, spam levels around the world instantly dropped by 70 percent.[5]

---

[1] Robert Lemos, "McColo Takedown Nets Massive Drop in Spam," *Security Focus*, November 13, 2008, http://www.securityfocus.com/brief/855 .

[2] Jaikumar Vijayan, "McColo Takedown: Internet Vigilantism or Online Neighborhood Watch?" *Computerworld*, November 17, 2008, http://www.computerworld.com/article/2529316/malware-vulnerabilities/mccolo-takedown--internet-vigilantism-or-online-neighborhood-watch-.html .

[3] Brian Krebs, "Host of Internet Spam Groups Is Cut Off," *Washington Post*, November 12, 2008; Jart Armin, ed., "McColo: Cyber Crime USA," HostExploit, 2008, http://hostexploit.com/downloads/view.download/4/14.html .

[4] Vijayan, "McColo Takedown."

[5] Brian Krebs, "A Closer Look at McColo," *Washington Post*, November 13, 2008, http://voices.washingtonpost.com/securityfix/2008/11/the_badness_that_was_mccolo.html ; see also

This story teaches us some lessons. First, because big money is involved, cybercriminals will not be easily deterred. They will find ways to abuse the Internet. The struggle is Sisyphean and considerable. Even if all American gates are closed before them, cybercriminals are likely to shift their activities offshore, where stopping the abuse is much more difficult. Second, although the challenge is great, cooperation between concerned citizens, responsible gatekeepers, and governments bears fruit in stifling cybercrime. In both instances, the spammers stopped operation when major intermediaries and concerned citizens cooperated. The responsible agents were successful because both companies were based in California. But international cybercrime necessitates international counteractivity and cooperation between governments, major ISPs, security companies, and the international community at large. With enhanced collaboration between responsible stakeholders, cybercriminals will be pushed to make more efforts and invest more resources to maintain their operations. The rising costs might force some to reconsider their business model or push others out of business.

TERRORISM

Another grave challenge we are facing today is terrorism. *Terrorism* is defined as the threat or use of violence against noncombatant targets for political, religious, or ideological purposes by subnational groups or clandestine individuals who are willing to justify all means to achieve their goals.[6] Terrorism is usually the work of a small number of committed individuals who strive for what they perceive as the "greater good" of a larger group with whom they identify.

In September 2014, the US Department of State Bureau of Counterterrorism listed 59 terrorist organizations.[7] More than 40 of them were established by radical Islamic groups. The majority of those terrorist groups use the Internet as a primary tool for their activities. The number of pro-terrorism websites is estimated to have increased from approximately 12 in 1998 to more than

P. W. Singer and Allan Friedman, *Cybersecurity and Cyberwar: What Everyone Needs to Know* (New York: Oxford University Press, 2014), 205.

[6] For a discussion on the complicated task of defining terrorism, see Alex P. Schmid, "The Definition of Terrorism," in *The Routledge Handbook of Terrorism Research: Research, Theories, and Concepts*, ed. Alex P. Schmid (London: Routledge, 2011), 39–98; Gus Martin, *Essentials of Terrorism* (Thousand Oaks, CA: Sage, 2011), 2–25.

[7] Bureau of Counterterrorism, "Foreign Terrorist Organizations," US Department of State, Washington, DC, September 2014, http://www.state.gov/j/ct/rls/other/des/123085 .

4,800 by 2010.[8] These websites use slogans to catch attention, often offering items for sale (such as T-shirts, badges, flags, and video or audio cassettes). Frequently the websites are designed to draw local supporters, providing information in a local language about the activities of a local cell as well as those of the larger organization. The website is thus a recruiting tool as well as a basic educational link for local sympathizers and supporters.[9]

Many password-protected forums refer to extensive literature on explosives. Tutorials provide insights on viruses, hacking stratagems, encryption methods, anonymity guidelines, and use of secret codes. Bomb-making knowledge is available on jihadi websites in the form of detailed, step-by-step video instructions showing how to build improvised explosive devices. Strong evidence indicates that such online instructions played a critical role in the March 2004 Madrid bombings, the April 2005 Khan al-Khalili bombings in Cairo, the July 2006 failed attempt to bomb trains in Germany, the June 2007 plot to bomb London's West End and Glasgow, and the April 2013 Boston Marathon bombing, in which Dzhokhar and Tamerlan Tsarnaev allegedly placed explosives-laden pressure cookers close to the finish line.[10]

Jihadism is a branch of political Islam that aims through armed conflict and war (*jihad*) to bring about the rule of Islam and the establishment of the caliphate (i.e., an Islamic state led by a religious and political leader known as the *caliph*) throughout the world on the basis of a specific interpretation of Salafist doctrine and the radical ideas of Sayyid Qutb. The term *salafi* is derived from the word *salaf*, which means to precede. Salafis believe in a strict return to the fundamentals of Islam and reject any idea that was not uttered by the Prophet Muhammad.[11] We should not overemphasize this radical challenge, but at the same time we should not underestimate it. The sheer

---

8   Luis Miguel Ariza, "Virtual Jihad: The Internet as the Ideal Terrorism Recruiting Tool," *Scientific American*, December 26, 2005, http://www.sciam.com/article.cfm?articleID=000 B5155-2077-13A8-9E4D83414B7F0101&sc=I100322 ; Gabriel Weimann, *Terror on the Internet: The New Arena, the New Challenges* (Washington, DC: US Institute of Peace Press, 2010).

9   Pauline Neville-Jones and Home Office, "Tackling Online Jihad," speech delivered at a Wilton Park conference, West Sussex, UK, January 31, 2011, https://www.gov.uk/government/speeches/tackling-online-jihad-pauline-neville-joness-speech ; Cindy C. Combs, "The Media as a Showcase for Terrorism," in *Teaching Terror: Strategic and Tactical Learning in the Terrorist World*, ed. James J. F. Forest (Lanham, MD: Rowman & Littlefield, 2006), 133–54, 139.

10 Marc Sageman, *Leaderless Jihad: Terror Networks in the Twenty-First Century* (Philadelphia: University of Pennsylvania Press, 2008), 113; Alia E. Dastagir, "Internet Has Extended Battlefield in War on Terror," *USA Today*, May 5, 2013, http://www.usatoday.com/story/news/nation/2013/05/05/boston-bombing-self-radicalization/2137191 ; Gabriel Weimann, "Virtual Packs of Lone Wolves," Woodrow Wilson International Center for Scholars, Washington, DC, https://medium.com/its-a-medium-world/virtual-packs-of-lone-wolves-17b12f8c455a .

11 On the Salafi doctrine, see Roel Meijer, ed., *Global Salafism: Islam's New Religious Movement* (New York: Columbia University Press, 2009); Quintan Wiktorowicz, "The New Global

accessibility of cyberwarfare capabilities to tens, perhaps hundreds, of millions of people is a development without historical precedent. Thus, the ethical dimensions of violence conducted by networks of individuals, operating via the virtual realm, are of significant importance.[12]

Many of the radical sites post the *Terrorist's Handbook*[13] and *The Anarchist Cookbook*,[14] two well-known manuals that offer detailed instructions on how to construct a wide range of bombs. These "classic" manuals are supplemented with al-Qaeda's own how-to books, whose writers have benefited from vast experience in the battlefields of Iraq, Afghanistan, and other bloody theaters of war. "The Mujahideen Poisons Handbook," written by Abdel-Aziz in 1996 and published on the official Hamas website, details how to prepare various homemade poisons, chemical poisons, poisonous gases, and other deadly materials for use in terrorist attacks. In the Introduction, Abdel-Aziz writes:

> *Warning: Be very careful when preparing poisons. It is much, much more dangerous than preparing explosives! I know several Mujahids whose bodies are finished due to poor protection etc.*
>
> On the positive side, you can be confident that the poisons have actually been tried and tested (successfully, he he!).[15]

"The Mujahideen Explosives Handbook," also written by Abdel-Aziz, teaches how to prepare a laboratory, how to cut glass pipe tube, and how to handle hot substances. The handbook explains the difference between electric and flame detonators, how to prepare different charges, how to test explosives, and how to mix dangerous materials. Interestingly, the handbook opens with a call to scan military books and send them to the Organization for the Preparation of Mujahideen through encrypted email. The public key of the Pretty Good

Threat: Transnational Salafis and Jihad," *Middle East Policy* 8, no. 4 (2001): 18–38, http://groups.colgate.edu/aarislam/wiktorow.htm . On Sayyid Qutb, see John Calvert, *Sayyid Qutb and the Origins of Radical Islamism* (New York: Columbia University Press, 2010); Albert J. Bergesen, ed., *The Sayyid Qutb Reader: Selected Writings on Politics, Religion, and Society* (New York: Routledge, 2007). See also E. S. M Akerboom, *Ideology and Strategy of Jihadism* (The Hague: National Coordinator for Counterterrorism, 2009).

12 Rik Coolsaet, ed., *Jihadi Terrorism and the Radicalisation Challenge* (Surrey, UK: Ashgate, 2011), especially part 1; John Arquilla, "Conflict, Security and Computer Ethics," in *The Cambridge Handbook of Information and Computer Ethics*, ed. Luciano Floridi (Cambridge: Cambridge University Press, 2010), 133–48; see also al-Qaeda's *Inspire* magazine on the Public Intelligence website, July 14, 2010, http://publicintelligence.net/complete-inspire-al-qaeda-in-the-arabian-peninsula-aqap-magazine/ .

13 *The Terrorist's Handbook*, http://www.dvc.org.uk/cygnet/tthb.pdf .

14 William Powell, *The Anarchist Cookbook*, http://www.scribd.com/doc/387846/The-Anarchist-Cook-Book .

15 Abdel-Aziz, "The Mujahideen Poisons Handbook," 1996, 1, http://cnqzu.com/library/Guerrilla%20Warfare/Poison%20Handbook-ACADEMIC%20USE%20ONLY.pdf .

Privacy encryption software (see chapter 2) is provided together with the key ID.[16] For some time, jihadists and terrorists used the encryption program Mujahideen Secrets, put out in January 2007 by the Global Islamic Media Front. It provided users with five encryption algorithms, advanced encryption keys, and data compression tools.[17] State-sponsored terrorists such as Hamas and Hezbollah are using products developed in-house. Many jihadists and terrorists are using one-time pad (OTP), an encryption technique that cannot be cracked if used correctly.[18] In this technique, the message is paired with a random, secret key (or *pad*). The encryption key must be at least the same length as the text. Provided that the key is random, that it is at least as long as the plaintext, that it never reused in whole or in part, and that it is kept completely secret, the resulting ciphertext will be impossible to decrypt or break.[19]

The so-called *Encyclopedia of Jihad*,[20] created by al-Qaeda, offers instructions on the manufacture of weapons and explosives, first aid instructions, analysis of intelligence information, information about American military training, information about establishing and operating an underground network, and detailed guidelines on carrying out violent attacks.

The *Al-Qaeda Manual* comprises 18 lessons.[21] The manual equips the modern Muslim warrior with the necessary information to carry out duties without being caught. Thus, for instance, lesson 3 concerns counterfeit currency and forged documents. Lesson 4 is about hiding places. Lesson 5 is concerned with communication and transportation. Lesson 7 discusses

[16] Abdel-Aziz, "The Mujahideen Explosives Handbook," Organization for the Preparation of Mujahideen, 1996, http://www.riskintel.com/wp-content/uploads/downloads/2011/06/Mujahideen-Explosive-Book.pdf . See also Asrar al-Mujahideen Tutorial (Software), *Jihad in the Cause of Allah with Wealth and Life* (blog), January 5, 2011, http://jihadfeesabilillah.blogspot.com/2012/08/asrar-al-mujahideen-tutorial-software.html .

[17] Yaakov Lappin, *Virtual Caliphate: Exposing the Islamist State on the Internet* (Washington, DC: Potomac Books, 2011), 32.

[18] I thank Aaron Weisburd, director, Internet Haganah, for this information via email communication, October 5, 2014.

[19] "One-Time Pad (OTP): The Unbreakable Code," Crypto Museum, http://www.cryptomuseum.com/crypto/otp.htm . See also "Cryptology and Data Secrecy: The Vernam Cipher," ProTechnix, Cornhill-on-Tweed, UK, http://www.pro-technix.com/information/crypto/pages/vernam_base.html .

[20] See the *Washington Post*'s description of the *Encyclopedia of Jihad* at http://www.washingtonpost.com/wp-dyn/content/custom/2005/08/05/CU2005080501351.html . For further discussion, see Alexei Malashenko, Stephen R. Bowers, and Valeria Ciobanu, "Encyclopedia of Jihad: Islamic Jihad," William R. Nelson Institute, James Madison University, Harrison, VA, 2001, http://www.c4ss.net/website/Web_site/RESEARCH/Islamic_Jihad.pdf .

[21] An English translation of the *Al-Qaeda Manual* is available on the US Department of Justice website at http://www.justice.gov/ag/manualpart1_1.pdf .

weapons, and lesson 8 is about members' safety. Lessons 11 and 12 are about espionage, and lesson 18 is about prisons and detention centers.

Mustafa Setmariam Nasar, known as Abu Mus'ab al-Suri, wrote *The Call for a Global Islamic Resistance*. This meticulous strategic and ideological 1,600-page magnum opus draws on the experience of prior conflicts to explain how global terror networks should organize, finance, recruit, train, and operate. The *Call* gives practical instructions for mass murder and urges wannabe jihadists to act independently. It is aimed at transforming al-Qaeda from a terrorist network into a truly global movement of individuals organized in small cells ("detachments"), thus making the movement impermeable to US and allied counterterrorism efforts. According to Nasar, the key to jihadi success is to make the global terror network looser, meaner, and more resilient. To defeat the United States, jihad should spread to all parts of the world and recruit far more warriors. The fighting should be conducted in three stages, each stage with the appropriate weapons and guerrilla warfare methods. The first stage is exhaustion, which includes assassinations, raids, ambushes, and explosion operations. It is followed by the balancing stage, where the jihadis move to great strategic attacks with great battles. Finally, the stage of decisiveness and liberation arrives after some sectors of the army join the guerrilla warriors and bring Muslim victory.[22] Should these manuals be perceived as protected speech, readily available on Western servers? Should the US Department of Justice post this information on its website? Granted that knowing the enemy should not be conflated with helping the enemy, does this public information help the enemy?

These manuals have been put into use. Mohammad Sidique Khan, Shehzad Tanweer, Hasib Hussain, and Germaine Lindsay, who were responsible for the deaths of 56 people and the injury of some 700 others during the London Underground bombing of July 7, 2005, downloaded the instructions on how to build their bombs from the Internet.[23]

In 2004, al-Qaeda issued a chilling manual directed at new volunteers who were "below the radar" of counterterrorist authorities and who could not undergo formal training in terrorist techniques. The manual encouraged the use of weapons of mass destruction.[24] *Al Battar* ("the sharp-edged sword") is an online al-Qaeda journal that has appeared regularly since 2003. It serves as a

---

[22] Abu Mus'ab al-Suri, *The Call for a Global Islamic Resistance*. Translated excerpts of the book are available at http://ia600503.us.archive.org/5/items/TheGlobalIslamicResistanceCall/The_ Global_Islamic_Resistance_Call_-_Chapter_8_sections_5_to_7_LIST_OF_TARGETS.pdf .

[23] Steve Hewitt, *The British War on Terror: Terrorism and Counter-terrorism on the Home Front since 9-11* (London: Continuum, 2008), 73.

[24] Jason Burke, "Al-Qaeda Launches Online Terrorist Manual," *Guardian*, January 17, 2004, http://www.guardian.co.uk/technology/2004/jan/18/alqaida.internationalnews .

virtual training camp, providing readers with instruction in weapons handling, explosives, kidnapping, poisoning, cell organization, guerrilla warfare, secure communications, physical fitness, construction of a suicide-bomb belt (even down to the correct thickness of the cloth), planting of land mines, detonation of a bomb remotely with a mobile phone, topography, orientation (including crossing a desert at night), map reading, and survival skills.[25] The courses are accompanied by statements and speeches of al-Qaeda leaders. Would-be jihadis are urged to follow the virtual courses at home or in groups, practice the instructions, obtain firearms, and maintain a high level of fitness in preparation for taking steps to join the mujahideen.

Some of the issues focus on single, important subjects. For instance, the September 2004 issue of *Al Battar* focused on how to properly conduct kidnapping operations. It provided detailed instructions as to requirements for conducting kidnapping, stages of public kidnapping, ways to deal with hostages, and security measures the kidnappers should take.[26] Al-Qaeda targeting guidance specifies types of targets inside cities, economic targets, the purpose of human targets, and the advantages and disadvantages of operations against cities.[27]

The Encyclopedia of Afghan Jihad is a detailed jihad manual that explains how to make and use explosives and firearms, and how to plan and carry out assassinations and terror attacks. The manual urges that plans be laid out to hit buildings such as skyscrapers, ports, airports, nuclear plants, and football stadiums, and suggests attacking large congregations at Christmas.[28]

In January 2008, the al-Qaeda-affiliated Global Islamic Media Front released "The Encyclopedia of Hacking the Zionist and Crusader Websites." This

[25] Abdel Bari Atwan, *The Secret History of al-Qaeda* (Berkeley: University of California Press, 2006), 144; Michael Kenney, "How Terrorists Learn," in *Teaching Terror: Strategic and Tactical Learning in the Terrorist World*, ed. James J. F. Forest (Lanham, MD: Rowman & Littlefield, 2006), 33–51, 39; Robert Spencer, "Al-Qaeda's Online Training Camp," Jihad Watch, January 6, 2004, http://www.jihadwatch.org/2004/01/al-qaedas-online-training-camp.html .

[26] Excerpts from *Al Battar*, Issue 10, September 4, 2004, are available at http://netwar04.blogspot.com/2004/09/al-battar-al-qaeda-manual-on.html .

[27] See "Al-Qaeda Targeting Guidance," IntelCenter/Tempest Publishing, Alexandria, VA, April 1, 2004, http://www.intelcenter.com/Qaeda-Targeting-Guidance-v1-0.pdf . The document includes an English translation of "Targets inside Cities," by Abdul Aziz al-Moqrin, an article from issue 7 of *Al-Battar*.

[28] Descriptions of *The Encyclopedia of the Afghan Jihad*, are available at http://www.unl.edu/eskridge/encyclopedia.html and http://emp.byui.edu/ANDERSONR/itc/Book%20_of_Mormon/08_helaman/helaman01/helaman01_08terrorism.htm . See also George Smith and Dick Destiny, "Download al Qaeda Manuals from the DoJ, Go to Prison?: In the U.K. It's All Down to Your Motivation," *Register*, May 30, 2008, http://www.theregister.co.uk/2008/05/30/notts_al_qaeda_manual_case/ ; Yaman Akdeniz, *Racism on the Internet* (Strasbourg, France: Council of Europe Publishing, December 2009), 12.

manual includes hacking instructions, a list of vulnerable websites, file-transfer programs, and the password cracker software John the Ripper.[29] The primary purpose of John the Ripper is to detect weak passwords. It conveniently combines a number of password crackers into one package.

Jihadists also benefit from other manuals that are freely available on the Net. For instance, *Bacteriological Warfare: A Major Threat to North America* was written by microbiologist Larry Wayne Harris. Its aim is to warn against the use of certain bacteria by terrorists. At the same time, it provides valuable data that might be exploited by those very terrorists.[30] Harris ends his book by saying that the public wants biological civil defense: "The public does not know that the government has failed to provide the Biological civil defense program which will preserve our society in the event of Terrorist Germ Warfare attacks."[31] However, the failure may be attributable in part to this very detailed book.

In addition to manuals and diagrams, training videos have become increasingly common on terrorist websites. For example, in early June 2005, a contributor to the militant Arabic-language Web forum Tajdid Al-Islami posted a series of training videos for beginner mujahideen that included discussions on basic fitness, ninja arts, proper uniform, and communication techniques. As a follow-up, in August 2005, the contributor posted a seven-part lesson on how to use a handheld, portable global positioning system receiver.[32]

Mohammad Zaki Amawi, an American citizen, tried to enter Iraq through Jordan to fight against the US and coalition forces. His failure to do so did not discourage him. In 2004, Amawi started to compile jihadist training manuals and videos through jihadist websites to build his own cell in Toledo, Ohio, and soon he recruited others. Among the materials Amawi collected were a "basic training" course for jihadists, videos on the production and use of improvised explosive devices, and an instructional video for building a suicide-bomb vest, titled "Martyrdom Operation Vest Preparation." Amawi and two others were arrested and indicted in February 2006.[33]

---

[29] The software is available on the Openwall website at http://www.openwall.com/john/ and the Help Net Security website at http://www.net-security.org/software.php?id=11 . An instructional video, "Using John the Ripper (JTR) to Crack a Hash," is available on YouTube at http://www.youtube.com/watch?v=wHxVT6u4kPU .

[30] Larry Wayne Harris, *Bacteriological Warfare: A Major Threat to North America, What You and Your Family Can do Before and After* (Indianapolis: Virtue International, 1995). See http://www.uhuh.com/reports/harris/book.htm for an edited and condensed version of Harris's *Bacteriological Warfare* by Richard L. Finke.

[31] Harris, *Bacteriological Warfare*, 130.

[32] James J. F. Forest, "Introduction," in *Teaching Terror: Strategic and Tactical Learning in the Terrorist World*, ed. James J. F. Forest (Lanham, MD: Rowman & Littlefield, 2006), 1–29, 10.

[33] Rita Katz and Josh Devon, "Web of Terror," Forbes, May 7, 2007, 184; prepared statement by Rita Katz, director, SITE Institute, and Josh Devon, senior analyst SITE Institute, "The

A Pakistani website, Mojihedun.com, contained a section called "How to strike a European city," which gave detailed instructions and suggestions for attacks. The forum al-Firdus provided highly detailed instructions for the preparation of explosive oils, recommending nitroglycerin because it is "more explosive than TNT."[34] To prevent intelligence officers from tracking their activities, terrorists sent training and recruitment videos over video game sites in Japan because traffic and file size at those sites are so large. Video games are used as recruitment and training tools.[35]

In 2010, al-Qaeda Global Islamic Media Front published *The Explosives Course*. This comprehensive and sophisticated manual, which is widely distributed online on extremist networks, aims to provide potential recruits and so-called lone wolves, who reside in the West and are unable to travel to terrorist camps, with the necessary knowledge to assemble explosives and make the utmost use of them.[36]

Just a month before the 2013 Boston Marathon bombings, the al-Qaeda branch in Yemen posted on the Web the *Lone Mujahid Pocketbook*,[37] a compilation of all the do-it-yourself articles with brisk English text, high-quality graphics, and teen-friendly shorthand, including a user-friendly manual on how to prepare a bomb from kitchen ingredients.[38]

Although I do not wish to evoke moral panic, the Internet is acknowledged as the single most important factor in transforming largely local jihadi concerns and activities into the global network that characterizes al-Qaeda

Internet: The Most Vital Tool for Terrorist Networks," in "Using the Web as a Weapon," Hearing before the Subcommittee on Intelligence, Information Sharing and Terrorism Risk Assessment of the Committee on Homeland Security, House of Representatives, 110th Cong., 1st sess., November 6, 2007; Federal Bureau of Investigation, Cleveland Division, "Three Sentenced for Conspiring to Commit Terrorist Acts against Americans Overseas," US Department of Justice, October 22, 2009, http://www.fbi.gov/cleveland/press-releases/2009/cl102209.htm ; *United States v. Mohammad Zaki Amawi et al.* 541 F.Supp.2d 955 (2008), http://www.leagle.com/decision/20081496541FSupp2d955_11402 .

34 Atwan, *The Secret History of al-Qaeda*, 144.
35 Sue Myrick, "Internet a Terrorist Haven," *Charlotte Observer*, May 17, 2007, 12A; United Nations Office on Drugs and Crime, *The Use of the Internet for Terrorist Purposes* (New York: United Nations, 2012), 5.
36 Alexander Meleagrou-Hitchens, "New English-Language al-Qaeda Explosive Manual Released Online," *ICSR Insight*, International Centre for the Study of Radicalization, King's College, London, December 31, 2010. For a discussion on lone wolves, see Gerry Gable and Paul Jackson, *Lone Wolves: Myth or Reality* (London: Searchlight, 2011); Weimann, "Virtual Packs of Lone Wolves."
37 An excerpt from *Lone Mujahid Pocketbook* can be viewed at http://worldanalysis.net/modules/news/article.php?storyid=2194 .
38 Scott Shane, "A Homemade Style of Terror: Jihadists Push New Tactics," *New York Times*, May 5, 2013, http://www.nytimes.com/2013/05/06/us/terrorists-find-online-education-for-attacks.html?nl=todaysheadlines&emc=edit_th_20130506&_r=0&pagewanted=print .

today.[39] Unfortunately, many of the terrorist websites are hosted by servers in the Western world. American ISPs hosted terrorist sites and helped the cause of jihad. Some did it knowingly while others did it inadvertently. Thus, for instance, InfoCom Corporation in Texas hosted websites for numerous clients in the Middle East. It served more than 500 Saudi Internet sites and notable Palestinian Hamas organizations, including the Islamic Association for Palestine and the Holy Land Foundation for Relief and Development. InfoCom was founded by Mousa Abu Marzook, one of the leaders of Hamas. InfoCom also served to launder money. Large amounts of money came from Saudi Arabia and the Gulf states to sponsor Hamas activities.[40]

Whereas InfoCom directors knew exactly what they were doing, Fortress ITX unwittingly hosted a jihadi site that urged attacks against American and Israeli targets. Among the webpages were "The Art of Kidnapping," "Military Instructions to the Mujahedeen," and "War inside the Cities." The informative site was shut down after Fortress learned about the content from a reporter.[41] Undoubtedly, not playing into the hands of terrorists requires exercising oversight and taking proactive steps. ISPs are reluctant to monitor their servers for economic reasons. American ISPs do not see monitoring as part of their service. They do not wish to compromise their customers' First Amendment rights.

Al-Qaeda has received funds from numerous social charities based in the United States. In the wake of the September 11, 2001, terror attack, the US government seized or froze the assets of several charities (among them the Benevolence International Foundation,[42] Elehssan

---

[39] Atwan, *The Secret History of al-Qaeda*, 124. See also Gabriel Weimann, "New Terrorism and New Media," Woodrow Wilson International Center for Scholars, Washington, DC, 2014.

[40] Anonymous [Rita Katz], *Terrorist Hunter* (New York: HarperCollins, 2003), 262–63; Brian Whitaker, "U.S. Pulled Plug on 500 Arab/Muslim Websites Day before Jetliner Attacks," Rense.com, September 10, 2001; "Mousa Abu Marzook," Investigative Project on Terrorism, April 20, 2012, http://www.investigativeproject.org/profile/106 .

[41] Eric Lipton and Eric Lichtblau, "Online and Even Near Home, a New Front Is Opening in the Global Terror Battle," *New York Times*, September 23, 2004, 12.

[42] The Benevolence International Foundation (BIF), based in Illinois, was founded in the 1980s by Saudi sheikh Adil Abdul Galil Batargy. BIF described itself as a humanitarian organization dedicated to helping those afflicted by wars. According to the Federal Bureau of Investigation, BIF raised millions of dollars for Osama bin Laden. It also reportedly sent $600,000 to Chechen extremists trained by al-Qaeda. In addition, BIF was linked to the 1993 World Trade Center bombing. Enaam Arnaout, executive director of BIF, helped bin Laden's al-Qaeda terrorism network move money and equipment around the world. See US Department of the Treasury, "Treasury Designates Benevolence International Foundation and Related Entities as Financiers of Terrorism," press release, November 19, 2002, http://www.treasury.gov/press-center/press-releases/Pages/po3632.aspx ; Kevin Johnson and Richard Willing, "Bosnian Evidence Makes FBI Case," *USA Today*, January 5, 2002, http://www.usatoday.com/news/sept11/2002/05/01/charity-terrorism.htm .

Society,[43] the Global Relief Foundation,[44] Goodwill Charitable Organization,[45] the Holy Land Foundation for Relief and Development,[46] Al Kifah Refugee Center,[47] and the Al-Haramain Foundation[48]) that allegedly used the Internet to raise money for al-Qaeda and Hamas.[49]

Some of the message boards and the information hubs where terrorists post texts, declarations, and recordings are often included in the "communities" sections of popular Western sites such as Yahoo!, Lycos, and others.[50] Jihadists use Twitter, YouTube, and Facebook extensively. Al-Qaeda, the Taliban, al-Shabaab (a militant Islamist group in Somalia), and Lashkar-e-Taiba (a Pakistani-based Islamic terrorist organization) have a noticeable presence on Facebook, filling hundreds of pages with information, photos, and videos.[51] An al-Qaeda website was registered in Toronto. Register.com is one of the most popular services for registering Internet domain names, operating from Nova Scotia. It offers anonymous registration service for a small extra fee, and was used by one of the world's largest pro-Hamas websites, where viewers could download martyrdom videos and watch terrorists bidding the world farewell before they launched deadly attacks.[52] The concept of content net neutrality, which rejects any responsibility for content, facilitates this phenomenon. However, the Internet is not outside the democratic realm. ISPs and WHSs are a necessary part of it. Terrorism is outside the democratic realm. A zero-sum game exists between democracy and terror. The victory of one comes at the expense of the other. Therefore, if the spirit and ideas of democracy are dear to ISPs and WHSs, and if they wish the democracy that enables their

[43] WMLR Newsdesk, "USA: Treasury Designates Charity as Conduit for Terrorism Funding," Banking Insurance Securities.com, May 5, 2005, http://www.bankinginsurancesecurities.com/aml_cft/USA-Treasury-designates-charity-as-conduit-for-terrorism-funding .

[44] "Protecting Charitable Organizations," US Department of the Treasury, Washington, DC, http://www.treasury.gov/resource-center/terrorist-illicit-finance/Pages/protecting-index.aspx .

[45] "U.S. Acts on Groups Aiding Hezbollah," *USA Today*, July 24, 2007, http://www.usatoday.com/news/washington/2007-07-24-treasury-hezbollah_N.htm .

[46] "Holy Land Foundation Shut Down," About.com, December 5, 2001, http://islam.about.com/library/weekly/aa120401a.htm .

[47] "Profile: Al-Kifah Refugee Center," History Commons, http://www.historycommons.org/entity.jsp?entity=al-kifah_refugee_center ; "Complete 911 Timeline," History Commons, http://www.historycommons.org/timeline.jsp?timeline=complete_911_timeline&financing_of_al-qaeda:_a_more_detailed_look=complete_911_timeline_al_kifah_mak .

[48] Bootie Cosgrove-Mather, "Al Qaeda Skimming Charity Money," CBS News, June 7, 2004, http://www.cbsnews.com/stories/2004/06/07/terror/main621621.shtml .

[49] "Jihad Online: Islamic Terrorists and the Internet," Anti-Defamation League, New York, 2002, http://www.adl.org/internet/jihad_online.pdf .

[50] Lawrence Wright, "The Terror Web," *New Yorker*, August 2, 2004.

[51] Abdel Bari Atwan, *After Bin Laden: Al Qaeda, the Next Generation* (London: Saqi, 2012), 240–43.

[52] Omar El Akkad, "Mind Field: Terror Goes Digital. With Canadian Help," *Globe and Mail*, August 18, 2007, F1.

operation to prevail, they cannot shield themselves under the concept of content net neutrality. They must take sides. They must distinguish good from evil and adopt the promotional approach (see chapter 6). ISPs and WHSs need to be proactive, scrutinizing their servers and ascertaining that they do not become terrorist hubs.

Many Internet experts believe that all they need to do is to provide the structure and the rest is up to the public. Such complacent neutrality is amoral at best and immoral at worst. Immoral conduct is conduct that is blatantly uncaring, disrespectful, and undignified. Amoral conduct is practiced when people are not aware that they transgress ethical boundaries. Some Internet intermediaries are oblivious to ethical standards because no one ever taught them the philosophical and ethical skills required for the job. I believe that a great deal of Internet professionals' conduct is amoral in nature. Many professionals are superficially familiar with their professional code of conduct because they received only limited training (which, one assumes, did not include the writings of Aristotle and Kant or Corporate Social Responsibility literature) and failed to fully comprehend the intricate philosophical layers underpinning their ethical code. The situation could be very different were Internet professionals to receive adequate training that pointed out the ethical issues at hand and made them aware of moral quandaries and ways to address and resolve such problems. High-tech companies should invest in teaching, planning, and developing ethical awareness and skills to shape new role models for the next generation of Internet professionals. The need is to create a culture of responsibility, trust, and accountability to the customers.

I said at the outset that I am not a relativist. I believe that people are able to discern between good and evil. Life is not a Michael Jackson song, where good is bad and vice versa. We can apply common sense and distinguish between that which is objectively right and that which is objectively wrong. Terrorism, child pornography, racism, and cyberbullying are objectively wrong, no matter what the people who are engaged in these activities say about them. Thus, all humane people perceive bombing civilian targets – be they buses, trains, airplanes, shopping malls, or buildings – as immoral, wrong, wicked, and odious. All humane people abhor beheadings and believe that such brutal conduct has no place in civilization. We also think that these views and beliefs are true and that people who justify terror are erroneous. We think, moreover, that our opinions are not just subjective reactions to the idea of the indiscriminate massacre of innocent lives but also opinions about its actual moral character. We think that as an objective matter – a matter of how things actually are – terrorism is manifestly wrong and immoral. Acknowledging the democratic catch (see chapter 3) – that unrestricted liberty is a prescription for dangerous and harmful conduct – all concerned citizens should address the terrorist challenge. At the same time, we should warn against government actions deemed unethical (collateral damage, indiscriminate killing) or illegal (stifling political opposition, undermining of critical yet legitimate speech).

Advancing content net neutrality at the expense of social responsibility serves wicked aims that undermine both the platform Internet professionals wish to protect and the society that promotes the democratic spirit in which they thrive.

CHILD PORNOGRAPHY

Adult pornography assumes consent: adults can voluntarily engage in sexual activity and may voluntarily agree to be photographed while engaging in that activity. Conversely, children cannot consent to their abuse – both physical abuse that engages them in sexual activity and pictorial abuse that photographs them in a sexual context. Thus, child pornography by definition is abusive and coercive. Every depiction of sexual intercourse with real children is considered molestation, a criminal act. Liberal democracies take upon themselves to protect vulnerable third parties, and children are perceived as worthy of protection against adult abuse.

The age of consent for sexual intercourse differs from one country to another. In the United Kingdom, the age of consent is 16. In the United States and Canada, it is between 16 and 18 years of age. In Australia, the criminal code speaks of young people under the age of 16. In Israel, it is between 14 and 16.[53]

The typical legal definition of *child pornography* involves sexual depictions of youth. In Europe, "audiovisual material which uses children in a sexual context" is restricted under the Council of Europe definition.[54] The International Criminal Police Organization (Interpol) definition of child pornography goes beyond visual representation to include a "child's sexual behavior or genitals."[55] In Victoria, Australia, child pornography means a film, photograph, publication, or computer game that describes or depicts a person

---

[53] "Age of Sexual Consent," AVERT, http://www.avert.org/age-of-consent.htm ; "Legal Age of Consent by State," Age-of-Consent.info, http://www.age-of-consent.info ; "Minimum Legal Age of Consent – Female," ChartsBin, http://chartsbin.com/view/hxj .

[54] Margaret A. Healy, "Child Pornography: An International Perspective," Computer Crime Research Center, August 2, 2004, http://www.crime-research.org/articles/536/ .

[55] Jeremy Harris Lipschultz, *Broadcast and Internet Indecency: Defining Free Speech* (New York: Routledge, 2008), 70–71.

who is – or appears to be – a minor engaging in sexual activity or depicted in an indecent sexual manner or context.[56]

US federal child pornography statutes make illegal the creation and distribution of photographs or video of sexual behavior between an adult and a minor.[57] For the purpose of this chapter, *online child sex offenders* are people who watch, download, store, and transmit images of children in a sexual context. *Child sex offenders* engage in these activities and also produce images of children in a sexual context.

A further distinction is between *child erotica* (sometimes termed soft-core child pornography) and *child exploitation material* (or child sexual abuse images, sometimes termed hard-core child pornography). Child erotica depicts children in the nude, in various positions. Some child erotica could be quite innocent, such as nude children who visit naturist camps with their families. One can easily access naturist magazines such as *H&E Naturist Magazine* on the Net.[58] European countries are far more tolerant than the United States, if not lax, in their attitude regarding the display of child nudity in a nonsexual context. The context and intentions are of utmost importance. Parents may take photos of their naked children on the beach, store the photos on their computer, and share the photos within their family. There is nothing wrong in doing so. But the moment they circulate those same pictures on the Internet for sexual motives, the photos are transformed from being innocent, private family photos into sexual, public child pornography.

Child exploitation material (CEM) depicts children engaging in sexual activity. If those images are of real children (as distinguished from computer-generated images), they are illegal. In most countries in the world, sexual activity with children violates applicable criminal codes.

Pedophilia is a clinical term and is used to refer to a psychosexual disorder involving a sexual preference for children. According to the *Diagnostic and Statistical Manual of Mental Disorders*, pedophilia involves recurrent, intense sexual urges and fantasies that last for at least six months and focus on prepubescent children.[59] The Pedophile Liberation Front, a mutual support group for those with a sexual interest in children, defines a *pedophile* as "an

---

[56] Crimes Act 1958, section 67A. The text of this provision is available at http://www.austlii.edu. au/au/legis/vic/consol_act/ca195882/s67a.html .

[57] 18 U.S.C. sections 2251–60. The text of this provision is available at http://www.law.cornell.edu/uscode/text/18/2251 .

[58] See the magazine's website at http://www.henaturist.net/ . NudistFun.com lists a number of online nudist magazines at http://www.nudistfun.com/magazines/ .

[59] Cited in "Child Protection Tips," LifeTips, 2014, http://childprotection.lifetips.com/cat/63575/pedophile-search-and-statistics/ .

adult that is sexually attracted to children."[60] As portrayed by organizations like the Pedophile Liberation Front, pedophilia is not only a sexual orientation but also a way of life. Within this subculture, distinctions are made between pedophiles and child molesters (sometimes called *predators*). From this perspective, *child molester* is a pejorative term and is generally used to refer to people who sexually abuse children, whereas *pedophilia* indicates a sexual interest in and love of children, which may never be acted on. Indeed, many pedophiles would argue that looking at photographs provides a safe outlet for feelings that might otherwise lead to a contact offense.[61] Detective Senior Constable Bruce McFarlane of the Victoria Police Force of Australia Global Terrorism Research Centre noted that most online child sex offenders who produce, watch, download, store, and transmit CEM images are not pedophiles.[62] Such online offenders carry on ordinary lives and maintain adult sexual relationships with their wives, girlfriends, and so on; however, they also use CEM as a means of sexual gratification. Hence, all offenders who use CEM are online child sex offenders, but not all online child sex offenders are pedophiles or molesters and predators.

The Internet facilitates online child sex offenders' and pedophiles' networking, allowing communication and access to unlimited numbers of like-minded people. The supportive environment offered by the Net involves both social consolidation and validation.[63] The Internet has allowed online child sex offenders to find entire online fraternities of like-minded people with whom to share experiences. Members of these fraternities transmit child pornography instantly and anonymously to one another and experience the comfort of a reassuring support group. Worse yet, modern encryption technology allows all these transmissions to be hidden.[64] Onion routers such as Tor enable Net abusers to hide their IP (Internet Protocol) addresses, enabling them to use Tor exit relay, which can be located anywhere in the world (see chapter 2).

Online child sex offenders like to collect images of child pornography and share those images. They trade images of child porn and abuse in the same way children swap sports cards or stamps. The collections are organized according to ages, gender, ethnic origin, hair color, and other characteristics, as well as the sexual act (e.g., assault, gross assault, sadism, bestiality), erotic

---

[60] Max Taylor and Ethel Quayle, *Child Pornography: An Internet Crime* (Hove, UK: Brunner-Routledge, 2003), 12.

[61] Ibid.

[62] Comments of Bruce McFarlane on a draft of this chapter, March 27, 2011.

[63] Taylor and Quayle, *Child Pornography*, 13.

[64] Terrence Berg, "The Internet Facilitates Crime," in, Mur *Does the Internet Benefit Society?*, ed. Cindy Mur (Farmington Hills, MI: Greenhaven, 2005), 35–39, 36–37.

poses, or nudism.[65] Often the collections are mixed: erotica images with graphic images. Online child sex offenders help one another to complete their collections. When online child sex offenders complete a series of images, many make the images freely available for others, as a comradeship service to other members of the community. They pride themselves on completing series of images (such as KG, KX, or Tiny Americans). KG and Tiny Americans are series of thousands of nude images of very young girls (in KG, many were between three and six years of age).[66] The KX series depicts these same girls in hard-core sexual situations with one or more men.

Some online child sex offenders use servers that host user-generated photos: hello.com, Yahoo!, and Google. They create their own webpages, and interested parties need a password to enter, but these sites are relatively easy to track. Online child sex offenders risk criminal accusations by creating such webpages, but some of them still do so. They want to share and exchange their photos, which gives them sexual gratification. Images have currency in terms of reputation. The ability to trade images and even to let people have free access to the collection is an important feature of the Internet. In the child porn community, having access to children facilitates climbing up the subculture hierarchy.[67] Offenders who get caught in the process pay the price for their lax security standards.

Before the Internet, trade had to be done either through magazines or by word of mouth, which meant a potential loss of anonymity. Equally, materials had to be sent through the post, which both restricted quantity and increased the likelihood of detection. Through the Internet, trade in child pornography not only becomes easier but also carries with it assumed anonymity.[68] Online child sex offenders have used peer-to-peer file-sharing programs such as Kazaa,[69] LimeWire,[70] and eMule. These programs are very useful for transferring large files of photos and video clips. Online child sex offenders who use IRC (Internet Relay Chat) can make available the contents of their

---

[65] Ethel Quayle and Max Taylor, "Child Pornography and the Internet: Perpetuating a Cycle of Abuse," *Deviant Behavior* 23, no. 4 (2002): 331–62; Max Taylor, Gemma Holland, and Ethel Quayle, "Typology of Pedophile Picture Collections," *Police Journal* 74, no. 2 (2001): 97–107.

[66] Philip Jenkins, *Beyond Tolerance: Child Pornography on the Internet* (New York: New York University Press, 2001), 62.

[67] Interviews with senior officials, National Center for Missing and Exploited Children, Alexandria, VA, April 2, 2008; discussion with a senior officer, UK Child Exploitation and Crime Protection Centre, Wilton Park, West Sussex, UK, February 1, 2011.

[68] Taylor and Quayle, *Child Pornography*, 160.

[69] Kazaa is no longer operating.

[70] LimeWire is under a court order dated October 26, 2010, to stop distributing its software. See http://www.limewire.com/ .

hard drive to other Netusers via FTP (File Transfer Protocol) sites. FTP sites remain the best method of transferring large files online. Another innovation that facilitates file sharing is TeamViewer, which enables Netusers to log into another's computer and download any file without leaving a digital trace (see chapter 2).

In the preceding Internet activities involving the use of images already available on the Net, no actual harm is inflicted by the present child pornographers on minors to participate in such images and films. So for now I have described the use of already available images, not the production of new ones through the use of real children.[71] However, some online child sex offenders and pedophiles use the Internet for tempting, seducing, grooming, and blackmailing children. According to GuardChild, 17 percent of teens surveyed said they received an email or online message with photos or words that made them feel uncomfortable. Moreover, 20 percent of teenage Internet users have been the target of an unwanted sexual solicitation (requests for sexual activities, chat, or information).[72] When asked whether child molesters use child pornography to seduce children, the answer of Michael J. Heimbach of the US Federal Bureau of Investigation (FBI) was undeniably yes.[73] FBI interviewees reiterated this statement in my meetings with them during 2008.

Child pornography is used as a learning instrument in the grooming process, whereby a child pornographer befriends a child, gains the child's trust, and then presses for sex. The virtual reality allows unlimited identity play, and manipulative child pornographers exploit this capacity, aiming to desensitize the child to sexual demands, reduce inhibition, and encourage the victim to normalize sexual activities.

Indeed, part of the grooming process is to "normalize" sexual activity with children and break down their inhibitions. Offenders use child pornography to teach children how to masturbate, perform oral sex, and engage in sexual intercourse. Sometimes blackmail is also involved, usually at the later stage

---

[71] Virtual child pornography, that is, computer-generated child pornography that does not involve real children, is beyond the scope of this discussion. According to the National Center for Missing and Exploited Children, the vast majority of the images deal with real children. On this issue, see *Ashcroft v. Free Speech Coalition*, 535 U.S. 234 (2002); Marie Eneman, "The New Face of Child Pornography," in *Human Rights in the Digital Age*, ed. Mathias Klang and Andrew Murray (London: GlassHouse, 2005), 27–40; Harry Henderson, *Internet Predators* (New York: Facts on File, 2005).

[72] "Internet Statistics," GuardChild, http://www.guardchild.com/statistics/. See also "Key Facts," National Center for Missing and Exploited Children, http://www.missingkids.com/KeyFacts.

[73] Testimony of Michael J. Heimbach, Criminal Investigative Division, Crimes against Children Unit, Federal Bureau of Investigation, before the US Senate, Subcommittee on Crime, Terrorism, and Homeland Security, Washington, DC, May 1, 2002, http://www.fbi.gov/news/testimony/supreme-courts-child-pornography-decision.

after the child has been exposed to some sort of pornography or after the child has performed sexual favors.[74] The saturation of the Internet with such material may serve to normalize this behavior and probably makes it easier to objectify children as sexual artifacts. Child pornography is also thought to reinforce a person's sexual attraction to children.[75]

Interactive communication on the Internet, as occurs in chat rooms, offers child sexual offenders easy ways to access children. In chat rooms, a child is usually alone and is therefore susceptible to influence and persuasion by the offender. Social networking sites have been shown to attract sexual predators. Such sites present ample opportunities for deviants to nurture their sexual interests by observing children online, interacting with them, and soliciting direct meetings offline.[76] Studies should continue to probe this new phenomenon of social networking abuse. Given our limited experience (the major social networking sites, MySpace and Facebook, were launched in 2003 and 2004, respectively), we need to explore the developing behavior of Netusers on such sites.

A number of Internet servers permit individuals to establish interest groups to which people can post images or messages. Because sites can be accessed without payment or subscription, users are *prima facie* anonymous. Yahoo! Groups is a collection of many thousands of groups on virtually every topic imaginable: business and computers, health and fitness, leisure, shopping, habits, sports, and so on. Opening a new group is free and straightforward; thus, unsurprisingly, we find thousands of sexually oriented groups, the vast majority dedicated to legal adult topics. Some, however, are involved with child pornography. The management of Yahoo!, MSN, Facebook, and other social networking Web providers is well aware of the problem. Offending groups are closed swiftly – sometimes within hours – when a Netuser informs the major Internet intermediaries of child porn activity.[77]

---

[74] Paul Peachey, "Paedophiles Blackmail Thousands of U.K. Teens into Online Sex Acts," *Independent* (London), September 20, 2013, http://www.independent.co.uk/news/uk/crime/paedophiles-blackmail-thousands-of-uk-teens-into-online-sex-acts-8827794.html?utm_source=indynewsletter&utm_medium=email20092013 .

[75] Janet Stanley, "The Internet Can Be Dangerous for Children," in Mur, *Does the Internet Benefit Society?*, 95–101, 99–100; Anneke Meyer, *The Child at Risk: Pedophiles, Media Responses, and Public Opinion* (Manchester, UK: Manchester University Press, 2007), 124.

[76] "Information on Child Monitoring Software," Awareness Technologies, Los Angeles, CA, http://www.awarenesstech.com/Parental/Articles/ ; Eric Schonfeld, "Thousands of MySpace Sex Offender Refugees Found on Facebook," *TechCrunch*, February 3, 2009, http://techcrunch.com/2009/02/03/thousands-of-myspace-sex-offender-refugees-found-on-facebook/ ; Child Exploitation and Online Protection Centre (CEOP), *Understanding Online Social Network Services and Risks to Youth* (London: CEOP, 2006).

[77] Jenkins, *Beyond Tolerance*, 63; interview with Ruth Allen, head of the Specialist Operational Support Child Exploitation and Online Protection Centre, London, April 18, 2011; interview

In the United Kingdom, ISPs have rejected a call by charities for children to implement the government's approved blocklist for images of child sexual abuse, arguing that the list does not stop anyone who wants to access such material. A warning was raised that 700,000 homes could access websites hosting images of abuse because small ISPs do not filter their networks. Small ISPs have resisted filtering their networks for several years on the grounds of both economics and principles. They would have to spend a lot of money on something that they do not deem absolutely necessary. Moreover, pedophiles with minimal technical knowledge are able to circumvent the blocklist. The policy, therefore, does not affect pedophiles' ability to access websites distributing child abuse images. Rather, the blocklist protects average Web users with no interest in such material from accidental exposure.[78]

The UK Internet Service Providers' Association (ISPA) has a code of practice that was adopted in January 1999 and amended in July 2007.[79] This code is intended to govern the conduct of ISP members. The ISPA code includes references to the work of the Internet Watch Foundation, with which ISPA cooperates in its efforts to remove illegal material from websites and newsgroups.[80] The relationship between the UK government and the formation of the Internet Watch Foundation provides an example of a co-regulatory approach.

In June 2013, Google, Yahoo!, Microsoft, Twitter, and Facebook announced that they would allow the Internet Watch Foundation to actively seek out abusive images, rather than just acting on reports they received. Prime Minister David Cameron said he wanted search companies to go even further and block certain search terms from providing results. Cameron was very explicit, saying that anyone searching for a word on a "blacklist" compiled by the Child Exploitation and Online Protection Centre should be made to view a webpage warning them of the consequences, "such as losing their job, their family, even access to their children."[81] Prime Minister Cameron explained, "There are some searches which are so abhorrent and where there can be no doubt whatsoever about the sick and malevolent

with Carolyn Atwell-Davis, director of legislative affairs, National Center for Missing and Exploited Children, Alexandria, VA, April 2, 2008.

[78] Christopher Williams, "Small ISPs Reject Call to Filter Out Child Abuse Sites," *Register*, February 25, 2009, http://www.theregister.co.uk/2009/02/25/iwf_small_isps/ .

[79] For more information about ISPA, see the organization's website at http://www.ispa.org.uk/ . The code of practice is available at http://www.ispa.org.uk/about-us/ispa-code-of-practice/ .

[80] Yaman Akdeniz, *Internet Child Pornography and the Law: National and International Responses* (Aldershot, UK: Ashgate, 2008), 248.

[81] "David Cameron Urges Internet Firms to Block Child Abuse Images," BBC News, July 21, 2013, http://www.bbc.co.uk/news/uk-23393851?print=true .

intent of the searcher."[82] Cameron urged the Internet companies to find solutions to technical obstacles if such solutions exist, but "don't just stand by and say nothing can be done."[83] A spokesman from Google said, "We have a zero tolerance attitude to child sexual abuse imagery. Whenever we discover it, we respond quickly to remove and report it."[84] Seven of the major UK ISPs (BT, Sky, O₂, Virgin, TalkTalk, Orange, and Plusnet) are cooperating with the Child Exploitation and Online Protection Centre. The majority of smaller ISPs also cooperate with the special law enforcement agency. An estimated 80 percent to 90 percent of ISPs cooperate. For the majority of ISPs, it is a reputation issue. They wish to engage in business and do not wish to clash with the law. The problem, however, lies with the 10 percent to 20 percent that do not cooperate.[85] Those ISPs fail to understand their responsibilities.

In Australia, the two major ISPs, Telstra and Optus, voluntarily decided in 2011 to block more than 500 child abuse websites.[86] Although this decision does not stop peer-to-peer sharing of abusive electronic material, it is a responsible step in the right direction.

In 2008, French Interior Minister Michèle Alliot-Marie announced that she planned to use the help of ISPs to block websites that disseminate child pornography. From September 2008 onward, Netusers could flag sites that carry child pornography, incitement to terrorism and racial hatred, or attempted fraud. This real-time information helped France draw up a blacklist of sites that disseminate child pornography and other illegal material, which it would transmit to ISPs that have agreed to block such sites.[87] AFA – the French ISP association (Association de Fournisseurs d'Accès et de Services Internet), established in 1997 – has been collaborating with the French law authorities since 1998 on a regular basis.[88]

---

[82] Ibid.

[83] Ibid.

[84] Ibid.

[85] Interview with Ruth Allen, head, Specialist Operational Support, Child Exploitation and Online Protection Centre, London, April 18, 2011.

[86] Jennifer Dudley-Nicholson, "Telstra, Optus to Start Censoring the Web Next Month," News Limited, June 22, 2011, http://www.news.com.au/technology/internet-filter/telstra-optus-to-begin-censoring-web-next-month/story-fn5j66db-1226079954138 .

[87] "France to Block Child Pornography Websites," Reuters, June 10, 2008, http://www.reuters.com/article/2008/06/10/us-france-internet-pornography-idUSL1077696620080610 .

[88] See AFA's website at http://www.afa-france.com . See also the website of the Defender of Rights, a French government office, at http://www.defenseurdesdroits.fr/connaitre-son-action/la-defense-des-droits-de-lenfant/actualites/lassociation-des-fournisseurs . The 2013 annual report of the Defender of Rights is available in French at http://www.defenseurdesdroits.fr/sites/default/files/upload/rapport_annuel_2013.pdf .

Also in 2008, the United States enacted the Protect Our Children Act, which requires ISPs to submit reports of online child pornography to the National Center for Missing and Exploited Children (NCMEC).[89] The NCMEC established the CyberTipline, which forwards reports involving child pornography–related incidents to law enforcement agencies designated by the US attorney general. An initial failure to report by an ISP could result in a fine of up to $50,000, and any subsequent failures could result in a fine of up to $100,000.[90] I am told two-way cooperation exists between the NCMEC and the major ISPs. The NCMEC provides queries and information to ISPs, and they reciprocate. There is no need to resort to coercive measures with ISPs for cooperation; they are willing to cooperate because it is good for their business. After all, they do not wish to serve online child sex offenders or to have the reputation of helping spread child pornography.[91]

As in the United Kingdom, the problem lies with small ISPs that do not report to the NCMEC. Senior officers at the NCMEC cannot say how many of these small servers exist.[92] According to a 2005 press release, only 142 of the more than 3,000 electronic communications service providers in the United States comply with the federal law and report to the NCMEC.[93] This lacuna should be addressed and rectified.

Another challenge is the possibility that online child sex offenders will establish their own servers. Interestingly, in 2008 NCMEC officials told me that this possibility was not a major problem, whereas FBI officials told me that online child sex offenders are setting up their own providers fairly regularly in an attempt to circumvent the need for subscribing to a large provider.[94] I asked for an update in 2014, and Michelle Collins, vice president of the Exploited Children Division of NCMEC, said that while some offenders may host their own servers, it is quite common for them to use servers belonging to others around the globe. Because there is no known list of US electronic service providers, there is no way for NCMEC to know how many companies are *not* reporting. Instead, 1,119 companies are capable of submitting reports to the

[89] Public Law 110-401, Protect Our Children Act (October 13, 2008), https://www.govtrack.us/congress/bills/110/s1738 .

[90] 18 U.S.C. section 2258A, Reporting Requirements of Electronic Communication Service Providers and Remote Computing Service Providers, http://www.law.cornell.edu/uscode/text/18/2258A . See also "Reporting Child Pornography," Cybertelecom Federal Internet Law and Policy, http://www.cybertelecom.org/cda/cppa.htm .

[91] Interviews with senior officials, NCMEC, Alexandria, VA, April 2, 2008. For further discussion, see the Perverted Justice website at http://www.perverted-justice.com/ .

[92] Interviews with senior officials, NCMEC, Alexandria, VA, April 2, 2008.

[93] Akdeniz, *Internet Child Pornography and the Law*, 262.

[94] Interview with Shawn Henry, deputy assistant director, Cyber Division, Federal Bureau of Investigation, Washington, DC, March 26, 2008.

CyberTipline when they become aware of "apparent child pornography" on their servers.[95] Only US companies are mandated to report to the NCMEC.

Some social networking sites review to some degree their own online spaces for inappropriate and illegal content, including child pornography, in addition to responding to Netcitizens' reports regarding such content. AOL, for instance, "has implemented technologies to identify and remove images of child pornography and to help eliminate the sending of known child pornography," including blocking transmissions of "apparent pornographic images."[96] AOL maintains a library of the signatures. When it identifies the transmission of one of the images, the transmission is blocked, and the image and user information are referred to the NCMEC for investigation. This procedure provides law enforcement with vital information necessary for the prosecution of child pornography purveyors.[97] Other social networking sites also take steps to address the challenge posed by online child sex offenders. Google Orkut, Google's online social network, which was officially shut down on September 30, 2014, launched image-scanning technology that aims to detect images of pornography (including child pornography) at the time of upload to the website, followed quickly by removal.[98] Community Connect[99] uses a "photo approval process for all social main photos to prevent inappropriate photos from appearing as the main photo on personal pages" and requires approval for "all main photos in Groups."[100] MTV Networks/Viacom screens uploads for inappropriate content using "human moderation and/or identity technologies."[101] MySpace "reviews images and videos that are

---

[95] Michelle Collins, email message to author, October 7, 2014.

[96] Internet Safety Technical Task Force, *Enhancing Child Safety and Online Technologies: Final Report of the Internet Safety Technical Task Force* (Boston: Berkman Center for Internet and Society, 2008), 25.

[97] Ibid., appendix E; discussion with Holly Hawkins, director, AOL Consumer Policy and Child Safety, New York, March 25, 2010.

[98] "Google Orkut," Berkman Center for Internet and Society, Boston, http://cyber.law.harvard. edu/sites/cyber.law.harvard.edu/files/Google_orkut%20feedback%202.pdf .

[99] See Community Connect's statement at http://cyber.law.harvard.edu/sites/cyber.law.harvard. edu/files/Community_Connect%20feedback.pdf . For further information about the company, see *Businessweek*'s profile at http://investing.businessweek.com/research/stocks/private/ snapshot.asp?privcapId=115659 .

[100] Internet Safety Technical Task Force, *Enhancing Child Safety and Online Technologies*. See also "Statement to the Technical Advisory Board from Community Connect Inc., on Member Safety Initiatives 2008," Berkman Center for Internet and Society, Boston, http://cyber.law. harvard.edu/sites/cyber.law.harvard.edu/files/Community_Connect%20feedback.pdf .

[101] Internet Safety Technical Task Force, *Enhancing Child Safety and Online Technologies*, 25. See also Elisabeth Staksrud, and Bojana Lobe, "Evaluation of the Implementation of the Safer Social Networking Principles for the EU Part I: General Report. European Commission Safer Internet Programme," European Commission, Luxembourg City, 2010, https://www.duo.uio. no/bitstream/handle/10852/27216/Safer-Social-Networking-part1.pdf?sequence=2 .

uploaded to the MySpace servers and photos deep-linked from third-party sites,"[102] and Facebook deploys a variety of technology tools, including easily available reporting links on photos and videos.[103] In February 2009, Facebook removed more than 5,500 accounts of convicted sex offenders. This action came after MySpace announced that it removed more than 90,000 accounts of sex offenders.[104] In 2012, Facebook alerted the police after its technology flagged an adult chatting about sex with a 13-year-old girl and planning to meet her the next day.[105]

Social networking sites restrict the ability of Netusers registered as adults to search for users registered as minors. On MySpace, profiles for Netusers under 18 are set to "private" upon account creation by default, and adults cannot add a Netuser under 16 as a friend without knowing that user's last name or email address.[106] MTV Networks/Viacom does not allow adults to search for minors, and adults can become "friends" with users under 16 only if they know the Netuser's last name, email address, or username.[107] Facebook does not allow minors and adults on the same regional network to see one another's profiles and does not allow adults to search for minors on the basis of profile attributes. Some of the privacy settings can be set to "Everyone," which, for adults, means anyone with access to the Internet. But for minors, "Everyone" has a different meaning: only the child's Friends, Friends of Friends, and people in any verified school or work networks they have joined on Facebook.[108]

In November 2009, Bebo became the first social networking site to install a panic button to protect children from predatory sex offenders who hunt for victims on the Web.[109] The panic button enables people to share concerns

[102] Internet Safety Technical Task Force, *Enhancing Child Safety and Online Technologies.*
[103] Ibid.
[104] "Facebook Removes 5,500 Sex Offenders," *ComputerWeekly*, February 20, 2009, http://www. computerweekly.com/Articles/2009/02/20/234945/facebook-removes-5500-sex-offenders.htm . See also "Facebook Blocked Some 5,500 Sex Criminals" [in Hebrew], *Haaretz*, February 20, 2009.
[105] Joseph Menn, "Social Networks Scan for Sexual Predators, with Uneven Results," Reuters, July 12, 2012, http://www.reuters.com/article/2012/07/12/us-usa-internet-predators-idUSBRE86B05 G20120712 .
[106] Internet Technical Safety Taskforce, *Enhancing Child Safety and Online Technologies*, 26; Tony Bradley, "Using MySpace.com Safely," About.com, http://netsecurity.about.com/od/news andeditorial1/p/myspace.htm .
[107] Internet Technical Safety Taskforce, *Enhancing Child Safety and Online Technologies*, 26. See also "Social Network Sites," Answers Investigation, http://www.answers.uk.com/services/social networkingsites.html .
[108] Anne Collier and Larry Magid, "A Parents' Guide to Facebook," ConnectSafely.org and iKeepSafe Coalition, 2012, http://www.connectsafely.org/pdfs/fbparents.pdf .
[109] Bebo was founded in 2005 by Michael and Xochi Birch. It grew into the third-largest social network in the world. In 2008, the Birches sold the company to AOL for $850 million. Over the years, the website faded away, and in 2013, the original founders bought Bebo back for $1 million

with qualified experts who can advise concerned Netusers about suspected deviant behavior.[110] Facebook, in contrast, was reluctant to install those buttons, thinking they could promote moral panics.[111] After long months of negotiations, in July 2010 Facebook announced it would allow a "panic button" application on its social networking site. Pressure mounted on Facebook following the rape and murder of 17-year-old Ashleigh Hall by a 33-year-old convicted sex offender, Peter Chapman. Chapman was posing as a teenage boy when Hall met him on Facebook.[112] Discussions were held between British law enforcement and Facebook to make such a report button available on every page. Disappointing, however, is that even at national levels, little cooperation or coordination seems to take place between ISPs to formulate coherent collective strategies to limit access to child pornography. As a first step in reducing the amount of child pornography available on the Internet, such coordination must become a high priority. It is difficult, for example, to see why newsgroups that are known to carry child pornography are allowed to survive, let alone made available to ISP subscribers.[113] It is unclear why ISPs do not follow their own codes of conduct, published on their own websites, which often prohibit child pornography. Codes of conduct should be enforced by the Internet content and service providers to ensure that they act in accordance with the law and with principles of social responsibility.

Both government officials and ISPs prefer self-regulation to state involvement. Self-regulation could involve the use of rating and filtering technology. To this end, content providers worldwide could mobilize to label their content voluntarily. Filters are available to empower guardians and all Netusers to make more effective choices about the content they wish to have enter their homes.[114] Voluntary self-rating would not harm anyone's right and, in fact,

and began reinventing it, aiming to relaunch it in three apps. See the company's website at http://www.bebo.com .

[110] "Bebo Installs Paedophile Panic Button as Police Warn Facebook and MySpace Are Failing to Protect Children," *Mail Online*, November 18, 2009, http://www.dailymail.co.uk/news/article-1228846/Bebo-installs-paedophile-panic-button-police-warn-Facebook-MySpace-failing-protect-children.html .

[111] The Political Issues Committee of the National Writers Union has also argued that there is a moral panic about the sexualization of children. See Bob Chatelle, "Kiddie Porn Panic," *PIC Newsletter* 2, no. 2 (November 1993).

[112] Daniel Emery, "Facebook Unveils Child Safety 'Panic Button,'" BBC News, July 12, 2010, http://www.bbc.co.uk/news/10572375 ; "Ashleigh Hall's Killer Had History of Sexual Violence," BBC News, March 8, 2010, http://news.bbc.co.uk/1/hi/england/wear/8555844.stm ; Haroon Siddique and Helen Carter, "Facebook Security Measures Criticised after Ashleigh Hall Murder," *Guardian*, March 9, 2010, http://www.guardian.co.uk/uk/2010/mar/09/ukcrime-facebook .

[113] Taylor and Quayle, *Child Pornography*, 202–3.

[114] Jens Waltermann and Marcel Machill, eds., *Protecting Our Children on the Internet: Towards a New Culture of Responsibility* (Gütersloh, Germany: Bertelsmann Foundation, 2000), 37.

represents the only form of channeling universally valid for cyberspace. A
<rating="">tag could be added to the next generation of HyperText Markup
Language (HTML). Ratings, which would be self-imposed by the creator of
each website, might approximate those assigned to movies: G, PG, R, and NC-
17. A parent installing a Web browser could follow simple instructions to set
the level of material that the browser would accept from the server. The
browser could also be set to refuse any pages lacking the <rating> tag. Alex
Stewart, one of the first to propose an HTML self-rating system, explained:

> We, as a community of informed, responsible people have a duty not only to
> the rest of the world but to ourselves to show that our community can be
> responsible and responsive to the concerns of parents, teachers, and others of
> the larger world in which we also live.[115]

The same ratings could be used for Usenet newsgroups or individual messages
and for telnet and FTP sites.[116]

Furthermore, a comprehensive self-regulatory system requires content
response and complaints systems for users, such as hotlines (see chapter 9).[117]
Awareness among Netusers of the means to filter and block content and to seek
redress if the rating level is not as promised by the industry is crucial to the
success of any self-regulatory framework. Education by public entities and
information distribution by private entities must be coordinated to raise this
awareness.[118]

Both terrorism and child pornography are illegal activities in the Western
world. To conceal their online activities, terrorists and online child sex
offenders invest in hiding their information from computer crawlers and
spiders. They resort to encryption, offshore ISPs, shifting proxies, private
online storage facilities, and other anonymizing techniques, all of which are
instrumental in protecting their privacy and make the life of law enforcement
difficult (see chapter 2). ISPs' cooperation is vital for the discovery of terrorist
networks and child porn rings.

CYBERBULLYING

Many social network providers also enable Netusers to premoderate any
comments left on their profile before they are visible by others. This opportunity can help Netusers prevent unwanted or hurtful comments from appearing

[115] Jonathan Wallace and Mark Mangan, *Sex, Laws, and Cyberspace: Freedom and Censorship on
the Frontiers of the Online Revolution* (New York: Henry Holt, 1996), 259.
[116] Ibid.
[117] Waltermann and Machill, *Protecting Our Children on the Internet*, 37.
[118] Ibid.

on their profiles for all to see. Netusers can also set their profiles to "private" so that only those authorized by the Netuser are able to access and see the profile. On receiving reports about cyberbullying, social networking sites investigate and can remove content that is illegal or violates their terms and conditions. They may issue warnings and they have the power to delete accounts of Netusers who have broken their rules. Here are some specific examples:

- *Facebook.* If Netusers notice abusive behavior from persons on their Friends List, they can remove the abuser from the list and report the individual, using the "Report/Block this Person" link that appears on the profile. If this action does not resolve the issue, abusers can be blocked by listing their name or email address in the Block List that appears on the Privacy Settings page. In the wake of several high-profile cases of cyberbullying, a Denver-based company has created a Facebook application that allows young people to report violations to Facebook officials and connect to safety and crisis-support organizations. SafetyWeb's application links teens to organizations including the US-based NCMEC CyberTipline, the National Suicide Prevention Hotline, and Facebook's abuse-reporting process. SafetyWeb cofounder Michael Clark likens it to an online version of the list of emergency phone numbers parents leave children when they are alone at home. Parents can encourage their children to bookmark the Find Help application to their Facebook page. When the children click Find Help, they are directed to phone numbers and links for reporting incidents.[119] In October 2014, under pressure to amend its research conduct on its 1.3 billion Facebook users, the company announced that it was looking to hire a manager to coordinate its research efforts and monitor ethics compliance.[120]
- *MySpace.* Following Megan Meier's tragic suicide (see chapter 4), MySpace issued a statement saying it "does not tolerate cyberbullying" and was cooperating fully with the US attorney.[121] Reports can be made via the "Site info" and then "Report abuse" links, which are accessible at the MySpace homepage (https://myspace.com/). When MySpace

[119] "Facebook App Tackles Cyber Bullying," *CBC News*, November 1, 2010, http://www.cbc.ca/technology/story/2010/11/01/tech-cyber-bullying-facebook-application.html .

[120] Vindu Goel, "Facebook Promises Deeper Review of User Research, but Is Short on the Particulars," *New York Times*, October 2, 2014, http://www.nytimes.com/2014/10/03/technology/facebook-promises-a-deeper-review-of-its-user-research.html?emc=edit_th_20141003&nl=todaysheadlines& nlid=33802468 .

[121] Jennifer Steinhauer, "Woman Indicted in MySpace Suicide Case," *New York Times*, May 16, 2008, http://www.nytimes.com/2008/05/16/us/16myspace.html .

receives notices or alerts about Netusers who contemplate suicide, the company contacts the National Suicide Prevention Hotline. According to Chief Security Officer Hemanshu Nigam, MySpace's intervention helped prevent 93 suicides in 2009.[122]

- *Ask.fm.* Following several suicide cases resulting from cyberbullying on this social networking site, cofounder Mark Terebin dismissed concern about trolling on the site, asserting that many teenagers were simply seeking attention not provided by their parents: "When they come to sites like these, they start trolling themselves so that their peers start protecting them. In this absurd way, they get the attention."[123] Ask.fm stated, "We are committed to ensuring that Ask.fm remains a safe, fun environment, and we have policies in place that empower our users to protect themselves and to invite our intervention when required."[124] Ask.fm explained that all users of the site can switch off anonymous questions in their privacy settings. If they receive a question that they do not like or find offensive, they can block the user and report the incident. All reports are read by an Ask.fm team of moderators to ensure that genuine concerns are heard and acted on immediately. Ask.fm said that it is "fully compliant with the Children's Online Privacy Protection Act."[125]

Posting messages on social networking sites is easy and instant. Because of the ease of posting statuses, users react spontaneously to statuses posted by other users without much reflection. One possible way to allow users more time to reflect before hitting the keyboard and posting their messages is to introduce settings by which users are asked to confirm that they wish to post the comment they typed. This additional step may filter some messages because further reflection may help users perceive when their instant typing is premature. This feature would be especially helpful in immediate Netuser discussions and exchanges – and more so when these discussions become inflamed and aggressive (see the discussion of the Megan Meier affair in chapter 4). Monitoring strings of words to prevent verbal abuse requires effort and resources, but this measure may save many lives.

[122] Danielle Keats Citron and Helen Norton, "Intermediaries and Hate Speech: Fostering Digital Citizenship for Our Information Age," *Boston University Law Review* 91, no. 4 (2011): 1435–84.

[123] "Hannah Smith, 14, Found Dead after Web Troll Abuse," *The Age* (Victoria, Australia), August 6, 2013, http://www.theage.com.au/digital-life/digital-life-news/hannah-smith-14-found-dead-after-web-troll-abuse-20130806-2rbm5.html .

[124] Ibid.

[125] Ibid. See also Jon Henley, "Ask.fm: Is There a Way to Make It Safe?," *Guardian*, August 6, 2013, http://www.theguardian.com/society/2013/aug/06/askfm-way-to-make-it-safe .

HATE SPEECH

Crime, terrorism, child pornography, and cyberbullying, I trust, are not contested issues. Hate speech, however, is contested and in the United States is protected under the First Amendment to the US Constitution. Morally speaking, hate speech is repugnant speech. Racist hate speech causes and is tantamount to harm.

Drawing a line as to what constitutes hate is not always simple. On the one hand, statements that assert "women should stay in the kitchen," "Jews are money hungry," "Islam equals terror," "gays are immoral," "blacks are chimps," or "Israel is an apartheid state" and calls to boycott Israel are all unpleasant yet legitimate speech. On the other hand, calls that provoke violence against individuals fall under the definition of incitement; here, the context is of harmful speech that is *directly linked* to harmful action. As explained in chapter 6, when I speak of hate speech, I refer to malicious speech aimed at victimizing and dehumanizing its target, often (but not always) members of vulnerable minorities. Hate speech is fuzzier than incitement and concretely more damaging than advocacy, which is speech designed to promote ideas. Hate speech creates a virulent atmosphere of "double victimization": the speakers are under attack, misunderstood, marginalized, or delegitimized by powerful forces (governments, conspiratorial organizations); the answer to their problem is the victimization of the target group. Their victimization is the speakers' salvation.

In 1996, the United States accounted for 66 percent of the world's Netusers, whereas in December 2013, the American market was reduced to 10.7 percent (see chapter 2). Still, the American influence on the Internet is highly significant. The United States takes the most liberal view in the world on the scope of freedom of expression, shielding hate speech under the First Amendment. No basic disagreement exists that hate speech is vile and offensive. Most Americans believe it is. Still, it is a price that Americans are willing to pay to preserve and protect free speech.

Hate speech, in its various forms, is harmful not only because it offends but also because it potentially silences the members of target groups, causes them to withdraw from community life, and interferes with their right to equal respect and treatment. Hateful remarks are so hurtful and intimidating that they might reduce target group members to speechlessness or shock them into silence. The notion of silencing and inequality suggests the great injury, emotional upset, fear, and insecurity that target group members might experience. Hate might undermine the individual's self-esteem and standing in the community.[126]

---

[126] See Richard Moon, *The Constitutional Protection of Freedom of Expression* (Toronto, ON: University of Toronto Press, 2000), 127; Richard Moon, "The Regulation of Racist Expression,"

Hate is a social evil that offends the two most basic Kantian and Millian principles that underlie any democratic society: respecting others and not harming others. In most cases, hate is derived from one form or another of racism, and modern racism has facilitated and caused untold suffering. It is an evil that has acquired catastrophic proportions in all parts of the world. Notorious examples include Europe under Nazism and since then Yugoslavia, Cambodia, South Africa, and Rwanda. Elsewhere, I argued that in hate messages, members of the targeted group are characterized as devoid of any redeeming qualities and innately evil. Banishment, segregation, and eradication of the targeted group are proposed to save others from the harm being done by this group. By using highly inflammatory and derogatory language, expressing extreme hatred and contempt, and making comparisons to and associations with animals, vermin, excrement, and other noxious substances, hate messages dehumanize the targeted groups.[127]

Hate messages undermine the dignity and self-worth of the targeted group members, and they erode the tolerance and open-mindedness that should flourish in democratic societies committed to the ideas of pluralism, justice, and equality. Hate messages undermine the equal status of targets in their community, their entitlement to basic justice, and their entitlement to the fundamentals of their reputation. Hate speech might lead to mental and emotional distress, racial discrimination, and political disenfranchisement.[128] Furthermore, hate speech might lead to hate crimes.

FROM HATE SPEECH TO HATE CRIME

Those who oppose hate speech regulation argue that allowing hatemongers and racists to release their pent-up emotions in the form of speech is better than their taking violent action. If they give vent to their feelings this way, their

in *Liberal Democracy and the Limits of Tolerance: Essays in Honor and Memory of Yitzhak Rabin*, ed. Raphael Cohen-Almagor (Ann Arbor: University of Michigan Press, 2000), 182–99; Raphael Cohen-Almagor, "Harm Principle, Offense Principle, and Hate Speech," in *Speech, Media, and Ethics: The Limits of Free Expression* (Houndmills, UK: Palgrave, 2005), 3–23.

[127] Raphael Cohen-Almagor, "In Internet's Way," in *Ethics and Evil in the Public Sphere: Media, Universal Values, and Global Development*, ed. Robert S. Fortner and P. Mark Fackler (Cresskill, NJ: Hampton Press, 2010), 93–115.

[128] Mari J. Matsuda, Charles R. Lawrence III, Richard Delgado, and Kimberly W. Crenshaw, eds., *Words That Wound: Critical Race Theory, Assaultive Speech, and the First Amendment* (Boulder, CO: Westview Press, 1993), 89–93; Ishani Maitra and Mary Kate McGowan, "On Racist Hate Speech and the Scope of Free Speech Principle," *Canadian Journal of Law and Jurisprudence* 23, no. 2 (2010): 343–72, 364; Jeremy Waldron, *The Harm in Hate Speech* (Cambridge, MA: Harvard University Press, 2012); Raphael Cohen-Almagor, "Is Law Appropriate to Regulate Hateful and Racist Speech? The Israeli Experience," *Israel Studies Review* 27, no. 2 (2012): 41–64.

targets will be much safer.[129] The trouble with this argument lies in its empirical assumption.

Social science evidence indicates that permitting someone to say or do hurtful things to another person increases, rather than decreases, the chances that he or she will do so again.[130] Just as important, observers may do likewise, thereby creating a climate in which the targets are at even greater risk. Once the speaker defines the category of the victim who deserves what he or she gets, the speaker's behavior is apt to continue and even escalate to include material discrimination and physical bullying.[131]

The Internet plays an instrumental role in spreading hate and in translating speech into action. Cyberhate produces a "permanent disfigurement" of group members.[132] *Hate crime* is defined as harm inflicted on a victim who is chosen because of the victim's actual or perceived race, color, religion, national origin, ethnicity, gender, disability, or sexual orientation. The impact of cruel websites should not be ignored or underestimated. More and more evidence shows that the issue is not mere words.

One of the rare incidents in which a hateful American website was shut down concerned Ryan Wilson, a white supremacist and former leader of the United States Nationalist Party, who in 1998 started a website for his racist organization, ALPHA HQ. Technical support for the site was provided by the godfather of hate sites, Don Black. The site depicted a bomb destroying the office of Bonnie Jouhari, a fair housing specialist in Pennsylvania who regularly organized antihate activities. Not only that, Jouhari was targeted as a "race mixer" because she had a biracial child, and she was considered a "race traitor" because she had had sexual relations with an African American man and because, as a fair housing advocate, she promoted integration. Next to her picture, the ALPHA HQ website stated, "Traitors like this should beware, for in our day, they will be hung from the neck from the nearest tree or lamp

---

[129] Nat Henthoff, *Free Speech for Me – But Not for Thee: How the American Left and Right Relentlessly Censor Each Other* (New York: Harper Collins, 1992), 134.

[130] Gordon W. Allport, *The Nature of Prejudice* (Cambridge, MA: Addison-Wesley, 1954); Leonard Berkowitz, "The Case for Bottling Up Rage," *Psychology Today* 7, no. 2 (1973): 24–31; Brad J. Bushman, "Does Venting Anger Feed or Extinguish the Flame? Catharsis, Rumination, Distraction, Anger, and Aggressive Responding," *Personality and Social Psychology Bulletin* 28, no. 6 (2002): 724–31; John F. Dovidio, Peter Glick, and Laurie A. Rudman, eds., *On the Nature of Prejudice: Fifty Years after Allport* (Oxford: Blackwell, 2005).

[131] William Peters, *A Class Divided: Then and Now* (New Haven, CT: Yale University Press, 1971); Richard Delgado and Jean Stefancic, *Understanding Words That Wound* (Boulder, CO: Westview, 2004), 205.

[132] Jeremy Waldron, "Dignity and Defamation: The Visibility of Hate," *Harvard Law Review* 123 (2010): 1596–667, 1601–7.

post."[133] Wilson reiterated the threat in a press interview. The website referred to Jouhari's daughter as a "mongrel," listed various types of guns, provided information on where to obtain various weapons, and provided a bomb recipe under the picture of Jouhari's office.

Following the Internet posting, Jouhari and her daughter began to receive numerous threatening phone calls. A known Ku Klux Klansman intimidated her by sitting long hours outside her office. Jouhari and her daughter were terrified. They decided to relocate to the West Coast, but a month after their arrival they began to receive the same kind of harassing and intimidating phone calls they had received in Pennsylvania. On one occasion, someone pounded on their door in the middle of the night. On another occasion, someone broke into their apartment. Jouhari's daughter, who was a minor at the time, was diagnosed by a forensic psychologist as "suffering from severe Post-Traumatic Stress Disorder with Delayed Onset."[134] Neither the state nor the federal government ever criminally charged anyone for these acts. I spoke to Bonnie Jouhari in 2010, and she was still traumatized by the ordeal. In a personal letter, Jouhari wrote to me in September 2011 that the impact of that ordeal "has been severe and ongoing. I no longer can comfortably return to my home and no longer [feel] like I have a home. The constant moving for safety reasons have left me with enormous [debt] and a destroyed previously great credit rating."[135]

Wilson was later charged by the Pennsylvania Commonwealth's attorney general with threats, harassment, and ethnic intimidation. The site was removed from the Internet, and the court issued an injunction against the defendant and his organization, barring them from displaying certain messages on the Internet. The chief administrative law judge said, "The website was nothing less than a transparent call to action. . . . When he published the ALPHA HQ website, Wilson created a situation that put Complainants in danger of harassment and serious bodily harm."[136]

In 1999, 21-year-old Benjamin Nathaniel Smith, an avowed Aryan supremacist, went on a racially motivated shooting spree in Illinois and Indiana over the July 4 weekend. Targeting African Americans, Jews, and Asian Americans,

[133] *The Secretary, United States Department of Housing and Urban Development, on behalf of Bonnie Jouhari and Pilar Horton v. Ryan Wilson and ALPHA HQ*, HUDALJ 03-98-0692-8, Initial Decision and Order (July 19, 2000), 5, http://portal.hud.gov/hudportal/documents/hud doc?id=Wilson7-19-00.pdf .

[134] Ibid., 23.

[135] Letter from Bonnie Jouhari to author, September 15, 2011.

[136] *HUD Secretary on behalf of Bonnie Jouhari and Pilar Horton v. Ryan Wilson and ALPHA HQ*, 24, 25. For further discussion, see "Cuomo Says Million Dollar Award Sends Clear Message against Racial Discrimination on the Internet," HUD News Release 00-165, Department of Housing and Urban Development, Washington, DC, July 20, 2000.

Smith killed two and wounded eight before taking his own life, just as law enforcement officers prepared to apprehend him.[137] Smith embarked on his killing spree after being exposed to Internet racial propaganda. He regularly visited the World Church of the Creator (WCOTC) website, a notorious racist and hateful organization founded in Florida in the early 1970s.[138] Smith was so consumed by the hate rhetoric of WCOTC that he was willing to murder and to take his own life in pursuit of his debased hate devotion. Smith said, "It wasn't really 'til I got on the Internet, read some literature of these groups that ... it really all came together." He maintained, "It's a slow, gradual process to become racially conscious."[139]

The same year two other hate-motivated murders occurred. Buford Furrow used to visit hate sites, including the Stormfront.org and Gore Gallery web-sites,[140] on which explicit photos of brutal murders were posted. Whether inspirational or instructional, the Internet supplied information that clearly helped fuel the explosion of a ticking human time bomb.[141] Furrow decided to act. He drove to the North Valley Jewish Community Center and shot an elderly receptionist and a teenage girl who cared for the young students attending the summer day school. He continued shooting, hitting three children, one as young as five years, before leaving the facility. Shortly thereafter Furrow fatally shot a Filipino American postal delivery worker because he worked for the federal government and was not white.

Likewise, Matthew Williams, a solitary student at the University of Idaho, turned to the Internet in search of a new spiritual path. Described as a "born fanatic" by acquaintances, Williams reportedly embraced a number of the

[137] Lorraine Tiven, *Hate on the Internet: A Response Guide for Educators and Families* (Washington, DC: Anti-Defamation League, 2003), 22.

[138] For information on World Church of the Creator (also known as the Creativity Movement), see "Creativity Movement," Anti-Defamation League, http://archive.adl.org/learn/ext_us/wcotc.html ; The Nizkor Project, http://www.nizkor.org/hweb/orgs/american/adl/cotc/ ; World Church of the Creator, http://www.wcotc.com/ ; World Church of the Creator: 'Racial Holy War' on the Web," Anti-Defamation League, 2001, http://www.adl.org/poisoning_web/wcotc.asp ; "The Creativity Movement," Apologetics Index, http://www.apologeticsindex.org/c171.html ; prepared statement of Howard Berkowitz, "Hate Crime on the Internet," Hearing before the Committee on the Judiciary, United States Senate, 106th Cong., 1st sess., September 14, 1999; Teal Greyhavens, "Creating Identity: The Fragmentation of White Racist Movements in America," *Spark Student Journal for Social Justice* (Fall 2007), http://www.whitman.edu/spark/rel355fa07_Greyhavens.html .

[139] Christopher Wolf, "Regulating Hate Speech Qua Speech Is Not the Solution to the Epidemic of Hate on the Internet," presented at the Organization for Security and Cooperation in Europe meeting on the Relationship between Racist, Xenophobic, and Anti-Semitic Propaganda on the Internet and Hate Crimes, Paris, June 16–17, 2004; personal communication with Christopher Wolf, Washington, DC, October 19, 2007.

[140] See, for example, "Murder at Best Gore," http://www.bestgore.com/category/murder/ and "Extreme Gore Gallery," http://www.zombiebloodbath.com/gore.html .

[141] Brian Levin, "Cyberhate," *American Behavioral Scientist* 45, no. 6 (2002): 958–88, 959.

radical right-wing, antisemitic, and antigovernment ideas he encountered online. He regularly downloaded pages from extremist sites and continually used printouts of these pages to convince his friends to also adopt these beliefs. At age 31, Matthew Williams and his 29-year-old brother, Tyler, were charged with murdering a gay couple, Gary Matson and Winfield Mowder, and with involvement in setting fire to three Sacramento-area synagogues. The police discovered boxes of hate literature at the home of the brothers.[142] It was argued that the Internet provided the theological justification for torching synagogues in Sacramento and the pseudointellectual basis for violent hate attacks in Illinois and Indiana.[143]

In early 2001, Richard Baumhammers, another Aryan supremacist, shot down six people, all members of minority groups, in suburban Philadelphia, inspired by material on the Internet. Tim Haney of the Allegheny County Police Department in Pennsylvania testified that computer records confiscated at the home of Baumhammers indicated his frequent visits to white supremacist Internet sites.[144]

Michael Brad Magleby burned a cross on an interracial couple's property. He also visited hate sites prior to transmitting this hateful message.[145] In 2002, Michael Kenneth Faust, a white supremacist who spent several hours a day on the Internet soliciting teens to take his classes on firearm use, shot and killed a black teenager.[146] Seven years later, a 22-year-old man, Keith Luke, murdered two black people and raped and nearly killed a third on the morning after Barack Obama was inaugurated as president (January 21, 2009). When he was captured, Luke told police that he intended to go to a synagogue that night and kill as many Orthodox Jews as possible. Luke apparently had had no direct contact with any extreme-right activists; he told the police that he had been reading "white power" websites for about six

---

[142] Tiven, *Hate on the Internet*, 22.

[143] Prepared statement of Abraham Cooper, "Hate Crime on the Internet," Hearing before the Committee on the Judiciary, US Senate, Washington, DC, September 14, 1999; discussion with Abraham Cooper, Jerusalem, December 17, 2009. For further discussion, see Raphael Cohen-Almagor, *The Scope of Tolerance: Studies on the Cost of Free Expression and Freedom of the Press* (London: Routledge, 2006), 238–63; Cohen-Almagor, "In Internet's Way."

[144] Michael A. Fuoco, "County Officer Specializes in Cyber Crime Cases," *Pittsburgh Post-Gazette*, September 4, 2001.

[145] See *United States v. Magleby*, 241 F.3d 1306, 1308 (10th Cir. 2001).

[146] Madeleine Gruen, "White Ethnonationalist and Political Islamist Methods of Fund-Raising and Propaganda on the Internet," in *The Changing Face of Terrorism*, ed. Rohan Gunaratna (Singapore: Marshall Cavendish, 2004), 127–45, 128; Gina Barton, "White Supremacist Arrested on Gun Charges," *Religion News Blog*, December 26, 2002, http://www.religionnews blog.com/1651/white-supremacist-arrested-on-gun-charges .

months (in other words, from about the time that Obama won the Democratic nomination) and had concluded that the white race was being subjected to genocide in America. Therefore, he told the police, he had to act.[147] This is a clear-cut case of propaganda translating directly into criminal violence.

Later the same year, on June 10, 2009, James von Brunn entered the US Holocaust Memorial Museum in Washington, D.C., and opened fire, killing security guard Stephen Tyrone Johns before other security guards stopped him. Von Brunn, a white supremacist and antisemite, spewed hate online for decades. He ran a hate website called holywesternempire.org and had a long history of associations with prominent neo-Nazis and Holocaust deniers. For a time, he was employed by Noontide Press, a part of the Holocaust-denying Institute for Historical Review, which was then run by Willis Carto, one of America's most prominent antisemites.[148]

In his self-published book, *Kill the Best Gentiles*, von Brunn railed against a Jewish conspiracy to destroy the white gene pool, offering a plan to remove "the cancer from our Cultural Organism."[149] A raging anti-Semite, von Brunn blamed "the Jews" for the destruction of the West. I do not intend to quote at length from this long, hateful tract. Suffice it to say that Jews, according to von Brunn, belong to "a dark and repulsive force." The Jews "are a nefarious and perverse sect." "Satan has prevailed upon them."[150] As a Holocaust denier, this angry, 88-year-old man, possessed with hatred, decided to wage an attack on the Holocaust Museum. He was not interested in visiting the museum to see the thousands of

---

[147] "DA Says Racism Drove Brockton Killings, Rape," Boston.com, January 23, 2009, http://www.boston.com/news/local/massachusetts/articles/2009/01/23/da_says_racism_drove_brockton_kil lings_rape/?page=1 ; "Killing 'Nonwhite People' Was Motive in Brockton Shooting Spree, Police Say," *Boston Globe*, January 22, 2009; David Holthouse, "Was Alleged Massachusetts Spree Killer a Neo-Nazi? Keith Luke Makes It Official," Southern Poverty Law Center, *Hatewatch* (blog), May 11, 2009, http://www.splcenter.org/blog/2009/05/11/was-alleged-massa chusetts-spree-killer-a-neo-nazi-keith-luke-makes-it-official/ . I thank Mark Potok, director of the Southern Poverty Law Center's Intelligence Project, for bringing this case to my attention.

[148] Heidi Beirich, "Holocaust Museum Shooter Had Close Ties to Prominent Neo-Nazis," Southern Poverty Law Center, *Hatewatch* (blog), June 10, 2009, http://www.splcenter.org/blog/2009/06/10/holocaust-museum-shooter-had-close-ties-to-prominent-neo-nazis/ ; Abraham H. Foxman and Christopher Wolf, *Viral Hate: Containing Its Spread on the Internet* (New York: Palgrave-Macmillan, 2013), 27–28.

[149] James W. von Brunn, *"Kill the Best Gentiles!" or "Tob Shebbe Goyim Harog!"* (Easton, MD: Holy Western Empire, 2002), 28.

[150] Ibid., 21–22. For further discussion, see Raphael Cohen-Almagor, "Holocaust Denial Is a Form of Hate Speech," *Amsterdam Law Forum* 2, no. 1 (2009): 33–42; Stephen L. Newman, "Should Hate Speech Be Allowed on the Internet? A Reply to Raphael Cohen-Almagor," *Amsterdam Law Forum* 2, no. 2 (2010): 119–23.

documents that reveal the magnitude of the horror. Von Brunn was beyond the point of deliberation, exchange of ideas, or speech. He was boiling inside with poisonous rage. In his mind, it was time for violent action, and the most appropriate place for the shooting was the museum that served the greatest hoax of all time.[151]

On April 13, 2014, 73-year-old American Nazi Frazier Glenn Miller murdered three people at two separate Jewish Community Centers in Overland Park, Kansas. Miller founded the Carolina Knights of the Ku Klux Klan and was its "grand dragon" in the 1980s. In 1985, he founded another white supremacist group, the White Patriot Party.[152] Miller had spouted his venomous hatred against Jews on hate websites, including his own, and in his self-published book, *A White Man Speaks Out*. On Vanguard News Network alone, Miller had more than 12,000 posts. The slogan of this antisemitic and white supremacist site is "No Jews, Just Right." Vanguard News Network founder Alex Linder has openly advocated "exterminating" Jews since December 2009.[153]

During his long career as an outspoken, blunt racist activist, Miller did not hide his disgust and hatred of Jews, whom he described as the greatest threat to white civilization. Jews are "swarthy, hairy, bow-legged, beady-eyed, parasitic midgets."[154] Adolf Hitler, in contrast, was "the greatest man who ever walked

---

[151] Anders Behring Breivik (see chapter 5) was motivated by Islamophobia and fear. In his twisted mind, he thought that killing members of the socialist Labor Party in Norway would generate a rethinking of immigration policies and eventually lead to a pan-European coups d'état, deportation of Muslims, and execution of traitors. See Neil Sears, "'Vlad the Impaler Was a Genius': The Crazed and Hate-Filled 'Manifesto' of the Mass Murderer," *Mail Online*, July 25, 2011, http://www.dailymail.co.uk/news/article-2018206/Norway-gunman-Anders-Behring-Breiviks-manifesto-Vlad-Impaler-genius.html ; Peter Beaumont, "Anders Behring Breivik: Profile of a Mass Murderer," *Guardian*, July 23, 2011, http://www.guardian.co.uk/world/2011/jul/23/anders-behring-breivik-norway-attacks ; Steven Erlanger and Scott Shane, "Oslo Suspect Wrote of Fear of Islam and Plan for War," *New York Times*, July 23, 2011, http://www.nytimes.com/2011/07/24/world/europe/24oslo.html?nl=todaysheadlines&emc=tha2 ; Scott Stewart, "Norway: Lessons from a Successful Lone Wolf Attacker," *Stratfor Global Intelligence*, July 28, 2011, http://www.stratfor.com/weekly/20110727-norway-lessons-successful-lone-wolf-attacker?utm_source=freelist-f&utm_medium=email&utm_campaign=20110728&utm_term=sweekly&utm_content=readmore&elq=b40a09dbfecc47849181377d509e22d8 .

[152] Heidi Beirich, "Frazier Glenn Miller, Longtime Anti-Semite, Arrested in Kansas Jewish Community Center Murders," Southern Poverty Law Center, *Hatewatch* (blog), April 13, 2014, http://www.splcenter.org/blog/2014/04/13/frazier-glenn-miller-longtime-anti-semite-arrested-in-kansas-jewish-community-center-murders/ .

[153] Ibid. See also John Avlon and Caitlin Dickson, "Hate – and Hitler – in the Heartland: The Arrest of Frazier Glenn Miller," *Daily Beast* April 14, 2014, http://www.thedailybeast.com/articles/2014/04/14/hate-and-hitler-in-the-heartland-the-arrest-of-frazier-glenn-miller.html .

[154] Beirich, "Frazier Glenn Miller, Longtime Anti-Semite, Arrested in Kansas Jewish Community Center Murders."

the earth.["155] Miller's website espoused views of white supremacy and virulent antisemitism and eschewed racial mixing.[156]

In his book, which was freely available on his website, Miller warned against Jewish domination of the media, art, music, literature, and culture of the Western world, "which has brought upon us the epidemics of drugs, venereal diseases, crime, pornography, ignorance, immorality, and yes, racial hatred."[157] Miller openly declared "total war" on ZOG (Zionist Occupation Government) because war is the only hope for the survival of the white race. "Together," Miller wrote, "we will cleanse the land of evil, corruption, and mongrels. And, we will build a glorious future and a nation in which all our people can scream proudly, 'This land is our land. This people is our people. This God is our God, and these we will defend – One God, One Race, One nation.'"[158] Miller called upon his fellow "Aryan warriors" to strike now:

> Strike for your homeland. Strike for your Southern honor. Strike for the little children. Strike for your wives and loved ones. Strike for the millions of innocent White babies murdered by Jew-legalized abortion, who cry out from their graves for vengeance. Strike for the millions of our people raped or assaulted or murdered by mongrels. Strike for the millions of our Race butchered in Jew wars.[159]

Miller was very explicit: "Let the blood of our enemies flood the streets, rivers, and fields of the nation in holy vengeance and justice."[160] Miller published this call in 1987 and repeated it frequently. For many years, he encouraged his followers to kill blacks, Jews, judges, and human rights activists.[161] Thus, no one should be surprised that Miller acted on his own call and went on a racially motivated killing spree.

I mention these incidents to show that dismissing such statements as just hyperbole and protecting them under the principle of free speech might entail significant costs that certainly not every society would find acceptable. Hate should not be taken lightly. The spiral of hatred can motivate and push bigots into action. Violent speech may lead to violent action. A two-year study by the

---

[155] Emma G. Fitzsimmons, "Man Kills 3 at Jewish Centers in Kansas City Suburb," *New York Times*, April 13, 2014, http://www.nytimes.com/2014/04/14/us/3-killed-in-shootings-at-jewish-center-and-retirement-home-in-kansas.html .

[156] The website is still active at http://www.whty.org/ .

[157] Glenn Miller, *A White Man Speaks Out* (1999), http://www.whty.org/book/awmso.pdf .

[158] Ibid.

[159] Ibid.

[160] Ibid.

[161] Steven Yaccino and Dan Barry, "Bullets, Blood, and Then Cry of 'Heil Hitler,'" *New York Times*, April 14, 2014, http://www.nytimes.com/2014/04/15/us/prosecutors-to-charge-suspect-with-hate-crime-in-kansas-shooting.html .

Southern Poverty Law Center shows that nearly 100 people in the past five years have been murdered by active users on Stormfront.org.[162] ISPs and WHSs should be aware of the connection between speech and action. Responsible organizations should always weigh the consequences of their conduct. They should not say, "I did not know." They should know. Ignorance cannot absolve them of responsibility. Society cannot take lightly calls for the murder of persons because of their race[163] or sexual orientation or for any other reason. A light and tolerant attitude constitutes clear-eyed and irresponsible akrasia. As Christopher Wolf, chair of the Internet Task Force of the Anti-Defamation League, argues while providing pertinent reports, "The evidence is clear that hate online inspires hate crimes."[164]

What ISPs and hosting companies could certainly do is to provide a uniform channel for user complaints. Such a channel (which could be as simple as a link to the CyberTipline) could easily be placed on the complaints or customer service page of the service provider.[165] Voluntary participation should be encouraged.

From an ethical perspective, ISPs and WHSs can and should have codes of conduct explicitly stating that they deny service to hatemongers. The promotional approach is in place, not content neutrality (see chapter 6). Sometimes, for whatever reasons (laziness, economic considerations, dogmatism, incuriosity, lack of care, contempt), we refrain from doing the right moral thing. But we should do the right thing. This is not a free speech issue, because we are not free to inflict harm on others. It is about taking responsibility for stopping those who abuse the Internet for their vile purposes.

In Canada, Fairview Technology Centre, an ISP owned by Bernard Klatt, whose server was located in Oliver, British Columbia, and was connected to the Internet via BC Tel, was identified as a host of a number of websites

[162] Beirich, "Frazier Glenn Miller, Longtime Anti-Semite, Arrested in Kansas Jewish Community Center Murders"; Caitlin Dickson, "Where White Supremacists Breed Online," *Daily Beast*, April 17, 2014, http://www.thedailybeast.com/articles/2014/04/17/where-white-supremacists-breed-online.html .

[163] International Convention on the Elimination of All Forms of Racial Discrimination, Office of the High Commissioner for Human Rights, United Nations, http://www.ohchr.org/EN/ProfessionalInterest/Pages/CERD.aspx .

[164] Christopher Wolf, "Needed: Diagnostic Tools to Gauge the Full Effect of Online Anti-Semitism and Hate," presented at the Organization for Security and Cooperation in Europe meeting on the Relationship between Racist, Xenophobic and Anti-Semitic Propaganda on the Internet and Hate Crimes, Paris, June 16, 2004; personal communication with Christopher Wolf, Berkeley, CA, June 5, 2009.

[165] Dick Thornburgh and Herbert S. Lin, eds., *Youth, Pornography, and the Internet* (Washington, DC: National Academies Press, 2002), 380; interview with Herb Lin, Washington, DC, May 15, 2008.

associated with hate speech and neo-Nazi organizations, including the Toronto-based Heritage Front, WCOTC, and the French Charlemagne Hammerhead Skinheads. The ISP was described as containing the most racist, hateful, fascist, antisemitic, Holocaust-denying websites in Canada by a wide margin. About a dozen neo-Nazi, white supremacist, and skinhead clients have used Klatt's server to publish material railing against immigration and "the homosexual agenda" while celebrating "Euro-Christianity" and Hitler's accomplishments.[166]

Klatt explained that Fairview did not control what subscribers upload to their websites. Klatt's business was that of selling computers and providing uncensored Internet access services. Klatt added:

> If you want to express a controversial viewpoint, I certainly don't have to agree with it, but I believe strongly that you have the right to express it.... If you don't agree with what you read or see – switch the topic. You have a choice. You don't have to read or watch any further. It won't come to you.[167]

Klatt was misleading. Unlike Stormfront, which specializes in hate, Fairview was more eclectic. The site provided access to hate sites as well as to local businesses, government agencies, and schools. Here the issue of trust was significant. Because Fairview served schools, students used it, and they could easily have been exposed to hateful messages and racial propaganda. Fairview entertained Holocaust denial. Differentiating between facts and bigot propaganda became a real challenge.

In 1998, Klatt announced that Fairview stopped serving as a provider. The announcement was the result of pressure from antiracists on BC Tel, the local Internet access provider. BC Tel demanded, on expiry of its contract with Fairview, that the new contract contain a term indemnifying BC Tel for any damages for which it might be liable as the result of the material hosted by Fairview. Fairview, rather than sign the BC Tel's proposed renewal contract, gave up providing Internet service.[168]

Judicial-Inc declares itself to be a pro-Christian website with approximately 110,000 monthly readers. In effect, it is an antisemitic site of the worst kind. All

---

[166] Robert Cribb, "Canadian Net Hate Debate Flares," *Wired*, March 25, 1998; Alan Dutton, Karen Mock, and Eliane Ellbogen, "Combatting Hate on the Internet," Stop Racism and Hate Collective, January 31, 2001, http://www.stopracism.ca/content/combatting-hate-internet .

[167] *Oliver Chronicle* (British Columbia), July 24, 1996.

[168] Ross Howard, "Notorious Internet Service Closes," *Globe and Mail*, April 28, 1998; David Matas, "Combating Hate on the Internet without Recourse to Law," panel contribution for the International Network against Cyber Hate Conference on Freedom of Speech versus Hate Speech, Amsterdam, November 9, 2009; personal interview with Mark J. Freiman, McCarthy Tetrault, Barristers & Solicitors, Toronto, October 6, 1998.

of Judicial's 4,200 Web pages were hosted by GoDaddy, which decided to suspend the Judicial account, apparently with zero notice.[169] Zionist Watch, another antisemitic site, was hosted by WordPress until WordPress suspended the site for a violation of the company's terms of service, which expressly prohibit hate content.[170]

Many other ISPs, WHSs, and social networks take a promotional approach – a responsible stance against hate – barring blatant expressions of bigotry, racism, and hate.[171] I have mentioned some responsible codes of conduct in chapter 6. In addition, Facebook prohibits posting content that is hateful or threatening.[172] In May 2010, Facebook took down a page titled "Kill a Jew Day," which urged Netusers to violence "anywhere you see a Jew" between July 4 and July 22. The company's spokesman, Andrew Noyes, explained:

> Unfortunately ignorant people exist and we absolutely feel a social responsibility to silence them on Facebook if their statements turn to direct hate. That's why we have policies that prohibit hateful content and we have built a robust reporting infrastructure and an expansive team to review reports and remove content quickly.[173]

MySpace prohibits content that "is patently offensive or promotes or otherwise incites racism, bigotry, hatred, or physical harm of any kind against any group or individual."[174] And XOOM.com of San Francisco, California, bans "hate propaganda" and "hate mongering."[175] However, having codes without adhering to them and enforcing them is meaningless. These responsible codes

---

[169] Some of the material posted by Judicial-Inc can still be viewed at the Judicial-Inc Archive at http://judicial-inc-archive.blogspot.com/ .

[170] The site was originally located at http://zionistwatch.wordpress.com/ . To read WordPress's terms of service, visit http://en.wordpress.com/tos/ . The Zionist Watch site is still available on other servers; see http://zionistwatch.blogspot.com/ and http://patrickgrimm.typepad.com/ .

[171] For instance, see Atlas Systems' use policy for its Aeon Software at http://www.atlas-sys.com/products/aeon/policy.html and Elluminate's agreement for software and services at http://www.elluminate.com/license_agreement.jsp .

[172] Facebook, "Statement of Rights and Responsibilities," November 15, 2013, http://www.facebook.com/terms.php?ref=pf .

[173] Yaakov Lappin, "'Kill a Jew' Page on Facebook Sparks Furor," *Jerusalem Post*, July 5, 2010, http://www.jpost.com/JewishWorld/JewishNews/Article.aspx?id=180456 ; Amanda Schwartz, "Anti-Semitism vs. Facebook," *Jewish Journal*, July 13, 2010, http://www.jewishjournal.com/community/article/anti-semitism_vs_facebook_20100713 . For similar reasons, Facebook also removed "Kick a Ginger Day" page (aimed at redheads).

[174] MySpace.com Terms of Use Agreement, https://myspace.com/pages/terms . MySpace's former chief safety officer explained that the company's approach stems from its sense of "what the company stands for and what would attract advertising and revenue." Quoted in Citron and Norton, "Intermediaries and Hate Speech."

[175] "Combating Extremism in Cyberspace: The Legal Issues Affecting Internet Hate Speech," Anti-Defamation League, 2000, 11, http://archive.adl.org/civil_rights/newcyber.pdf .

should be steadfast. I reiterate that we expect ISPs, web-hosting companies, and social networks to abide by their own terms of conduct.

I stress this point because Facebook, despite what is said above, still hosts the National Association for the Advancement of White People.[176] Facebook does not perceive Holocaust denial as a form of hate speech and upholds what it calls the people's right to be factually wrong about historic events. In response to pleas to remove those pages, Facebook said, "We think it's important to maintain consistency in our policies, which don't generally prohibit people from making statements about historical events, no matter how ignorant the statement or how awful the event."[177] How this stance can be reconciled with Facebook's prohibition on posting content that is hateful or threatening is something for the Facebook managers to reckon with and answer.

In another publication, I argued that Holocaust denial is a form of hate. Those who deny the Holocaust deny history, reality, and suffering. Holocaust denial might create a climate of xenophobia that is detrimental to democracy. It generates hate through the rewriting of history in a vicious way that portrays Jews as the anti-Christ and as destructive forces that work against civilization. Hateful messages desensitize members of the public on very important issues. They build a sense of possible acceptability of hate and resentment of the other that might be more costly than the cost of curtailing speech.[178] At best, those who help hatemongers disseminate their vile ideas show a strong form of akrasia. At worst, they intend to express bigotry and hate.[179] Other intermediaries, such as MySpace, had no difficulty in removing Holocaust-denial sites.[180]

---

[176] View the site at http://www.facebook.com/pages/The-National-Association-for-the-Advancement-of-White-People/102208269835141 .

[177] Miriam Grossman, "Facebook Firm on Holocaust Denial Pages, Despite Survivors' Letter," *JTA*, July 28, 2011, http://www.jta.org/news/article/2011/07/28/3088748/facebook-firm-on-holocaust-denial-pages-despite-survivors-letter . See also Andre Oboler, "It's Time Facebook Repents," Online Hate Prevention Institute, October 3, 2014, http://ohpi.org.au/its-time-facebook-repents/ .

[178] Cohen-Almagor, "Holocaust Denial Is a Form of Hate Speech"; Newman, "Should Hate Speech Be Allowed on the Internet?," 125–32. See also Brian Cuban, "Facebook at Odds with Obama on Holocaust Denial," *Cuban Revolution* (blog), http://www.briancuban.com/facebook-at-odds-with-obama-on-holocaust-denial/ ; Ian Paul, "Facebook Boots Holocaust Denial Groups," *PC World*, May 12, 2009, http://www.pcworld.com/article/164765/facebook_boots_holocaust_denial_groups.html ; Chris Matyszczyk, "Facebook: Holocaust Denial Repulsive and Ignorant," CNET News, May 6, 2009, http://news.cnet.com/8301-17852_3-10234760-71.html .

[179] Some examples are Aaargh, "Holocaust Revisionism in English," http://vho.org/aaargh/engl/engl.html ; Adelaide Institute, http://www.adelaideinstitute.org/ ; Toben Review, http://www.toben.biz/ ; Royalty Free Photos, http://www.air-photo.org/index.php .

[180] Citron and Norton, "Intermediaries and Hate Speech." For legal analysis of Holocaust denial in Canada, see Irwin Cotler, "Holocaust Denial, Equality, and Harm: Boundaries of Liberty and Tolerance in a Liberal Democracy," in *Liberal Democracy and the Limits of Tolerance:*

Anti–hate speech advocates should explain to ISP managers the nature of the contested hate, its potential harms, and why corporate responsibility means taking the content off their servers. Such actions may lead ISPs to take proactive steps to avoid hosting hate sites on their servers.

In this context, let me mention that the US Congress included a "Good Samaritan" provision, in the 1996 Communications Decency Act (section 230(c)(2)), which protects ISPs that voluntarily take action to restrict access to problematic material:

> No provider or user of an interactive computer service shall be held liable on account of –
> (A) any action voluntarily taken in good faith to restrict access to or availability of material that the provider or user considers to be obscene, lewd, lascivious, filthy, excessively violent, harassing, or otherwise objection-able, whether or not such material is constitutionally protected.[181]

One may ask how we decide whether something on the Internet is terroristic or hateful. Many cases are quite straightforward, but other cases might be obscure or contested. One solution is to approach law enforcement agencies or the courts. This process, however, might be long and costly. An alternative is to establish a Netcitizens Committee to study the problematic websites and to issue recommendations (see the Conclusion). A third option is to seek online arbitra-tion. Online arbitration is a private dispute resolution process that involves the intervention of a neutral decision maker – namely, the arbitrator – who listens to the arguments of both parties and renders a decision that is binding on them. Compared to a court procedure, arbitration is faster, cheaper, and confidential.

Online arbitration is increasingly appreciated by companies active on the Internet because it is more rapid and less costly than legal proceedings or classical arbitration. Everything is done online: the claimant fills out a form on the Cyber Tribunal site, and the form is then sent to the other party. If the other party agrees to participate in arbitration, he or she is asked to respond to the claim. When they undertake arbitration, the parties agree to comply with the award, no matter what the decision is. In case of noncompliance and in accordance with applicable laws and treaties, the injured party can obtain enforcement of the award.[182] It is necessary that the arbitration between the

---

Essays in Honor and Memory of Yitzhak Rabin, ed. Raphael Cohen-Almagor (Ann Arbor: University of Michigan Press, 2000), 151–81.

[181] 47 U.S.C. section 230(c)(2)(A), Protection for private blocking and screening of offensive material, http://www4.law.cornell.edu/uscode/47/230.html .

[182] See, for instance, net-ARB at http://www.net-arb.com/ ; World Intellectual Property Organization On-Line Arbitration at http://www.wipo.int/amc/en/arbitration/online/ . For further discussion, see Ethan Katsh and Janet Rifkin, *Online Dispute Resolution: Resolving*

complainant and the ISP be conducted swiftly, within a matter of days. During the process, the problematic speech remains on the site.

ISPs and Web-hosting companies should develop standards for responsible and acceptable practices for Netusers. ISPs' terms of service usually grant ISPs the unilateral right and ability to block service to clients that violate the terms. ISPs are reluctant to do so because they wish to maintain business. They are for profit. However, ISPs have denied service in some instances, commonly because of violation of copyrights. Following complaints about copyright violation, ISPs took the material off their servers. In addition, Colin Nederkoorn shows that Verizon violated Net neutrality when it throttled its Netflix connection to prevent smooth streaming,[183] and the scary-news department of the Net Neutrality website argues that ISPs are actively trying to block the use of encryption.[184]

An example of cooperation between an Internet monitoring organization and an ISP concerns the Anti-Defamation League (ADL). This nongovernmental organization, founded in 1913, has 30 regional offices in the United States and three overseas offices in Israel, Italy, and Russia. ADL fights all forms of bigotry, defends democratic ideals, and protects civil rights for all, through information, education, legislation, and advocacy.[185] Brian Marcus, who headed ADL's Internet division, explained that private companies may decide not to post messages containing hate speech if they perceived that doing so would be bad for their business. The ADL approached a chief executive officer (CEO) of a Texas Web-hosting company and asked him where he would draw the line between legitimate and illegitimate speech. The CEO answered that hate speech is protected speech – but threats are not. Marcus indicated that one of the sites the CEO's company hosted claimed that all members of minority groups should be hanged from street lamps. The CEO was surprised. Although such language is not considered a threat according to the American law,[186] it was too much for the CEO. Marcus

Conflicts in Cyberspace (San Francisco: Jossey-Bass, 2001); Kanzlei Hellinger, ODR – Online Dispute Resolutions (database), http://hellinger.eu/en/law/odr-online-dispute-resolution/ .

[183] Colin Nederkoorn, "Verizon Fios Throttles Netflix," YouTube, http://www.youtube.com/watch?v=5vs3QhEx_3w .

[184] Mike Masnick, "Revealed: ISPs Already Violating Net Neutrality to Block Encryption and Make Everyone Less Safe Online," Net Neutrality, October 13, 2014, https://www.techdirt.com/blog/netneutrality/articles/20141012/06344928801/revealed-isps-already-violating-net-neutrality-to-block-encryption-make-everyone-less-safe-online.shtml .

[185] For more information, see the Anti-Defamation League website at http://www.adl.org/ .

[186] Anna S. Andrews, "When Is a Threat 'Truly' a Threat Lacking First Amendment Protection? A Proposed True Threats Test to Safeguard Free Speech Rights in the Age of the Internet," *UCLA Online Institute for Cyberspace Law and Policy*, May 1999; Cohen-Almagor, *Scope of Tolerance*, 256–58.

then showed him some 150 such sites. After deliberation, the company closed approximately 120 of the problematic sites.[187]

As technology facilitates freedom of expression, it can also facilitate the delineation of the boundaries of free expression. As of October 2012, YouTube claimed 280 million members who uploaded an average of 72 hours of video content to the site every minute of every day.[188] This amount of information is staggering, yet YouTube is very efficient in filtering material that violates copyrights and all forms of pornography. The YouTube business model factors in precautionary measures to ensure its video streaming does not violate copyright law and is free of pornography. The combination of detecting technology, a dedicated team of technicians, and answerability to users' complaints proves quite effective in addressing the hefty challenge.

In December 2008, YouTube contacted the ADL for its expertise in dealing with Net hate. As a result of that partnership, the ADL is a contributor to YouTube's Policy Center, where Netcitizens are empowered to identify and confront hate and to report abuses.[189] The YouTube Policy Center features information and links to resources developed by ADL to help Netcitizens respond to and report offensive material and extremist content that violates YouTube's Community Guidelines on hate speech.[190] However, quite a few questionable video clips on YouTube are still available.[191] After receiving complaints, YouTube decided to include this content warning before people enter the sites: "The following content has been identified by the YouTube community as being potentially offensive or graphic. Viewer discretion is advised."[192]

Like ISPs, connectivity providers that supply ISPs and domain name registrars should insist as a condition for the provision of their service that it will not

---

[187] Interview with Brian Marcus, former ADL director of Internet monitoring, Washington, DC, April 16, 2008.

[188] Foxman and Wolf, *Viral Hate*, 91.

[189] See the Policy Center's webpage at http://www.youtube.com/yt/policyandsafety/policy.html . See also YouTube's policy on hate speech at https://support.google.com/youtube/answer/ 2801939 .

[190] "YouTube Taps ADL as Partner in Fight against Hate," press release, Anti-Defamation League, December 11, 2008, http://www.adl.org/PresRele/Internet_75/5416_75.htm .

[191] See, for example, "Hate Society – Hail Blood and Honour," http://www.youtube.com/watch?v=s SGXQmfHzsU ; "skrewdriver – blood and honour," http://www.youtube.com/watch?v=YoUuqF aoMiM&NR=1 ; "Hate Society and Razors Edge – Teenage Rebellion," http://www.youtube.com/ watch?v=lzE5Z3SZ1Jg; "Svastika flies again!," http://www.youtube.com/watch?v=Fu3XvQv_IXE ; "SS-Sturmführer," http://www.youtube.com/watch?v=dBhFgnADkpY&NR=1 ; "No remorse- Oi monkey," http://www.youtube.com/watch?v=dw2oMnHFCwI&feature=PlayList&p=1AFED9 C2D71B446B&index=14 .

[192] See, for example, http://www.youtube.com/verify_controversy?next_url=/watch%3Fv%3DdBhF gnADkpY%26NR%3D1 .

be used for the promotion of hate. Every socially responsible access provider should insert into its subscription contract an antihate provision.

Whatever responsible steps corporations may take, these steps should be transparent and communicated to the public. In 2002, Google quietly deleted more than 100 controversial sites from some search result listings. It did so secretly, without public discussion or explanation and, as a result, was subjected to intense criticism. Most of the sites were removed from Google.fr (France) and Google.de (Germany) and were antisemitic, pro-Nazi, or related to white supremacy.[193] However, the removed sites continue to appear in listings on the main Google.com site.

In 2005, Google spokesman Steve Langdon announced that Google News does not allow hate content. "If we are made aware of articles that contain hate content, we will remove them," he said.[194] Among the removed news publications was *National Vanguard*, then a publication of the National Alliance, an organization for "people of European descent" that aimed to achieve a new consciousness, a new order, and a new people. *National Vanguard* describes itself as "fearless – uncompromising – brilliant – witty – educational. *National Vanguard* provides the information and the insights that White America's future leaders will need to guide our nation through the dangerous, revolutionary times ahead."[195] Indeed, news organizations have editorial discretion over what they run and do not run. Google made a conscious decision not to help in the spread of racism and bigotry. In this case, Google proactively adopted ethical norms of Corporate Social Responsibility.[196]

Conversely, in another instance concerning a notorious hate site, Jew Watch,[197] Google refused to change its search results to lower the scale of the site's popularity. Instead, Google published an explanation of its search results, saying that searching for the word *Jew* may result in referring to disturbing sites and that Google does not endorse antisemitism.[198] Google also inserts a link to "Offensive Search Results" on the page where the link to

---

[193] Declan McCullagh, "Google Excluding Controversial Sites," CNET News, October 23, 2002.

[194] Susan Kuchinskas, "Google Axes Hate News," InternetNews.com, March 23, 2005, http://www.internetnews.com/xSP/article.php/3492361 .

[195] See the organization's website at http://www.natvan.com/national-vanguard/ . See also http://nationalvanguard.org/ .

[196] For critique of Google for its lax attitude on human rights in China, see Gary Elijah Dann and Neil Haddow, "Just Doing Business or Doing Just Business: Google, Microsoft, Yahoo!, and the Business of Censoring China's Internet," *Journal of Business Ethics* 79, no. 3 (2008): 219–34.

[197] The organization's Web address is http://www.jewwatch.com/ .

[198] Google's algorithm depends on hundreds of factors. Google Team, "An Explanation of Our Search Results," Google, 2011, http://www.google.com/explanation.html . On March 13, 2011, I Googled the word *Jew*. Jew Watch came second after Wikipedia. On July 3, 2011, Jew Watch came fourth after Google's explanation on offensive search results and two Wikipedia entries. On

Jew Watch appears, stating, "We're disturbed about these results as well." But Google is evidently not disturbed enough to stop assisting the spread of vile bigotry on its pages.[199]

In October 2011, I met Yoram Elkaim, head of Google Legal – Southern and Eastern Europe, Middle East, and Africa. He explained that Google does not think that it should be the judge of free speech online. Google believes in free speech and tolerance. It is mobilized to fight against ignorance by informing people, providing them with information, and digitizing world-leading libraries. Google brought the Harvard library to Africa. Google also supported the digitization of the Yad Vashem archives, dedicated to educating people about the Holocaust. Moreover, Google sponsors events against violence and extremism. The company certainly strives to exclude illegal content from the Net. Elkaim emphasized that Google's job is to provide *relevant* information. Google's job is to inform – not to misinform. Google aims to provide correct information and does its best not to play into the hands of people who try to game the system by excessively affecting the Google ranking algorithm. I asked Elkaim why Google ranks Jew Watch in fourth place if it aims to provide relevant information about Jews. Does he think that the Jew Watch information is relevant to those who seek information about Jews or that this information is true or correct in one way or another? To my surprise, Elkaim was not familiar with the Jew Watch controversy.[200]

I object to search engine optimization that is designed to promote narrow commercial interests; however, in accordance with the promotional approach, I welcome search engine optimization stemming from principles of moral and social responsibility. In practice, the major companies and search engines are engaged in online profiling that is designed to target individual online conduct and direct users to relevant advertisements. Google, Facebook, and Yahoo! are heavily biased to serve the interests of big business.[201] The neutrality claim is just a hypocritical facade.

Search engines surely aspire to be efficient. Google reached the position it enjoys today as the leading search engine in the world because it was better at referring Netusers to the information they were looking for. To keep its position,

---

October 3, 2014, Jew Watch came fifth. For further discussion, see Brian Leiter, "Cleaning Cyber-Cesspools: Google and Free Speech," in *The Offensive Internet: Speech, Privacy, and Reputation*, ed. Saul Levmore and Martha C. Nussbaum (Cambridge, MA: Harvard University Press, 2010), 155–73.

[199] In May 2011, a court in Buenos Aires ordered Google to stop recommending 76 antisemitic and racist websites. Judge Carlos Molina Portela also ordered that advertisements be removed from those sites.

[200] Interview with Yoram Elkaim, Paris, October 10, 2011.

[201] Paul Bernal, "Web Spies," *Index on Censorship* 40, no. 2 (2011): 108–14.

Google should not flood Netusers with irrelevant and misleading information. The results should be transparent and relevant. Search engines should be attentive to words and string of words, and the results should be as free from bias and manipulation as possible. Thus, Jew Watch should not appear prominently on the search for the term *Jew*, but it should appear prominently when one searches *Jew hatred*. Similarly, Holocaust denial sites should not feature in a prominent place when one searches the term *Holocaust*, but they should be the first to appear when one searches *Holocaust denial*.

In other words, I am not suggesting that search engines should necessarily omit pertinent results about racism. People who seek racist and bias-motivated speech may use Google to get it. But Google should do a better job in catering to the interests of Netusers and optimizing its search results.

Furthermore, considerations of moral and social responsibility may come into play when search engines rank results for other antisocial activities. For instance, they may adopt a promotional approach in their ranking of the results for the term *jihad*, giving preference to nonviolent sites that explain what jihad means and object to the use of terror over those that promote violence and terror.

Moreover, information on dangerous activities that might harm others, especially vulnerable third parties such as children, should not be made readily available for potential abusers. This is indeed the case when child abuse is concerned. Thus, when one Googles terms such as *pedophilia, child pornography, child sex, child abuse*, or *sex with children*, the prominent results do not lend advice and support for child abusers. In addition, prominent search results for the terms *rape* and *cyberbullying* do not condone and promote those practices. On these issues, Google upholds the promotional approach, a nonneutral approach, unwilling to assist lawbreakers and other violent abusers to bluntly undermine the safety of vulnerable third parties.

In 2011, Google Ideas (Google's think tank) launched a global network called AVE (Against Violent Extremism). The network brings together former extremists, victims, private businesses, foundations, and experts to share knowledge, experience, and resources in the fight against extremism. AVE is privately funded, although government officials have attended the network's launch and provided advice when needed.[202]

CONCLUSION

The gatekeepers of the large information technology companies – Google, Facebook, Yahoo!, and Twitter – are young Americans who were brought up

---

[202] National Security Program, "Countering Online Radicalization in America," Bipartisan Policy Center, Washington, DC, December 2012, 34.

on the values of the First Amendment. For them, freedom of expression is the most important principle that guides their actions – so much so that Facebook at first did not have rules on what speech violated its terms of service,[203] and Twitter's rules are designed to provide freedom of expression the widest possible scope.[204] Consequently, hate speech is a legitimate, protected speech. But the role of gatekeeper, which invests great power, also requires great responsibility. Gatekeepers need to strike a balance between freedom of expression and social responsibility, between rowdiness and civility, between creating a wide-open marketplace of ideas and ensuring that the marketplace does not facilitate violence and lawlessness.

Luciano Floridi argues that the ethical use of information and communication technologies and the sustainable development of an equitable information society need a safe and public infosphere for all, where communication and collaboration can flourish coherently with the application of human rights and the fundamental freedoms in the media. Sustainable development means that our interest in the sound construction of the infosphere must be associated with an equally important, ethical concern for the way in which the latter affects and interacts with the physical environment, the biosphere, and human life in general, both positively and negatively.[205] Ethical behavior considers the consequences of one's actions, and it is about being accountable for them. Information professionals cannot be neutral regarding content because such behavior is irresponsible and unprofessional. They have a *prima facie* moral duty to provide stakeholders with a certain level of security.

Ethics, Floridi rightly notes, is not only a question of dealing morally with a given world. It is also a question of shaping the world for the better. This suggests a proactive approach that perceives agents as world owners, creators, game designers, producers of moral goods and evils, providers, and hosts.[206] Accordingly, ISPs should be able to plan and initiate action responsibly, in anticipation of future events, in an attempt to control their course by making something happen or by preventing something from happening.

Moreover, I have argued that the Internet is a form of new media, but it is still a medium of mass communication. It is not reasonable to prohibit certain expressions in print and allow the same objectionable expression electronically. We cannot be neutral with regard to certain conduct that falls within the

[203] Jeffrey Rosen, "The Delete Squad," *New Republic*, April 29, 2013, http://www.newrepublic.com/article/113045/free-speech-internet-silicon-valley-making-rules .

[204] See Twitter's policy at https://support.twitter.com/articles/18311-the-twitter-rules .

[205] Luciano Floridi, "Ethics in the Infosphere," *Philosophers' Magazine* 6 (2001): 18–19.

[206] Luciano Floridi, "Ethics after the Information Revolution," in *The Cambridge Handbook of Information and Computer Ethics* (Cambridge: Cambridge University Press, 2010), 3–19. See also Luciano Floridi, *The Ethics of Information* (Oxford: Oxford University Press, 2013).

parameter of harming others; then the dangers to democracy, to our fellow citizens, to the moral basis of society, and to the values that we hold dear might be too grave.

Netcitizens, when acting collectively, have power. They are able to change companies' policies and conduct. Yahoo! had adult items on its shopping pages since 1999. Then, in 2001 it quietly expanded its offering of hard-core videos and DVDs in search of new revenue. When the *Los Angeles Times* reported this, the company was swamped with angry calls and emails. Under pressure, Yahoo! announced that it would stop selling X-rated videos and other pornographic material on its Web pages. It would also stop entering into new contracts for banner advertisements for adult merchandise.[207] Yahoo! also has come under fire for serving as a host to online chats by hate groups. It began donating ad space in the chat rooms to Tolerance.org, set up by the Southern Poverty Law Center. The ads also appear when users enter words such as "Nazi" or "hate" on the Yahoo! search engine.[208] Agents' collective action, driven by moral and social responsibility considerations, may affect business to better their Net conduct in fear of revenue losses. At the same time, businesses should adopt guidelines for moral conduct that are driven by considerations of moral and social responsibility.

Furthermore, in chapter 5, I mentioned VampireFreaks.com, a busy site with hundreds of thousands of postings. Some effort is needed for moderators of such large sites to monitor the heavy traffic. Two practical questions arise: one related to cost, the other related to ability. As for costs, web experts should be hired to devise a research algorithm for identifying what is regarded as problematic material. The costs can be shared among the website's members. Cost alone is an insufficient reason to relieve ISPs from their moral and social responsibilities. As for ability, among laypersons it is a contested issue whether monitoring websites – especially very large and voluminous websites with heavy traffic – is technologically possible. The issue is far less contested among experts. A Web expert who monitored groups for Yahoo! verified that a small number of experts who specialize in social networking could devise batches of programs to look for illegal material and remove it. This expert did so for Yahoo! in its struggle against child pornography.[209] Microsoft had created a system to help monitor and track down online child pornography.[210]

[207] Brian Bergstein, "Yahoo Takes the xxx out of Its www," *Philadelphia Inquirer*, April 14, 2001, A1.

[208] Ibid.

[209] Personal communication with a research specialist, George Washington University, Washington, DC, June 12, 2008. See also Steve Silberman, "The United States of America v. Adam Vaughn," *Wired*, October 10, 2002, http://www.wired.com/wired/archive/10.10/kidporn.html .

[210] Yaakov Lapin, "Police Arrest Internet Pedophile Ring," *Jerusalem Post*, January 22, 2009.

Among the tools Microsoft uses in the fight against these illegal images is PhotoDNA, developed by Microsoft Research in collaboration with Dartmouth College. PhotoDNA helps to refine and automate the search for child pornography among the billions of photos on the Internet.[211]

David Corchia founded Concileo, which develops and manages community and participatory strategies by providing community platforms forums and internal teams for moderating user content submitted to its clients' user participation areas. Among his clients are some major French newspapers and magazines, including *Figaro* and *Elle*. His team of 30 people monitors some 100,000 texts a day.[212] Marc Rotenberg, president of the Electronic Privacy Information Center, said that the capability to monitor the Internet is greater than what most people assume. It is a question of will, not of ability.[213]

National security organizations have developed mechanisms to scrutinize large parts of the Internet susceptible to criminal activity. The University of Florida has created a software tool called ICARUS that monitors traffic over its network, identifies traffic that appears to be characteristic of peer-to-peer file sharing, and then suspends network service to the computer generating the traffic for 30 minutes. Users may regain network access only if they complete a 10-minute interactive presentation on copyright law.[214] I mentioned in chapter 6 companies such as Check Point, Symantec Gateway Security, and Allot Communications that provide traffic management facilities and security management tools. Edward Snowden's revelations about the American government surveillance capabilities on phone and Internet communications affirm the existence of comprehensive mechanisms to monitor and analyze large data.[215]

Thus, it is possible to monitor traffic on large websites. It is a question of will and of priorities in allocating resources for monitoring. At present,

[211] Microsoft, "Combating Child Exploitation Online," February 2013, download.microsoft.com/download/B/8/2/B8282D75-433C-4B7E-B0A0-FFA413E20060/combating_child_exploitation.pdf; Microsoft, "PhotoDNA Newsroom," http://news.microsoft.com/presskits/photodna/; Microsoft, "Microsoft and National Center for Missing & Exploited Children Push for Action to Fight Child Pornography," press release, December 15, 2009, http://news.microsoft.com/2009/12/15/microsoft-and-national-center-for-missing-exploited-children-push-for-action-to-fight-child-pornography/.

[212] Discussion with David Corchia, Paris, October 10, 2011. Concileo's website address is http://www.concileo.com/.

[213] Interview with Marc Rotenberg, president of the Electronic Privacy Information Center, Washington, DC, May 2, 2008.

[214] Eric Evans, "From the Cluetrain to the Panopticon: ISP Activity Characterization and Control of Internet Communications," *Michigan Telecommunications and Technology Law Review* 10, no. 2 (2004): 445–99, 498.

[215] Luke Harding, *The Snowden Files: The Inside Story of the World's Most Wanted Man* (London: Guardian Faber Publishing, 2014); Glenn Greenwald, *No Place to Hide: Edward Snowden, the NSA, and the U.S. Surveillance State* (New York: Macmillan, 2014).

VampireFreaks is not exceptional in its reluctance to monitor sites and relieve itself of responsibility. Most ISPs and hosting companies shy away from assuming such responsibility because avoidance is the easiest and most profitable path to pursue. But this attitude may change. It is already changing in the sphere of child pornography.

To be sure, ISPs cannot go too far in their monitoring. Adopting an overzealous monitoring policy might itself create a negative reputation for an ISP, which in turn would encourage its users to experiment with other service providers.[216] My argument is very limited in scope, focusing on a relatively small number of potentially problematic websites where criminals look for appreciative company, venting their hostile aspirations and plans.

Furthermore, search engines may play a role in optimizing search results. Search engines, asserts James Grimmelmann, are attention lenses: "they bring the online world into focus. They can redirect, reveal, magnify, and distort. They have immense power to help and to hide."[217] Netusers who are looking for data rarely explore beyond the first few pages of results. Optimizing search results has been influenced by business considerations, and it can be influenced by standards of social responsibility. In the United Kingdom, when one searches *suicide*, topping the Google list is "Need help? In the United Kingdom, call 08457 90 90 90 Samaritans." The same message, with a different phone number, appears in the United States. In addition, Google ads appear that offer counseling, help, the opportunity to chat with others about depression, and the chance to find hope and life instead of despair and death.

Another idea to push Internet intermediaries to take proactive steps against antisocial forms of expression is to introduce league tables naming and shaming ISPs who fail to take down offensive material. In the United Kingdom, the former culture minister, Barbara Follett, and her successor, Ed Vaizey, backed the idea that Web providers must be embarrassed into dealing with violent, sexually explicit Web content. Follett said she wanted to see the prescreening of material on sites such as YouTube, as occurs at present on MySpace. She thought that the Internet represented growing chaos and order that needed to be established: "Many people have said that the Internet is like the Wild West

---

[216] Assaf Hamdani, "Who's Liable for Cyberwrongs?," *Cornell Law Review* 87, no. 4 (May 2002): 901–57, 929–30.

[217] James Grimmelmann, "Some Skepticism about Search Neutrality," in *The Next Digital Decade: Essays on the Future of the Internet*, ed. Berin Szoka and Adam Marcus (Washington, DC: TechFreedom, 2010), 435–59, 435. See also Geoffrey A. Manne, "The Problem of Search Engines as Essential Facilities: An Economic and Legal Assessment," in *The Next Digital Decade: Essays on the Future of the Internet*, ed. Berin Szoka and Adam Marcus (Washington, DC: TechFreedom, 2010), 419–34.

in the gold rush and that sooner or later it will be regulated. What we need is for it to be regulated sooner rather than later."[218] She was clear about her expectations: Service providers need to come forward and show that they are responsible organizations.[219]

If all the preceding does not help, human rights organizations should seek to ban services of hateful ISPs. If ISPs fear that they might lose business, they may change their ways and assume responsibility.

Lasting social change requires a combination of solid governmental support and committed corporate action. A comprehensive look at the movement for Corporate Social Responsibility shows that market forces often jumpstart responsibility. Consumer demand for responsibility may push companies to produce certain products and abandon others; actual (or threatened) consumer boycotts can influence decision-making processes; "naming and shaming" practices by nongovernmental organizations, pressure from socially responsible investors, and values held by employees and management are all influential. Yet there is no guarantee that a company will sustain its efforts past a marketing campaign if practices and standards are not enshrined in law. And corporations will only participate in Corporate Social Responsibility over the long term if it is good for their bottom line. Although profitability may not be the only reason corporations will or should behave virtuously, it has become the most influential one. Corporate social responsibility is sustainable only if virtue pays off.[220]

Economically speaking, a concern might be raised that enforcing liability on Internet intermediaries will significantly raise the cost of doing business. One way to offset the incurred costs might be for companies to charge higher subscription fees. Although some subscribers who are unable to afford the service might leave the market, there are several ways to address this concern. One is for companies to display more advertising – or the same amount of advertising at a higher price, accompanied by an explanation for the reason the price has risen. Another is to provide government tax incentives to Internet intermediaries on the basis of number of subscribers. A third way is for companies to impose financial penalties against those who systematically violate their

[218] Patrick Wintour, "Web Providers to Be Named and Shamed over Offensive Content," *Guardian*, November 15, 2008.

[219] Ibid.

[220] David Vogel, *The Market for Virtue: The Potential Limits of Corporate Social Responsibility* (Washington, DC: Brookings Institution Press, 2005); Philip Kotler and Nancy Lee, *Corporate Social Responsibility: Doing the Most Good for Your Company and Your Cause* (Hoboken, NJ: Wiley, 2005); Tom Campbell and Seumas Miller, eds., *Human Rights and the Moral Responsibilities of Corporate and Public Sector Organisations* (Dordrecht, Netherlands: Kluwer, 2004).

own terms and conditions of service, thus penalizing people who abuse the service for their misconduct.

In the next chapter, I turn to the issue of state responsibility in tackling antisocial and violent behavior on the Net. Governments have an important role to play when individuals and corporations violate national laws and act irresponsibly for partisan, selfish interests.

8

State Responsibility

Without commonly shared and widely entrenched moral values and obligations, neither the law, nor democratic government, nor even the market economy will function properly.

–Václav Havel

Simon Guy Sheppard and Stephen Whittle do not like Jews and black people, and they believe that the Holocaust is a hoax. Both were charged and convicted in Britain for possessing, publishing, and distributing racially inflammatory material. Sheppard operated and contributed to a number of websites, including heretical.com, klan.org, nazi.org, and whitepower.co.uk. Having edited material written by Whittle, Sheppard posted it to a website in Torrance, California, thinking that by doing so he was shielded by the First Amendment to the US Constitution. The British Crown Court decided that it had jurisdiction to try the appellants for their conduct because a substantial measure of the activities constituting the crime took place in England. Sheppard was sentenced to 3 years and 10 months in jail, and Whittle was sentenced to 1 year and 10 months imprisonment.[1] The message is clear: England does not tolerate hate speech on the Internet. It sees it as its responsibility to protect vulnerable minorities from bigoted, inflammatory, hateful expression.

In his germinal cyber law scholarship, Lawrence Lessig distinguishes between two claims. One claim is that, given the Internet's architecture, governments find it difficult to *regulate behavior* on the Net. The other claim is that it is difficult for governments to *regulate the architecture* of the Net. The first claim is true. The second is not. It is not hard for governments to take steps to alter Net architecture and, in so doing, help regulate Net behavior.[2]

---

[1] *Regina v. Simon Guy Sheppard and Stephen Whittle* (2010), EWCA Crim 65.

[2] Lawrence Lessig, *Code and Other Laws of Cyberspace* (New York: Basic Books, 1999), 43–44.

In the mid 1990s, the Internet seemed a perfect medium for business: it was supranational and diffusive, had wide distribution and little regulation, and offered enormous opportunities to investors. In his famous "Declaration of the Independence of Cyberspace," the Internet theorist John Perry Barlow wrote:

> Governments of the Industrial World, you weary giants of flesh and steel. . . . You have no sovereignty where we gather. . . . You have no moral right to rule us nor do you possess any methods of enforcement we have true reason to fear. . . . Cyberspace does not lie within your borders.[3]

I beg to differ on both principled and practical grounds. On principled grounds, governments have a moral duty to intervene to enforce their customary moral norms of the day. Because liberal democracies are founded on the ideas of respect for others (treating citizens as equals) and not harming others, they need to adhere to the promotional approach (as explained in chapter 6) advancing these underpinning ideas. Being cognizant of the democratic catch (see chapter 3), governments have the responsibility to set boundaries to liberty and tolerance both offline and online. On practical grounds, cases like Yahoo! and Google, to be discussed subsequently, may mark the beginning of the end of the no-sovereignty illusion. They demonstrate that Internet service providers (ISPs) have to maintain societal responsibilities and to respect domestic state legislation to avoid legal risks. The Internet is international in character, but it cannot be abused to override law. There is not one law for the Internet and another for all other forms of communication.

The following discussion is concerned with several challenges that democracies are facing today on the Net. It opens with the challenge of racism and hate, focusing attention on the Yahoo! controversy in France and the United States. The controversy juxtaposes two contrasting views: Yahoo!'s Internet-separatist view aimed at being able to determine its own rules and regulation worldwide, notwithstanding national laws and morals, in contrast to the view that holds that countries have the right and ability to assert their sovereignty on the Net. The discussion continues by an analysis of other forms of antisocial speech: child pornography, cyberbullying, suicide, terror, and crime. Finally, I call for state monitoring of dangerous, antisocial websites to avert violent activities.

---

[3] John Perry Barlow, "A Declaration of the Independence of Cyberspace," Electronic Frontier Foundation, February 8, 1996, https://projects.eff.org/~barlow/Declaration-Final.html . See also "The Internet's New Borders," *Economist*, August 9, 2001; Christopher Shea, "Sovereignty in Cyberspace," *Boston Globe*, January 15, 2006, K4.

HATE AND RACISM

First Amendment scholars and free speech advocates promote the idea of tolerance. Respecting our fellow citizens entails that we should see them, in Kantian terms, as ends rather than means, appreciate diversity and differences, and not be quick to judge others as "strange" or "peculiar" only because they adhere to a different way of life or conception of the good. In a democracy, government and people are said to tolerate others, thus providing others with scope to develop themselves and their respective ways of life.

However, the important proviso is that democracy should protect itself against those who exploit tolerance to harm others. Tolerance should not mean enabling extreme, intolerant groups to destroy the very foundations that prescribe tolerance. Karl Popper explained that because of the strong belief in toleration on the one hand and the fear of being intolerant on the other hand, people are inclined to extend toleration to those who spread intolerant ideologies that aim to destroy the very foundations of toleration. Many see themselves as committed to treating every individual as a moral agent and to allowing any person the opportunity to practice freedom – even if this attitude might prove conducive to promoting intolerance. However, the moral ideal of toleration does not require that we put up with anything and everything. Popper asserts that to allow freedom of speech to those who would use it to eliminate the very principle on which they rely is paradoxical. He does not imply that we *always* should suppress utterances of intolerant philosophies but that we should claim the *right* to suppress them if necessary.[4] It is absurd to assume that we should tolerate the intolerant no matter what. The delicate task is to maintain a balance between tolerance and intolerance; otherwise the very foundation of tolerance might provide the intolerant with the tools to destroy democracy (the *democratic catch*).

The liberal state has an interest in limiting the scope and spread of hate and bigotry. Europe, a continent that suffered the horrors of the Nazi dictatorship, is far more sensitive to Nazism and racism than is the United States. Hate speech and propaganda led not only to the destruction of European Jewry but also to mass murders of horrific scale of all "inferior races" and "undesired elements."[5] Thus, Spain passed legislation authorizing judges to shut down

[4] Karl R. Popper, *The Open Society and Its Enemies*, vol. 5 (London: Routledge and Kegan Paul, 1962), 265; Karl R. Popper, "Toleration and Intellectual Responsibility," in *On Toleration*, ed. Susan Mendus and David Edwards (Oxford: Clarendon Press, 1987), 17–34.

[5] On the horrors of World War II, their root causes, and their justifications, see Raul Hilberg, *The Destruction of the European Jews* (New York: Holmes and Meier, 1985); George L. Mosse, *Toward the Final Solution: A History of European Racism* (New York: Howard Fertig, 1997); Michael Burleigh and Wolfgang Wippermann, *The Racial State: Germany 1933–1945* (Cambridge: Cambridge University Press, 1993); Karola Fings, *In the Shadow of the Swastika: Volume 2: The Gypsies during the Second World War* (Hatfield, UK: University of Hertfordshire

Spanish sites and block access to US Web pages that do not comply with national laws; Spain did so to address threats to its national defense and public order.[6] In the Netherlands, section 137 of the criminal code dictates that it is a criminal offense to "deliberately give public expression to views insulting to a group of persons on account of their race, religion or conviction, or sexual preference."[7] In Sweden, the Freedom of the Press Act (chapter 7, article 4) prohibits the expression of contempt for a population group "with allusion to its race, skin color, national or ethnic origin, or religious faith."[8]

Britain recognized dangers associated with the fascist uses of hate speech. Its earliest attempt to curb such speech was article 5 of the Public Order Act (1936).[9] This legislation was aimed to counter the verbal attacks on the Jewish community made by Oswald Mosley and his fellow members in the British Union of Fascists, which led to outbreaks of violence in Britain. The law was further bolstered in 1965, when section 6 of the British Race Relations Act made it an offense to stir up hatred against a racial group.[10] Section 17 of the

Press, 1999); Gotz Aly, Susanne Heim, and A. G. Blunden, *Architects of Annihilation: Auschwitz and the Logic of Destruction* (Princeton, NJ: Princeton University Press, 2003); Clarence Lusane, *Hitler's Black Victims: The Historical Experiences of European Blacks, Africans, and African Americans during the Nazi Era* (New York: Routledge, 2002); William I. Brustein, *Roots of Hate: Anti-Semitism in Europe before the Holocaust* (Cambridge: Cambridge University Press, 2003); Eric A. Johnson and Karl-Heinz Reuband, *What We Knew: Terror, Mass Murder, and Everyday Life in Nazi Germany* (New York: Basic Books, 2006); Eric Ehrenreich, *The Nazi Ancestral Proof: Genealogy, Racial Science, and the Final Solution* (Bloomington: Indiana University Press, 2007); Christopher R. Browning, *The Origins of the Final Solution: The Evolution of Nazi Jewish Policy, September 1939–March 1942* (Winnipeg, MB: Bison Books, 2007); Daniel Jonah Goldhagen, *Worse Than War: Genocide, Eliminationism, and the Ongoing Assault on Humanity* (New York: Public Affairs, 2009); Ian Kershaw, *Hitler, the Germans, and the Final Solution* (New Haven, CT: Yale University Press, 2009); "Nazi Racism," *Holocaust Encyclopedia*, United States Holocaust Memorial Museum, Washington, DC, http://www.ushmm.org/outreach/en/article.php?ModuleId=10007679 ; "Racism: An Overview," *Holocaust Encyclopedia*, United States Holocaust Memorial Museum, Washington, DC, http://www.ushmm.org/wlc/en/article.php?ModuleId=10005184 ; Robert S. Wistrich, *A Lethal Obsession: Anti-Semitism from Antiquity to the Global Jihad* (New York: Random House, 2010).

6   Julia Scheeres, "European Outlaw Net Hate Speech," *Wired News*, November 9, 2002; Anita Ramasastry, "Can Europe Block Racist Web Sites from Its Borders?," CNN, February 6, 2003.

7   Wojciech Sadurski, *Freedom of Speech and Its Limits* (Dordrecht, Netherlands: Kluwer, 1999), 179.

8   Ibid.

9   See the United Kingdom's legislation website for Public Order Act, 1936, 1 Edw. 8 & 1 Geo. 6, c. 6, § 5, http://www.statutelaw.gov.uk/content.aspx?activeTextDocId=2236942 .

10   Eric Barendt, *Freedom of Speech* (New York: Oxford University Press, 2007); Michael Supperstone, *Brownlie's Law of Public Order and National Security* (London: Butterworths, 1981), 15; Home Office, *Racial Discrimination*, White Paper, Commd. 6234, September 1975; Commission for Racial Equality, *Reviews of the Race Relations Act* (1985,

Public Order Act of 1986 defines "racial hatred" as hatred against a group of persons by reference to color, race, nationality (including citizenship), or ethnic or national origins.[11] Other countries that have enacted laws penalizing the distribution of hate propaganda include Austria, Belgium, Cyprus, Hungary, Italy, and Switzerland, as well as Brazil and India.[12]

In November 2000, the Internet bookseller Amazon.com responded to German government complaints by deciding to stop shipping Adolf Hitler's *Mein Kampf* to customers in Germany, where the book is banned.[13] In December 2000, Germany's supreme court said that German laws against Nazi incitement could apply to people who put Nazi material "on the Internet on a foreign server ... that is accessible to Internet users in Germany."[14]

The case involved Frederick Töben, an Australian, who was prosecuted for violating a post–World War II German law outlawing the Nazi Party and forbidding any glorification of Nazism. Töben sent leaflets through the mail to Germany denying that the Nazis had ever perpetrated the Holocaust. He also posted the same information on an Australian Internet website. On visiting Germany in April 1999, Töben was arrested and charged with inciting racial hatred and "defaming a segment of the national population."[15] A trial court found Töben guilty of sending leaflets by mail to Germany. It dismissed the charges of posting materials on the Internet, however, holding that German law could not be applied to content on a foreign website.

1992); Her Majesty's Stationery Office, *Race Relations Act 1976* (1976); Her Majesty's Stationery Office, *Race Relations (Amendment) Act 2000*.

11   See the United Kingdom's legislation website for the Public Order Act 1986, http://www.opsi. gov.uk/RevisedStatutes/Acts/ukpga/1986/cukpga_19860064_en_4 .

12   Thomas David Jones, *Human Rights: Group Defamation, Freedom of Expression, and the Law of Nations* (Boston: Martinus Nijhoff, 1998), 189–24, 259–313; Alexander Tsesis, "Prohibiting Incitement on the Internet," *Virginia Journal of Law and Technology* 7, no. 2 (2002): 1–41, 5, http://vjolt.net/vol7/issue2/v7i2_a05-Tsesis.pdf ; Alexander Tsesis, "Dignity and Speech: The Regulation of Hate Speech in a Democracy," *Wake Forest Law Review* 44 (2009): 499–501; Raphael Cohen-Almagor, "Countering Hate on the Internet: A Rejoinder," *Amsterdam Law Forum* 2, no. 2 (2010): 125–32.

13   Angela Doland, "Nazi Staff on Yahoo Offensive to France," *Sun-Sentinel* (Fort Lauderdale, FL), July 25, 2000, 7A. There are also implications to libel law. In 2002, the Australian High Court ordered that a businessman, Joseph Gutnick, could sue Dow Jones for defamation in Australia on the basis of a *Barron's* magazine story that emanated from the company's computer servers in New Jersey. See *Dow Jones v. Gutnick* [2002] HCA 56, 2002 AUST HIGHCT LEXIS 61, 34 (High Court of Australia).

14   Peter Finn, "Neo-Nazis Sheltering Web Sites in the U.S.; German Courts Begin International Pursuit," *Washington Post*, December 21, 2000, A1. For further discussion, see *Jones v. Töben* [2002] Federal Court of Australia (FCA) 1150 (September 17, 2002).

15   Ray August, "International Cyber-Jurisdiction: A Comparative Analysis," *American Business Law Journal* 39, no. 4 (2002): 531–74.

Both Töben and the prosecution appealed. On appeal, Germany's supreme court, the Federal Court of Justice (Der Bundesgerichtshof), upheld Töben's conviction for mailing leaflets to Germany and reversed the trial court's ruling regarding the Internet. The fact that material glorifying the Nazi Party – material that Germany considers highly offensive to its national interest – could be accessed from within Germany was sufficient to give German courts jurisdiction. Germany's legislation that banned the Nazi Party and any glorification of it – including "Auschwitz lies" – can be applied to Internet content that originates outside of the country's borders.[16] *Auschwitz-Lüge* ("Auschwitz Lie") held that a statement denying the Holocaust was not protected speech under article 5 of the Basic Law.[17] According to the Constitutional Court, the fundamental right protected under article 5 is freedom of opinion.[18] However, freedom of opinion is not unconditionally guaranteed. General laws may restrict the right to freedom as well as other Basic Rights, such as personal honor.[19] In interpreting a statute limiting freedom of opinion, the court must balance the value of freedom of opinion along with the legal interest that the statute restricting the basic right serves.[20]

In 2009, Germany called for foreign ISPs to remove neo-Nazi images, texts, and other content that can be viewed in the country in violation of laws prohibiting any Nazi symbols. Stefan Glaser of Jugendschutz.net, a youth protection group, said that the Internet has become the number one propaganda platform for far-right extremists. In 2008, his organization catalogued 1,600 such sites, and the number is growing.[21]

HATE COMMERCIALIZATION: THE YAHOO! SAGA

Sales of Nazi merchandise are against the law in France, which strictly prohibits the selling or displaying of anything that incites racism. The French criminal code prohibits the display of Nazi symbols.[22] The Yahoo! saga started in February 2000 when Marc Knobel, director of the International League against Racism and Antisemitism (Ligue Internationale Contre le

---

[16] Alexander Tsesis, *Destructive Messages: How Hate Speech Paves the Way for Harmful Social Movements* (New York: New York University Press, 2002), 188; Donald P. Kommers, *Constitutional Jurisprudence of the Federal Republic of Germany* (Durham, NC: Duke University Press, 1997), 382–87.

[17] *Auschwitzlüge*, BVerfGE 90, 241–55 (1994).

[18] Ibid., B(II)(1).

[19] Ibid., B(II)(1)(c).

[20] Eric T. Eberwine, "Sound and Fury Signifying Nothing: Jürgen Büssow's Battle against Hate-Speech on the Internet," *New York Law Review* 49 (2004–2005): 353–410.

[21] "Germany Urges Ban of Neo-Nazi Sites Abroad," Associated Press, July 10, 2009.

[22] See section R645–1 of the French Criminal Code.

Racisme et l'Antisémitisme, or LICRA) and a member of the Observatory of Antisemitism, went to the Yahoo! auction site and saw pages of Nazi-related paraphernalia. The site featured swastika armbands, SS (Shutzstaffel) daggers, concentration camp photos, striped uniforms once worn by Nazi camp prisoners, and replicas of the Zyklon B gas canisters.[23] Knobel acknowledged that the auctions might be legal in the United States but believed them to be absolutely illegal within the borders of France.[24]

In April 2000, LICRA, together with two other organizations, the Union of French Jewish Students (Union des Étudiants Juifs de France, or UEJF) and the Movement against Racism and for Friendship between Peoples (Mouvement contre le Racisme et pour l'Amitié entre les Peuples, or MRAP), asked Yahoo! to either remove the Nazi memorabilia from its American websites or make all such auctions inaccessible to Web surfers in France and its overseas territories, such as Martinique and French Guyana, in accordance with its own terms of service agreement, which prohibited Netusers from posting content that was "hateful, or racially, ethnically, or otherwise objectionable."[25] If it did not, the organizations asked that the California-based company be fined $96,000 for each day of noncompliance. Ronald Katz, a lawyer representing the French groups, asserted, "There is this naive idea that the Internet changes everything. It doesn't change everything. It doesn't change the laws in France."[26]

Yahoo! did not respond to the demands, and legal action commenced. At the first hearing in the Tribunal de Grande Instance de Paris on May 15, 2000, defending counsel Christophe Pecard noted that Yahoo! maintained a French-language website (Yahoo.fr) that complied with French law. He argued that "Internet users who go to Yahoo.com undertake a virtual voyage to the U.S.," so no offense could be said to take place in France.[27] In any case, he said, it would be technically impossible for Yahoo! to block all access to its sites from France. Yahoo! claimed that it had no power to identify the national

---

[23] David Crumm and Alexa Capeloto, "Hate Is up for Bid on Some Web Sites: Sellers Hawk Hitler Trinkets, KKK Knives," *Detroit Free Press*, December 11, 2000, A1; "Yahoo Headed for Trial in France," Reuters, February 27, 2002. I am grateful to Marc Knobel and Philippe A. Schmidt for their reflections on a draft of this chapter.

[24] "Anti-racism Group Sues Yahoo for Hosting Auctions of Nazi-Related Items," Associated Press, April 12, 2000, http://www.nytimes.com/2000/04/12/technology/12yahoo-france.html .

[25] See Yahoo!'s terms of service at http://uk.docs.yahoo.com/info/terms.html/ .

[26] Lisa Guernsey, "Welcome to the Web: Passport, Please?," *New York Times*, March 15, 2001, http://www.nytimes.com/2001/03/15/technology/welcome-to-the-web-passport-please.html .

[27] Marc Le Menestrel, Mark Hunter, and Henri-Claude de Bettignies, "Internet E-ethics in Confrontation with an Activists' Agenda: Yahoo! on Trial," *Journal of Business Ethics* 39, nos. 1–2 (2002): 135–44.

origins of its customers and thus no control over where in the world its digital products went. Were Yahoo! forced to comply with French law, it would need to remove the Nazi items from all its servers, thereby depriving Yahoo! users everywhere from buying them and making French law the effective rule for the world.[28] In response, the plaintiffs' lawyer, Stéphane Lilti, asserted that France had the sovereign right to prohibit the sale of Nazi merchandise within its borders and argued that Yahoo! should not be exempt from French law.[29]

The Court Orders

On May 22, 2000, Judge Jean-Jacques Gomez ruled that Yahoo!'s sales were "an offense to the collective memory of a nation profoundly wounded by the atrocities committed in the name of the Nazi criminal enterprise."[30] He rejected all of Yahoo! Inc.'s jurisdiction-related arguments, holding that although the Yahoo.com site was located on a server in California and perhaps intended for an American audience, harm was suffered in French territory, and Yahoo! auctions were not protected in France by the First Amendment to the US Constitution. The judge ordered Yahoo! Inc. "to take all measures such as would dissuade and render impossible all consultations on yahoo.com of the service of auctioning of Nazi objects as well as any other site or service which constitute an apology of Nazism or which contest the nazi crimes."[31] In other words, the French court said that there could be no apology for Nazi crime and that it was impossible to contest or downplay the horrific magnitude of evildoing that had happened.[32] Yahoo! Inc. was ordered to prevent access from French territory to the Nazi objects and hate speech sites in question or face a penalty of FF 100,000 (about $13,600) per day for noncompliance.[33]

[28] Jack Goldsmith and Tim Wu, *Who Controls the Internet? Illusions of a Borderless World* (New York: Oxford University Press, 2006), 5.

[29] Le Menestrel, Hunter, and de Bettignies, "Internet E-ethics in Confrontation."

[30] *LICRA v. Yahoo! Inc. and Yahoo! France* (TGI, Paris, May 22, 2000), affirmed in *LICRA and UEJF v. Yahoo! Inc. and Yahoo! France* (TGI, Paris, November 20, 2000).

[31] *La Ligue Contre le Racisme at l'Antisémitisme (LICRA) and l'Union des Etudiants Juifs de France (UEJF) v. Yahoo! Inc. and Yahoo! France*, Interim Court Order, The County Court of Paris 6 (May 22, 2000). The Superior Court of Paris reiterated this in its November 20, 2000, order. The original and the English translation are provided in the appendix to the Complaint for Declaratory Relief in *Yahoo! Inc. v. LICRA and UEJF*, 169 F. Supp. 2d 1181 (N.D. Cal. 2001) (No. 00-21275).

[32] *Apology* in French means *discours visant à prendre la défense de quelqu'un ou de quelque chose*, or speech to stand up for someone or something. It may also mean *apologia* as in exposition and analysis in Latin. In this context, it indicates that there is no place to try to understand or to downplay the Nazi crimes. I thank Agnès Lefranc for clarifying this usage.

[33] Uta Kohl, *Jurisdiction and the Internet: A Study of Regulatory Competence over Online Activity* (Cambridge: Cambridge University Press, 2007), 201–2.

In reaction to this court judgment, Heather Killen, a Yahoo! vice president, commented, "It's very difficult to do business if you have to wake up every day and say 'OK, whose laws do I follow?' . . . We have many countries and many laws and just one Internet."[34] Yahoo! argued that even if French officials identified and blocked the offending offshore website, the same information could be posted on mirror sites outside France;[35] to keep out the Nazi pages, Yahoo! said, France would need to shut down every single Internet access point within its borders. Furthermore, even this action would not be completely effective because determined users in France could access the Net by a telephone call to an Internet access provider in another country.[36] Yahoo! co-founder Jerry Yang did not believe that one country's laws should regulate the Internet in other parts of the world and asserted that asking Yahoo! to filter access to its sites according to the nationality of Web surfers was "very naive."[37] Later it became clear that it was Yang who was naive. France instantiated new legal standards for the Internet that came into conflict with Yahoo!'s business practices.

Marc Levy, who represented the International League against Racism and Anti-Semitism, observed that "freedom of expression is not unlimited. . . . The law does not permit racism in writing, on television, or on the radio, and I see no reason to have an exception for the Internet."[38] Yahoo!, he said, should not be exempt from laws in the countries where it does business. Levy expressed "great satisfaction" with the ruling, saying that the judge had "rendered a service to the Internet," which otherwise ran the risk of becoming a "no-law zone."[39]

Judge Gomez gave Yahoo! two months to work out how to block French surfers from the disputed auction sites. During this interval, Cyril Houri, the founder of a fledgling American firm called InfoSplit, contacted the plaintiff's lawyer, Stéphane Lilti, and told Lilti that he had developed a new technology that could identify and screen Internet content on the basis of its geographic source.[40] Using this technology, Houri learned the Yahoo! servers accessed by Netusers in France, which the firm had claimed were protected by the First

[34] Joelle Tessler, "Online Auction of Nazi Items Sparks Debate Issue: National Laws on Global Web," *San Jose Mercury News*, July 25, 2000.

[35] Goldsmith and Wu, *Who Controls the Internet?*, 2.

[36] Ibid., 3.

[37] Brian Love, "Auctions of Nazi Gear May Yet Cost Yahoo!," *Seattle Times*, August 11, 2000, C6.

[38] Lee Dembart, "Boundaries on Nazi Sites Remain Unsettled in Internet's Global Village," *International Herald Tribune*, May 29, 2000.

[39] "French Court Says Yahoo Broke Racial Law," *New York Times*, May 23, 2000, http://www.nytimes.com/2000/05/23/business/french-court-says-yahoo-broke-racial-law.html .

[40] "Method and Systems for Locating Geographical Locations of Online Users," patent application, http://www.google.com/patents/US8239510 .

Amendment to the US Constitution, were actually located on a website in Stockholm, Sweden. Yahoo! placed constantly updated "mirror" copies of its US site on servers in Sweden to make access to it in Europe faster.[41]

When the trial resumed on July 24, 2000, Yahoo! lawyers again asserted that it was technically impossible to identify and filter out French visitors to the firm's US-based websites. "It's technically not in Yahoo!'s power to do this," said Armando Fox, a computer science professor at Stanford University. He added, "All Yahoo! sees is an IP [Internet Protocol] address, and anyone can set up a tunneling proxy to change an IP address. There's no way to reliably map an incoming connection."[42] Attorneys for the company said they had pulled Third Reich paraphernalia from Yahoo.fr and added warnings to pages with sensitive material, alerting French Netusers that they risked breaking French law by viewing them.[43]

But this solution was not acceptable to the plaintiffs because it was still possible to buy the illegal items via the US server. In addition, Lilti raised Houri's geographic location technology with the court, alleging that Yahoo! auctions in France were not, in fact, coming from American servers and that the assumption that every Web page was equally accessible to every Netuser everywhere in the world was simply wrong. If Yahoo! could target French users from Swedish servers, Lilti argued, it could potentially identify Netusers by geographic location and screen them out.[44]

In August 2000, Judge Gomez appointed three Internet experts – Vinton Cerf, the American Internet pioneer known as one of the "fathers of the Internet"; Ben Laurie, a British software engineer and visiting fellow at Cambridge University's computer laboratory; and François Wallon, a French technologist – to assess the extent to which Yahoo! could block transmissions into France. The experts concluded that it was possible to locate 70 percent of Netusers, a figure that could be increased by 20 percent if Yahoo! asked the users who requested the illegal contents to declare their nationality.[45] The three experts also suggested that Netusers could be forced to declare their

[41] Cyril Houri to Jack Goldsmith, personal communication (September 7, 2004), quoted in Goldsmith and Wu, *Who Controls the Internet?*, 7.

[42] Tom Perotta, "Yahoo! Ruling Exposes Risks of Being Global," *Internet World*, July 1, 2000.

[43] "Groups Sue Yahoo! over Sale of Nazi Objects in France," Associated Press, July 25, 2000. Although Yahoo! removed the items from the commercial auction site, it still continued to allow them to be sold in chat rooms accessible worldwide. See Bob Egelko, "Yahoo Getting New Hearing on Posting Nazi Items," *San Francisco Chronicle*, February 11, 2005, C3.

[44] Goldsmith and Wu, *Who Controls the Internet?*, 7; discussion with Philippe A. Schmidt, LICRA vice-president for international affairs, New York, March 26, 2010.

[45] Union des Étudiants Juifs de France, TGI Paris, November 20, 2000, Ord. ref., J.C.P. 2000, Actu., 2219; Peter Piazza, "Jurisdictional Comity Is No Joke: Yahoo! Must Prevent French from Accessing Nazi Memorabilia Auction Sites," *Security Management* 45, no. 2 (2001): 38.

geographic location and to answer certain questions if they used key words such as "Nazi" in their searches. Although Cerf expressed some philosophical reservations about the proposals, he admitted they were technically feasible.[46]

In November 2000, Judge Gomez reaffirmed his May 22 order and ruled that Yahoo! was avoiding a moral and ethical exigency that all democratic societies share.[47] In his final ruling on November 20, Gomez said Yahoo! had already prohibited the sale of human organs, cigarettes, live animals, drugs, and used underwear. He observed that it would cost the company very little to extend the list of banned goods to include Nazi symbols and that doing so "would have the merit of satisfying an ethical and moral standard shared by all democratic societies."[48] He noted that Yahoo! welcomed French visitors to its US website with French-language advertisements, which showed that Yahoo! was tailoring content for France and that it could, at least to some extent, identify and screen Netusers by geography.[49] Marc Knobel said in reaction, "The French justice system has heard us.... It is no longer OK for online retailers to say they are not affected by existing laws."[50] He maintained that if global Internet companies were not willing to put "ethics and morals" first themselves, they would be forced to do so.[51]

Yahoo! Inc. was again directed to satisfy the terms of Gomez's previous order within three months or pay a fine of FF 100,000 per day thereafter if it failed to comply with its legal obligations.[52] Yahoo! representatives responded by saying that France had no jurisdiction in the case and indicated that the

[46] Eduardo Cue, "National Boundaries: Latest Frontier in Cyberspace," *Christian Science Monitor*, January 10, 2001, 1.

[47] "Conclusions pour la Société Yahoo! Inc., A Monsieur le Président du Tribunal de Grande Instance de Paris," referred hearing of May 15, 2000, 8.

[48] Cue, "National Boundaries," 1.

[49] *La Ligue Contre le Racisme et l'Antisémitisme (LICRA) and L'Union des Etudiants Juifs de France (UEJF.) v. Yahoo! ! Inc. and Yahoo! France*, Interim Court Order, County Court, Paris, May 22, 2000. The original and English translation are provided in the appendix to the Complaint for Declaratory Relief in *Yahoo! Inc. v. LICRA and UEJF*, 169 F. Supp. 2d 1181 (N.D. Cal. 2001) (No. 00-21275). For critical discussion, see Robert Corn-Revere, "Caught in the Seamless Web: Does the Internet's Global Reach Justify Less Freedom of Speech?," in *Who Rules the Net? Internet Governance and Jurisdiction*, ed. Adam Thierer and Clyde Wayne Crews Jr. (Washington, DC: Cato Institute, 2003), 219–38.

[50] *LICRA et UEJF v. Yahoo! Inc.*, Ordonnance Référé, TGI Paris (November 20, 2000), Ord. ref., J.C.P. 2000, Actu., 2219 (English translation available at http://www.lapres.net/yahen11.html). See also Kristi Essick, "Judge to Yahoo: Block Nazi Goods from French," *PCWorld*, November 21, 2000,

[51] *LICRA et UEJF v. Yahoo! Inc.*, Ordonnance Référé, TGI Paris (November 20, 2000).

[52] Two days after Judge Gomez decided the *Yahoo!* case another judge rendered his verdict on similar facts and issues against UEJF. In this case, Multimania hosted a website titled "nsdap" (an

company would ignore the decision unless a US court were to enforce it. Yahoo! said it would refuse to pay the fine.[53] Considerations of moral and social responsibility, so it seems, were not at the forefront for the Yahoo! executives. They were interested only in the legal aspects, which they thought were on their side.

This was a case of clear-eyed akrasia. Yahoo! knew it violated French law. Its managers knew that Nazi paraphernalia was highly offensive to the French people, especially the Jews, after the Holocaust. Yet the company did not care. For Yahoo!, it seems, the issue was a matter of principle, integral to the company's business philosophy. Yahoo! did not wish to abide by the French law. Notwithstanding different moral and social norms, Yahoo! wished to implement American legal norms globally. The French did not like that. Yahoo! should have made a bona fide effort to keep users in France off the Nazi paraphernalia sites, but it had dragged its feet.

American Salvation?

After realizing that Yahoo! could not win in France, the company directors decided to seek help on its home soil, thinking that an American court would decline to enforce the French judgment against the company and would grant Yahoo! permission to continue its unfettered business practices on the global Internet. Yahoo! filed suit against LICRA and UEJF in federal district court, seeking a declaratory judgment that the interim orders of the French court were not enforceable in the United States.[54]

acronym for the Nazi party) whose content related to Adolf Hitler, the Nazi ideology, Nazi texts, and symbols. Once on notice, Multimania removed access to the website. Multimania had also supervised the websites it hosted by use of a search engine and keywords relating to usual illegal content found on the Internet. But Multimania had not used the acronym "nsdap" for its search. The court found that Multimania acted reasonably and promptly given its competence and the technical means available to detect illegal content. Unlike Yahoo!, Multimania acted in good faith, and the court held that it was not liable. See *Ass'n Union des Étudiants Juifs de France v. SA Multimania Prod.*, TGI Nanterre (May 24, 2000). See also Xavier Amadei, "Standards of Liability for Internet Service Providers: A Comparative Study of France and the United States with a Specific Focus on Copyright, Defamation, and Illicit Content," *Cornell International Law Journal* 35 (November 2001–February 2002): 189–230. In *National Football League v. TVRadioNow Corp.*, 53 U.S.P.Q.2d 1831 (2000), a Canadian website was enjoined from transmitting copyrighted programming material into the United States.

[53] "Yahoo Loses Court Ruling in France," Associated Press, November 21, 2000; Douglas W. Vick, "Regulating Hatred," in *Human Rights in the Digital Age*, ed. Mathias Klang and Andrew Murray (London: GlassHouse, 2005), 41–54, 41–42.

[54] See Joel R. Reidenberg, "Technology and Internet Jurisdiction," *University of Pennsylvania Law Review* 153 (2005): 1951–74, 1959. Reidenberg argues that Yahoo! introduced a misleading translation of the French decision at the district court. See also Joel Reidenberg, "Yahoo and Democracy on the Internet," *Jurimetrics* 42 (2002): 261–80.

In November 2001, the US District Court for the Northern District of California considered the important differences between the French legal norms and the US First Amendment and ruled that the Yahoo! order could not be enforced in the United States. Judge Jeremy Fogel concluded that the French ruling was inconsistent with the First Amendment and held that, while France could regulate speech in its territory, "this court" would not enforce a foreign order that violated the protections granted under the US Constitution. Yahoo! showed that the threat to its constitutional rights was real and immediate.[55]

The litigation culminated in January 2006 with a lengthy and fractured opinion by an en banc panel of the US Court of Appeals for the Ninth Circuit.[56] Eight of the 11 judges concluded that the District Court had personal jurisdiction over the French organizations, but notwithstanding the court's view of the jurisdictional issue, 3 of the 8 judges also concluded that Yahoo!'s claim was not "ripe for adjudication" and should be dismissed on those grounds. They pointed out that LICRA and UEJF had not sought enforcement of the French court's orders in the United States, that the French court could not impose a fine even if it did ask for one, and that it was unlikely that a US court would enforce such a fine even if a French court imposed one. Enforcement was unlikely "not because of the First Amendment, but rather because of the general principle of comity under which American courts do not enforce monetary fines or penalties awarded by foreign courts."[57] Although 3 of those 8 judges did think Yahoo!'s case was "ripe," 3 of the court's 11 judges concluded that the District Court did not have personal jurisdiction over the French organizations. The court could not lend support to Yahoo! to uphold the First Amendment to violate French criminal law and to facilitate the violation of French law by other Netusers. Since a majority of judges (6 of the 11) voted to dismiss the case for one reason or another, dismissed it was. Yahoo! did not receive the judicial support it was hoping for.

THE LEGAL DIMENSION

France was pushing forward new legal standards for the Internet. The *Yahoo!* case opened a renewed discussion about the national boundaries of the

---

[55] *Yahoo! Inc. v. La Ligue Contre le Racisme et l'Antisémitisme*, 169 F. Supp. 2d 1181; 2001 US Dist. Lexis 18378 (November 7, 2001).

[56] *Yahoo! Inc. v. LICRA and UEJF*, 433 F 3d 1199 (9th Cir. 2006). See also *Yahoo! Inc. v. LICRA and UEJF*, 379 F 3d 1120 (9th Cir. 2004).

[57] *Yahoo! Inc. v. LICRA and UEJF*, 433 F 3d 1199 (9th Cir. 2006), http://law.justia.com/cases/federal/appellate-courts/F3/433/1199/546158/ .

Internet.[58] The case prompts us to reconsider the international aspects of the Internet as a global phenomenon, with revamped responsiveness to national sovereignty and laws.

Yahoo!'s managers are able to grasp and apply moral norms. Yet they saw the norms of respecting others and not harming others as secondary to US law and challenged the authority of the French court. As a US company committed to the First Amendment, Yahoo!'s managers thought that freedom of expression was the single most important aspect of its business. The First Amendment is enshrined in the US legal and political culture. It explicitly instructs:

> Congress shall make no law respecting an establishment of religion, or prohibiting the free exercise thereof; or abridging the freedom of speech, or of the press; or the right of the people peaceably to assemble, and to petition the Government for a redress of grievances.[59]

This statement is sharp and uncompromising. No law, leading American scholars and judges argue, means no law. One of the preeminent justices of the US Supreme Court, Hugo L. Black, asserted in a classic article his belief that the US Constitution "with its absolute guarantees of individual rights, is the best hope for the aspirations of freedom which men share everywhere."[60] Another iconic legal authority, Alexander Meiklejohn, asserts that the First Amendment declares that with respect to belief, political discussion, political advocacy, and political planning, the citizens are the sovereign and the Congress is their subordinate agent.[61] The First Amendment condemns with its absolute disapproval any suppression of ideas. Meiklejohn coined the saying that "to be afraid of any idea is to be unfit for self-government."[62] According to this view, the public responsibilities of citizenship in the free world are in a vital sense beyond the reach of any legislative control. Consequently, freedom of expression in the American tradition occupies an especially protected position. Generally speaking, expression is perceived as doing less injury to other social goals than action. It has less immediate

---

[58] On June 21, 2004, prevention and repression measures against online antisocial expressions were enhanced by Law 2004–575. Internet hosts and access providers have the obligation of contributing to preventing the distribution of pedophile, revisionist, and racist information. The same year, the law on the digital economy expressly confirmed the ability of judges in chambers, apart from any other criteria of competence, to prescribe filtering of racist and antisemitic websites (article 6-I.8). See Marc Knobel, *The Law and the Internet* (Paris: CRIF, 2013).

[59] See the text of the First Amendment and annotations at http://constitution.findlaw.com/amendment1.html .

[60] Hugo L. Black, "The Bill of Rights," *New York University Law Review* 35 (1960), 865–81, 879.

[61] Alexander Meiklejohn, *Political Freedom* (New York: Oxford University Press, 1965), 107.

[62] Ibid., 124.

consequences and is less irremediable in its impact.[63] Only when expression might immediately translate into harmful action or when one is able to prove a clear link between the harmful speech and the resulting harmful action is it possible to consider restrictions on freedom of expression. This approach sets a very high threshold to satisfy. Only in clear and exceptional cases are there grounds to limit expression.[64] Only hate crimes are criminalized.

The northern neighbor of the United States, Canada, has one of the most progressive laws against hate speech. Canadian criminal law is far more extensive on prevention of hate speech than is US law. In both cultures, diversity is believed to be good. In both cultures, minorities are encouraged to speak and express opinions. But Canada recognizes that hate speech builds on differences and targets minorities for hatred. Hate speech destroys the mosaic that is so important to the Canadian identity.[65] Section 13(1) of the Canadian Human Rights Act holds:

> It is a discriminatory practice for a person or a group of persons acting in concert to communicate telephonically or to cause to be so communicated, repeatedly, in whole or in part by means of the facilities of a telecommunication undertaking within the legislative authority of Parliament, any matter that is likely to expose a person or persons to hatred or contempt by reason of the fact that that person or those persons are identifiable on the basis of a prohibited ground of discrimination.[66]

[63] Thomas I. Emerson, *The System of Freedom of Expression* (New York: Random House, 1970), 9, 292. See also Lillian R. BeVier, "The First Amendment and Political Speech: An Inquiry into the Substance and Limits of Principle," *Stanford Law Review* 30, no. 2 (1978): 299–358; Raphael Cohen-Almagor, *The Boundaries of Liberty and Tolerance: The Struggle against Kahanism in Israel* (Gainesville: University Press of Florida, 1994), especially chapter 5; Owen Fiss, "Freedom of Speech and Political Violence," in *Liberal Democracy and the Limits of Tolerance: Essays in Honor and Memory of Yitzhak Rabin*, ed. Raphael Cohen-Almagor (Ann Arbor: University of Michigan Press, 2000), 70–78.

[64] Jeremy Waldron, "Dignity and Defamation: The Visibility of Hate," *Harvard Law Review* 123, no. 7 (2010), 1596–657; Steven J. Heyman, *Free Speech and Human Dignity* (New Haven, CT: Yale University Press, 2008), 164–83; Frederick M. Lawrence, "The Hate Crime Project and Its Limitations: Evaluating the Societal Gains and Risk in Bias Crime Law Enforcement," Public Law Research Paper 216, George Washington University, Washington, DC, 2006.

[65] For further discussion of the social and legal aspects of hate propaganda, see Frederick M. Lawrence, *Punishing Hate* (Cambridge, MA: Harvard University Press, 1999); James Weinstein, "An American's View of the Canadian Hate Speech Decisions," in *Free Expression: Essays in Law and Philosophy*, ed. W. J. Waluchow (Oxford: Clarendon Press, 1994), 175–221; Richard L. Abel, *Speaking Respect, Respecting Speech* (Chicago and London: University of Chicago Press, 1998), 14–21; William B. Rubenstein, "The Real Story of U.S. Hate Crimes Statistics: An Empirical Analysis," *Tulane Law Review* 78 (2004): 1213–46.

[66] Shirish Pundit Chotalia, "Canadian Human Rights Act," in *Human Rights Law in Canada* (Toronto, ON: Carswell, 1998). The act supplements the hate provisions of the *Criminal Code*, R.S.C. 1985, chapter C-46. For further discussion, see Raphael Cohen-Almagor, "Ethical

Canada has been quick to act against the problem that hate speech on the Internet poses in its society. There is a particular risk of Internet sites transmitting neo-Nazi messages from the United States, where such speech is not punishable, to Canada, where its expression is understood to be socially inimical.[67] Furthermore, British and Canadian statutory documents affirm a corollary proposition about the effect of hateful speech on the community at large.

Being cognizant of the democratic catch, Canada constantly balances freedom of expression with considerations of its multicultural mosaic. It asserted its jurisdiction in prosecuting Ernst Zündel for violating the law by operating a hateful website in the United States from his Toronto office. In August 1996, a complaint was brought before the Canadian Human Rights Commission against Zündel for posting material on his website, www.zundelsite.org, that was likely to expose Jews to hatred or contempt. Deputy Attorney General Mark Freiman explained that the prosecution had to show that Zündel was using telephone lines to operate the Californian site. This issue was, of course, very contentious. Zündel contested the view that he was operating the site from Toronto. Freiman was able to prove that Zündel was the one who was communicating, that he was communicating "telephonically," that he was using Canadian telecommunications facilities in whole or in part, and that his communications were likely to expose an individual or a group to hatred or contempt based on their membership in an ethnic group.[68] On January 18, 2002, the Canadian Human Rights Tribunal ordered that Zündel, as well as any other individuals who act in his name or in concert with him, cease the discriminatory practice of communicating telephonically or causing to be communicated telephonically by means of the facilities of a telecommunication undertaking

Considerations in Media Coverage of Hate Speech in Canada," *Review of Constitutional Studies* 6, no. 1 (2001): 79–100; Richard Moon, "Report to the Canadian Human Rights Commission Concerning Section 13 of the Canadian Human Rights Act and the Regulation of Hate Speech on the Internet," Canadian Human Rights Commission, Ottawa, October 2008, http://www.safs.ca/moonreport.pdf ; Heather Belle Guest, ed., *Underneath the Golden Boy*, vol. 6 (Winnipeg: University of Manitoba, 2010).

[67] Neil Weinstock Netanel, "Cyberspace Self-Governance: A Skeptical View from Liberal Democratic Theory," *California Law Review* 88 (2000): 395. See also *Warman v. Harrison*, Canadian Human Rights Tribunal (August 15, 2006).

[68] Interview with Mark J. Freiman, deputy attorney general and deputy minister responsible for native affairs for the Province of Ontario, July 11, 2002. I also benefited from conversations with Richard G. Dearden, expert on media law, Gowling, Strathy & Henderson, Barristers & Solicitors, Ottawa, September 29, 1998; Martin Freeman, director and general counsel, Department of Justice, Constitutional and Administrative Law Section, Ottawa, July 17, 2002; Ian Binnie, justice of the Supreme Court of Canada, July 18, 2002; Frank Iacobucci, former justice of the Supreme Court of Canada, September 14, 2006; Irwin Cotler, member of the Canadian Parliament, December 17, 2009; and Peter Cory, former justice of the Supreme Court of Canada, May 16, 2010.

within the legislative authority of the Canadian Parliament.[69] The ruling was unenforceable, however, because Zündelsite was and still is hosted by an American company. Zündel, however, decided to emigrate from Canada. He moved to the United States and later was deported back to Canada, which, in turn, deported him to Germany, where he stood trial for Holocaust denial.

In New Zealand, section 61 of the Human Rights Act, Racial Disharmony, makes it unlawful for any person (a) to publish or distribute written matter that is threatening, abusive, or insulting; (b) to broadcast by means of radio or television words that are threatening, abusive, or insulting; or (c) to use in any place words that are threatening, abusive, or insulting if the person using the words knew or ought to have known that those words were reasonably likely to be published in a newspaper, magazine, or periodical or broadcast by means of radio or television.[70]

In Australia, section 3 of the Racial Hatred Act 1995 prohibits public behavior that is likely "to offend, insult, humiliate, or intimidate another person or group of people" if the act is done because of the race, color, or national or ethnic origin of the other person or a group.[71]

In Israel, amendment no. 20 (1986) of the penal code makes "incitement to racism" a criminal offense. Anyone who publishes anything with the purpose of inciting to racism is liable for five years of imprisonment (section 144B), and anyone who has racist publications in his or her possession for distribution is liable for imprisonment of one year (section 144D). The term *racism* is defined as "persecution, humiliation, degradation, manifestation of enmity, hostility or violence, or causing strife toward a group of people or segments of the population – because of color or affiliation with a race or a national-ethnic origin" (section 144A). The penal code also addresses the issue of hate. Section 144F(a) holds:

> If a person commits an offense out of a racist motive ... or out of enmity toward a public because of their religion, religious group, community of origin, sexual inclination or because they are foreign workers, then he shall be liable to double the penalty set for that offense or to ten years imprisonment, whichever is the lesser penalty.[72]

---

[69] *Citron and Toronto Mayor's Committee on Community and Race Relations v. Zündel*, Canadian Human Rights Tribunal (January 18, 2002). See also Yaman Akdeniz, *Racism on the Internet* (Strasbourg, France: Council of Europe Publishing, December 2009), 68.

[70] The full text of the Human Rights Act 1993 is available at http://www.dredf.org/international/NewZconst.html .

[71] Sadurski, *Freedom of Speech*, 179.

[72] Chapter 8, article 1, section 144F(a) of the *Penal Law*, in *Penal Law*, 5737–1977 (Aryeh Greenfield: A.G. Publications, March 2005). For further discussion, see Raphael Cohen-Almagor,

The implications of the *Yahoo!* case are particularly relevant for Germany, where many racist and hate groups post messages that are illegal in Germany on US sites. German radicals are accessing the Internet for those vile purposes from Germany, in clear violation of German law. Children are targeted in an attempt to lure them to racist, radical ideology. According to a recent study, the number of right-wing extremist contributions from Germany to Internet platforms aimed at schoolchildren or music fans – Facebook, YouTube, Twitter, and other social networking tools – rose from 750 in 2007 to about 6,000 in 2010.[73] Until recently, Germany did not do enough to address this issue.[74] In July 2009, Germany's minister of justice, Brigitte Zypries, said her office would appeal to foreign Internet providers to use their own terms of service as grounds for eliminating content promoting fascism. She called for ISPs in the United States and elsewhere to remove neo-Nazi images, text, and other content that could be viewed inside the country in violation of laws forbidding any Nazi symbols.[75] To date, US companies are not rushing to remove material that is protected under the First Amendment.[76]

"Regulating Hate and Racial Speech in Israel," *Cardozo Journal of International and Comparative Law* 17 (2009): 101–10.

[73] Toby Axelrod, "Internet Extremism Growing in Germany," *JTA*, August 26, 2010, http://www.jta.org/2010/08/26/news-opinion/world/internet-extremism-growing-in-germany .

[74] In January 1996, Deutsche Telekom blocked access to Internet sites that were spreading antisemitic propaganda, a crime in Germany. This action followed a request by Mannheim prosecutors, who were investigating one of the most prominent Holocaust deniers in the world, Ernst Zündel. Deutsche Telekom also blocked access to a Californian company, Web Communications, because it provided access to Zündel's site. However, although Deutsche Telekom blocked access, the site was still available through CompuServe. See Internet Law and Policy Forum Working Group on Content Blocking, "Material That Vilifies People on the Basis of Race, Gender, Sexual Preference, or Disability, or Incites or Promotes Hatred on those Bases," *Internet Law and Policy Forum*, http://www.ilpf.org/groups/content/vilify.htm .

[75] Patrick McGroarty, "Germany Calls for Ban of Neo-Nazi Sites Abroad," *Sydney Morning Herald*, July 10, 2009, http://news.smh.com.au/breaking-news-technology/germany-calls-for-ban-of-neo nazi-sites-abroad-20090710-devv.html .

[76] In August 2000, the Düsseldorf District Authority president, Jürgen Büssow, wrote to four American ISPs, requesting that they prevent access to four websites containing racist, neo-Nazi material. This action was unsuccessful. See Yaman Akdeniz, *Internet Child Pornography and the Law: National and International Responses* (Aldershot, UK: Ashgate, 2008), 236. On February 8, 2002, Büssow ordered all ISPs in the German state of Nordrhein-Westfalen to block user access to two specific US–based hate sites, Stormfront and Nazi-Lauck. See Düsseldorf District Authority, "Düsseldorf District Authority Issues Blockage Orders of Extreme Right Offers on the Internet" [in German], press release, February 8, 2002, http://www.brd.nrw.de/presse/pressemitteilungen/2002/02Februar/Bezirksregierung_Duesseldorf_erlaesst_S422002.html . More than 30 of the 76 ISPs in Nordrhein-Westfalen lost various court battles, which may be found in Oberverwaltungsgericht Münster, 2003 Multimedia und Recht (MMR) 348; Verwaltungsgericht Düsseldorf 2003 MMR 305; Verwaltungsgericht Arnsberg 2003 Zeitschrift für Urheber- und Medienrecht Rechtsprechungsreport 222. However, this blocking directive is local and does not compel

THE BUSINESS DIMENSION

The business dimension of the Yahoo! saga was implicitly acknowledged in November 2000, when Yahoo!'s management decided that auctions would henceforth be a paying service and Yahoo! would decide what was proper for sale. Yahoo! realized that it had assets in France, including a French subsidiary, that might be at risk of seizure if the company failed to comply with the French court's ruling. Shares of Yahoo! Inc. fell nearly 15 percent on New York's NASDAQ Stock Market after the Gomez verdict. This was their lowest level in two years.[77] Yahoo! was sensitive to the wishes of its foreign customers because 40 percent of its traffic at that time was outside the United States. Analysts noted that legal issues facing the company were not likely the main cause of the stock's weakness but feared that any additional successful suits could hurt international revenues.[78]

In January 2001, after both interim orders had been entered by the French court and after Yahoo! had filed suit in US federal district court, Yahoo! adopted a new policy prohibiting use of auctions or classified advertisements on Yahoo.com to offer or trade in items associated with groups principally known for hateful and violent positions based on race or similar factors. Yahoo! pulled all Nazi, Ku Klux Klan, and similar items associated with hatred and violence from its auction sites, announcing that it would "no longer allow items that are associated with groups which promoted or glorify hatred and violence, to be listed on any of Yahoo!'s commerce properties."[79] Yahoo! also said it would start a new policy that included having trained representatives monitoring the site regularly. In addition, Yahoo! would use software to identify potentially objectionable items.[80] Because these actions brought Yahoo! into substantial compliance with French law, the fines were not imposed.

Sometimes, the tension between freedom of information on the one hand and moral and social responsibility on the other can have significant business implications not only for the company at hand but also for other information and communication companies in the future. Corporations are careful to abide by legal responsibilities, but they also have other responsibilities. These

the other German states to follow suit. Moreover, it relates to only two hate sites. See Eberwine, "Sound and Fury Signifying Nothing."

77 "French Court Tells Yahoo to Block Nazi Auction Sites," *Reuters*, November 21, 2000.

78 "Yahoo! Stock Plunges to Lowest Level Since '98," *Reuters*, November 22, 2000.

79 Troy Wolverton and Jeff Pelline, "Yahoo! to Charge Auction Fees, Ban Hate Materials," *CNET News*, January 2, 2001, http://news.cnet.com/2100-1017-250452.html . See also "A Web of Thought Control," *Chicago Tribune*, January 13, 2001, 22.

80 Wolverton and Pelline, "Yahoo! to Charge Auction Fees, Ban Hate Materials."

are customary responsibilities, subject to negotiation according to the interests and balance of power of the parties involved and dictated by competitive necessities. Corporate Social Responsibility (CSR) scholar Keith Davis asserts that it is a firm's obligation to consider the effects of its decisions on society in a manner that will accomplish social benefits as well as traditional economic benefits. This obligation means that "social responsibility begins where the law ends. A firm is not being socially responsible if it merely complies with the minimum requirements of the law, because this is what any good citizen would do."[81] Yahoo! focused on the legal aspects of its auction site by merely complying with the requirements of the US law that champions freedom of expression.

According to Archie Carroll's formulation (see chapter 6), Yahoo! did not behave responsibly because its conduct lacked ethical perspective. Instead, the Yahoo! officials exhibited amoral management. "Amoral managers," Carroll explains, "are neither immoral nor moral but are not sensitive to the fact that their everyday business decisions may have deleterious effects on others."[82] In my assessment, the Yahoo! managers ignored the ethical dimension of their business when they were unresponsive to local laws and were inattentive to the implications of their conduct on stakeholders.

CSR on the Internet may prompt information and communication technology (ICT) professionals to adopt different modes of operation. The *first* mode is to follow the law of the land. The *second* mode to adopt is to do more than following the laws. I believe my analysis implies that multinational companies must do more than follow the laws in the home country of the head office; they need to integrate the laws of the host country into their decision making by, for instance, banning hate speech even where they are not required to do so, as is the case in the United States. And the *third* mode of operation ICT professionals need to adopt is to ignore the laws of the host country if the laws are manifestly unethical. ICT professionals should use this approach, for instance, when the law is aimed to censor political speech or certain groups because of their race, culture, religion, or sexual orientation.

David Radin, president of a technology startup, asserts that removing Nazi memorabilia from auction sites is the right business decision. First,

---

[81] Keith Davis, "The Case for and against Business Assumption of Social Responsibilities," *Academy of Management Journal* 16, no. 2 (1973): 312–22, 313. See also Philip Kotler and Nancy Lee, *Corporate Social Responsibility: Doing the Most Good for Your Company and Your Cause* (Hoboken, NJ: Wiley, 2005).

[82] Archie B. Carroll, "The Pyramid of Corporate Social Responsibility: Toward the Moral Management of Organizational Stakeholders," *Business Horizons* 34, no. 4 (1991): 39–48. See also Øyvind Ihlen, Jennifer L. Bartlett, and Steve May, eds., *The Handbook of Communication and Corporate Social Responsibility* (Chichester, UK: Wiley, 2011).

removing the objectionable materials will probably not harm the company's sales dramatically but will have a goodwill effect throughout the world, possibly attracting more users. Second, from a business standpoint, it is easier to implement restrictions worldwide than to create business and technical processes that treat citizens of different countries differently in an effort to universally stay on the right side of the law. Third, Radin rightly notes, it is the morally responsible thing to do.[83] Yahoo!'s commercial image would not have gained much by condoning the sale of Nazi memorabilia on its websites. Even in the absence of the ability to enforce such laws, factors such as market forces, moral beliefs, or a combination of them do so.[84]

Yahoo! has been complying with the court orders. On October 3, 2014, a search of the term *Nazi* on the Yahoo! Shopping site yielded 906 results, mostly books and documentary material. A search on another auction site, eBay, yielded 26,825 results. Most of the items appeared to be items that are not problematic – coins and stamps. This finding is in line with eBay's revised policy adopted in May 2001, which does not let people sell items associated with Nazi Germany, the Ku Klux Klan, or "notorious individuals who have committed murderous crimes."[85]

The *Yahoo!* court case marked that sovereign states may affirm their laws on the Internet. In February 2010, three Google executives were convicted of violating Italy's privacy laws after a three-minute video of a boy with disabilities who was being bullied was posted on Google. The young boy, who had Down syndrome, was shown as he was punched and kicked by four teenagers at a Turin school. Google removed the video and cooperated with the authorities on investigating the clip. However, the Italian prosecutor claimed that the video had been viewed 5,500 times over a period of two months. It reached the top of Google Italy's "most entertaining" video list, and the company ignored Netcitizens' appeals to remove it. Only after Google was notified by the authorities did the company take active steps.

Google has long been subject to criticism over how long it takes to respond to complaints and requests to remove inappropriate content. Prosecutor Alfredo Robledo said, "We are very satisfied because we have dealt with a serious problem, that is, protecting a person, and that should always come

---

[83] David Radin, "Yahoo! Auction Is Right to Ban Nazi Goods," *Pittsburgh Post-Gazette*, January 11, 2001, F3. See also Irene Lynch Fannon, *Working Within Two Kinds of Capitalism* (Portland, OR: Hart, 2003), 93–103.

[84] Uta Kohl, *Jurisdiction and the Internet: A Study of Regulatory Competence over Online Activity* (Cambridge: Cambridge University Press, 2007), 207.

[85] Todd Pack, "Ebay Bans Products Linked to Klan, Nazis," *Orlando Sentinel*, May 17, 2001, C1.

above business freedom."[86] Whereas Google's "launch first, correct later" approach was protected by the First Amendment to the US Constitution, in Italy freedom of expression is bounded by respect for other human rights, among which stands out the right to privacy. Although Google celebrated its approach as ultracreative and flexible as well as vital to maintaining an expanding dynamic business, the prosecution perceived this approach as problematic because it might violate privacy and human dignity. For the prosecution, the child was the weak party who deserved protection. Because Google handled user data and used content to generate advertising revenue, it was a content provider, not a service provider, and therefore broke Italian privacy law. Google, Robledo stressed, could not continue to ignore Italian laws if it wished to operate in the country.

Google argued that it was technically impossible to check all content on its server. Robledo, however, maintained that Google could easily find ways to monitor its content, and that it should not profit from advertising revenue generated from content that violated privacy laws. If Google had found a way to create filters in China, Robledo said, it could do the same in Italy, not to monitor political content "but to protect human dignity."[87]

On February 24, 2010, Judge Oscar Magi of the Milan court convicted three Google executives (David Drummond, senior vice president and chief legal officer; George De Los Reyes, former Google Italy board member; and Peter Fleischer, global privacy counsel) for violating Italian privacy laws and sentenced them in absentia to six-month suspended sentences.[88] As in the *Yahoo!*

[86] "Google to Appeal Italy's Privacy Convictions," video, http://www.youtube.com/watch?v=UK VOfbYAfYk ; "Italy Convicts 3 Google Execs in Abuse Case," video, http://www.youtube.com/ watch?v=8PlxJ4UW4Zw ; "Google Executives Jailed in Italy," video, http://www.youtube.com/ watch?v=xo94EraVlt4 ; Adrian Shaw, "Google Bosses Convicted over YouTube Bully Video," *Mirror Online*, February 25, 2010, http://www.mirror.co.uk/news/top-stories/2010/02/25/google-bosses-convicted-over-youtube-bully-video-115875-22068321/ .

[87] Stephen Shankland, "Execs Convicted in Google Video Case in Italy," CNET, February 24, 2010, http://news.cnet.com/8301-30685_3-20000092-264.html ; Colleen Barry, "Google Privacy Violation Conviction: Three Executives Found Guilty in Italy," *Huffington Post*, February 24, 2010, http://www.unz.org/Pub/HuffingtonPost-2010feb-08631 ; Rachel Donadio, "Larger Threat Is Seen in Google Case," *New York Times*, February 24, 2010, http://www.ny times.com/2010/02/25/technology/companies/25google.html ; Lionel Barber and Maija Palmer, "Google Chief Prizes Creativity," FT.com, June 3, 2010, http://www.ft.com/cms/s/ 2/bdecoee8-6f4f-11df-9f43-00144feabdco.html ; Massimo Mucchetti, "The Prosecutor, Privacy, and Google: They Want the Far West" [in Italian], *Corriere della Sera*, June 6, 2010, http://www.corriere.it/cronache/10_giugno_06/mucchetti-privacy-google_ec489b1c-71 3e-11df-82e2-00144fo2aabe.shtml .

[88] The decision can be found at http://speciali.espresso.repubblica.it//pdf/Motivazioni_sentenza _Google.pdf . Interestingly, in an interview, Peter Fleischer said that he and his colleagues were sentenced to "six months in jail for content." See Peter Fleischer, "Private Lives," *Index on Censorship* 40, no. 2 (2011): 78–89, 87.

case, whereas the Internet intermediary perceived the case as an attack on the principles of free expression on which the Internet is built, the court emphasized that the Internet is not a borderless entity where everything is allowed. The ruling means, in essence, that hosting platforms are criminally responsible for illegal content that Netusers upload. Internet intermediaries such as YouTube, Facebook, and Blogger are required to abide by the laws of the countries in which they operate. This ruling, similar to the *Yahoo!* decision, asserted the nation's local law and may force Internet intermediaries to change their business models.

However, these are still contentious issues, and the Web companies are doing their best to resist efforts compelling them to take active measures to scrutinize content. Google appealed the conviction, and, in December 2012, the appeals court overturned the decision.[89] Italy's Court of Appeals held that Google could not be subject to a monitoring obligation on all content uploaded by users before posting, not only because that obligation was effectively impossible to carry out but also because placing such an obligation would alter the nature of the service that Google provides, would affect the functioning of the platform, and would ultimately conflict with the right to freedom of expression.[90]

Finally, I should mention that in 2010, Google faced legal proceedings yet again for privacy violation, this time in Canada. On October 19 of that year, a court ruled that Google violated the privacy of citizens when it introduced Street View without their consent. The court ordered Google to delete all confidential data.[91] Similarly, in 2011, the French privacy watchdog fined Google for accidentally collecting and storing data through its Street View cars, and the same year, in Spain, Google was instructed to delete links to websites that contained personal information.[92]

[89] Jon Brodkin, "Italy Finally Acquits Google Execs Convicted over User-Uploaded Video," Ars Technica, December 21, 2012, http://arstechnica.com/tech-policy/2012/12/italy-finally-acquits-google-execs-convicted-over-user-uploaded-video/ .

[90] Flavio Monfrini, "Court of Appeals Overturns Conviction of Google Italy Executives, Redefines Liability of Hosting Providers under Privacy Legislation," Italy Legal Focus, Studio Legale Bernascone & Soci, March 26, 2013, http://www.lexology.com/library/detail. aspx?g=b36ffdc4-ee2b-4dfb-ae83-01bcb15ff5f7 .

[91] Office of the Privacy Commissioner of Canada, "Google Contravened Canadian Privacy Law, Investigation Finds," news release, October 19, 2010, http://www.priv.gc.ca/media/nr-c/ 2010/nr-c_101019_e.cfm ; Josh Halliday, "Google Street View Broke Canada's Privacy Law with Wi-Fi Capture," *Guardian*, October 20, 2010, http://www.guardian.co.uk/technology/ 2010/oct/19/google-street-view-privacy-canada ; "Index Index," *Index on Censorship*, 40, no. 2 (2011): 140–44, 140.

[92] Fleischer, "Private Lives," 79.

CHILD PORNOGRAPHY

Countries have affirmed their responsibility in the realm of child pornography. The Netherlands has established a telephone hotline and a website available for Netusers to complain about Internet child porn sites. Complaints are relayed to the ISP, which is required to withdraw the sites to forestall police intervention. At the same time, German authorities attempted to make ISPs liable for illegal content. On this principle, the government tried to force the CompuServe Information Service to bar its 4 million Netusers from accessing 200 newsgroups with sexual content. This action affected many Netusers in the alt.binaries.pictures.erotica group. A senior CompuServe official was then criminally charged for failing to ensure that his company suppressed such traffic. The conviction was subsequently overturned.[93]

The German government established an official administrative authority called the Federal Department for Media Harmful to Young Persons (Bundesprüfstelle für jugendgefährdende Medien, or BPjM). Its task is to protect children and adolescents from any media that might contain harmful or dangerous contents. This work is authorized by the Youth Protection Law (Jugendschutzgesetz). Objects are considered harmful or dangerous to minors if they tend to endanger their process of developing a socially responsible and self-reliant personality. In general, this restriction applies to objects that contain indecent, extremely violent, crime-facilitating speech as well as racist material. Completely prohibited are depictions of sexual acts involving children, animals, or violence.[94]

United Kingdom

In the United Kingdom, the government tackled the problem by establishing in 1996 a self-regulatory industry body, the Internet Watch Foundation (IWF). The UK Home Office, the Department for Trade and Industry, the Metropolitan Police, and ISPs all played a crucial role in the process. The result was a Safety Net proposal that required UK ISPs to implement reasonable, practicable, and proportionate measures to hinder the use of the Internet for illegal purposes and to provide a response mechanism in cases where illegal material or activity is identified.[95] This proposal led to the formation

---

[93] Philip Jenkins, *Beyond Tolerance: Child Pornography on the Internet* (New York: New York University Press, 2001), 191.

[94] For general information, see the BPjM website at http://www.bundespruefstelle.de/ .

[95] See Executive Committee of the Internet Service Providers, London Internet Exchange, and Safety-Net Foundation, "R3 Safety-Net: Rating, Reporting, Responsibility for Child

of the IWF Hotline.[96] The hotline enables Netusers to report potentially illegal content on websites, newsgroups, and online groups.[97] The IWF undertook to inform all British ISPs once it had located the supposed illegal content. The ISPs could not claim that they were unaware of the illegal content, and the UK police would be entitled to take action against any ISP that did not remove the relevant content. The hotline is mainly concerned with child pornography.[98]

The IWF is the only authorized organization in the United Kingdom that provides an Internet hotline for Netusers to report their exposure to potentially illegal online images (e.g., child abuse images and criminally racist and obscene content hosted in the country).[99] After the IWF traces the child abuse content complained of, it judges the potential illegality according to British law. If the offending content is hosted in the United Kingdom, the IWF issues a notice to ISPs to take down the content and advises the National Crime Squads Paedophile Online Investigation Team. If the content is hosted abroad, the IWF passes that information to the relevant national hotline, if one exists, and, in any event, to the National Police Intelligence Service for onward transmission to Interpol.

The British telecommunication service company BT, in partnership with IWF, developed the Cleanfeed Project in late 2003. The Cleanfeed Project is aimed toward blocking access to any images or websites that contain child pornography within the IWF database. Customers of BT are prevented from accessing the blocked content and websites. In July 2004, BT claimed that within the first three weeks of its launch, the Cleanfeed system had blocked 230,000 attempts to access child abuse websites.[100] The project attempts to prevent people from deliberately or accidentally accessing illegal sites in the United Kingdom and abroad. UK security officials think that Cleanfeed is able to block 90 percent of all child pornography. This estimate is optimistic. Cleanfeed fails to capture dynamic content posted in chat rooms and instant

Pornography and Illegal Material on the Internet," industry proposal, September 23, 1996, http://www.mit.edu/activities/safe/labeling/r3.htm .

[96] See the IWF website at http://www.iwf.org.uk/ . Also see Akdeniz, *Internet Child Pornography*, 252–53. I will elaborate on hotlines in chapter 9.

[97] For more on the IWF Hotline, see https://www.iwf.org.uk/hotline .

[98] See Akdeniz, *Internet Child Pornography*, 254.

[99] See IWF's reporting page at http://www.iwf.org.uk/report . See also Ben Wagner, "The Politics of Internet Filtering: The United Kingdom and Germany in a Comparative Perspective," *Politics* 34, no. 1 (2013): 58–71.

[100] See Akdeniz, *Internet Child Pornography*, 258. See also Catharine Lumby, Lelia Green, and John Hartley, *Untangling the Net: The Scope of Content Caught by Mandatory Internet Filtering* (Sydney: Australian Law Reform Commission, 2011), http://www.alrc.gov.au/sites/default/files/pdfs/ci_2522_l_green_w_attachments.pdf .

messaging. Online child sex offenders and pedophiles make great efforts to circumvent regulation. Often they have their own obscure providers. But at least content-filtering systems such as Cleanfeed are able to stop new seekers of child pornography, the majority being people who have not found their way into pedophile rings.

Operation PIN of the Virtual Global Taskforce involved the creation of a website that purported to contain images of child abuse. In fact, it was a law enforcement site. Anyone who entered the site and attempted to download images was confronted online by a law enforcement officer. The person was informed that he or she had entered a law enforcement website and had committed an offense and that his or her details had become known to the relevant national authorities. Since its launch in December 2003, Operation PIN has captured the details of individuals from a number of countries who were looking for images of child abuse. However, data gathering is not the primary aim of this initiative. Operation PIN was designed to reduce crime, but its real success has been in undermining the confidence of Net abusers who think that the Internet is an anonymous place where online child sex offenders, pedophiles, and other criminals can operate without fear of being caught.[101]

In 2003 and 2004, the British government also funded major media advertising campaigns against Internet pedophiles. The campaigns were supported by an official website, Thinkuknow.[102] Advertisements and the website were designed to inform children and parents about the danger of Net pedophiles, and particularly about the risks of being "groomed" online to become a victim, about pedophiles who assume false identities, and about ways to devise protective practices and promote social responsibility.[103]

In 2005, a law enforcement agency called the Child Exploitation and Online Protection Centre (CEOP) was established. It is dedicated to eradicating the sexual abuse of children. It does research and intelligence on how offenders operate and think, how children and youth behave, and how technological advances are developing. A specialist policing focus is placed on (a) keeping organized crime from profiteering from publishing or distributing images, (b) supporting local forces in computer forensics and covert

---

[101] "Police to Trap Online Paedophiles," BBC News, December 18, 2003, http://news.bbc.co.uk/1/ hi/uk/3329567.stm ; For more information, visit the Virtual Global Taskforce's website at http:// www.virtualglobaltaskforce.com/what-we-do/ .

[102] You can visit the website at http://www.thinkyouknow.co.uk/ .

[103] Anneke Meyer, *The Child at Risk: Pedophiles, Media Responses, and Public Opinion* (Manchester, UK: Manchester University Press, 2007), 135. See also Anne-Marie McAlinden, *"Grooming" and the Sexual Abuse of Children: Institutional, Internet, and Familial Dimensions* (Oxford: Oxford University Press, 2012).

investigations, and (c) bolstering international cooperation. As child abusers tend to have a lifelong predilection for this kind of offense, the CEOP researches the suspect's environment and career history and the ways and means for the suspect to have access to children.[104] The CEOP also incorporates the United Kingdom's national victim identification program through which child victims can be identified using a comprehensive database.

On September 29, 2008, British Prime Minister Gordon Brown launched the "Click Clever Click Safe" Campaign.[105] The campaign, designed to promote children's Net safety, was the result of work done by the UK Council for Child Internet Safety, which brought together more than 140 organizations and individuals. Its composition includes companies, government departments and agencies (including those in Scotland, Wales, and Northern Ireland), law enforcement officers, academic experts, charities, and parenting groups.

Between March 1, 2009, and February 28, 2010, CEOP received 6,291 reports about children at risk. Many of the reports came through the ClickCEOP button,[106] which is on the UK authoritative website for online safety. During 2006 to 2010, 624 children were safeguarded or protected from sexual abuse as the result of CEOP activity. During that period, 1,131 suspected child sexual offenders were arrested, and 262 high-risk sexual offender networks were disrupted and dismantled.[107] In 2010 and 2011 alone, 132 high-risk sex offender networks were disrupted and dismantled, and 414 children were safeguarded or protected from sexual abuse as the result of CEOP activity.[108] In 2012 to 2013, CEOP received an unprecedented total of 18,887 reports of child sexual abuse, showing an overall increase of 14 percent compared to the previous year.[109] Greater awareness of the problem and of the work of the CEOP leads to a greater number of reports.

---

[104] See the CEOP website at http://ceop.police.uk/ . See also National Policing Improvement Agency, *Guidance on Investigating Child Abuse and Safeguarding Children* (Wyboston, UK: Association of Chief Police Officers, 2009); discussion with Ruth Allen, head of Specialist Operational Support, CEOP, February 1, 2011.

[105] UK Council for Child Internet Safety, "Click Clever, Click Safe: The First U.K. Child Internet Safety Strategy," http://ceop.police.uk/Documents/UKCCIS_Strategy_Report.pdf . See also *Get Safe Online* (blog) at http://www.getsafeonlineblog.org/ .

[106] See the CEOP website at http://ceop.police.uk/ . See also http://www.facebook.com/clickceop .

[107] CEOP Centre, *Annual Review: 2009–10*, CEOP, London, 2010, http://ceop.police.uk/Docu ments/CEOP_AnnualReview_09-10.pdf .

[108] CEOP Centre, *Annual Review 2010–11 and Centre Plan 2011–12*, CEOP, London, 2011, http:// www.ceop.police.uk/Documents/ceopdocs/Annual%20Rev2011_FINAL.pdf .

[109] CEOP Centre, *Annual Review 2012–2013 and Centre Plan 2013–2014*, CEOP, London, 2013, http://ceop.police.uk/Documents/ceopdocs/AnnualReviewCentrePlan2013.pdf .

Canada

In Canada, the National Child Exploitation Coordination Centre (NCECC) was established in 2004 as part of Canada's National Strategy to Protect Children from Internet Sexual Exploitation and in response to recognition of the growing and disturbing crime of Internet-facilitated child sexual exploitation. The NCECC's mandate has been refined to reduce the vulnerability of children to Internet-facilitated sexual exploitation by identifying victimized children; investigating and assisting in the prosecution of sexual offenders; and strengthening the capacity of municipal, territorial, provincial, federal, and international police agencies through training and investigative support.[110] Since its inception, the NCECC has been a valuable part of the Canadian response to Internet-facilitated child sexual exploitation. It has contributed significantly to Canadian law enforcement's ability to investigate these offenses and coordinate national and international investigative files.

The NCECC is a partner of the Canadian Coalition against Internet Child Exploitation (CCAICE). CCAICE comprises ISPs, their industry associations, Justice Canada, Industry Canada, Cybertip.ca, and Canadian law enforcement.[111] The mandate of CCAICE is to provide solutions to reduce children sexual exploitation.

NCECC's efforts through CCAICE have led to the development of the Child Exploitation Tracking System (CETS). CETS is software developed by NCECC, international law enforcement officials, and Microsoft Corporation to help battle online child abuse.[112] This tool helps law enforcement officials collaborate and share information with other police services. CETS increases the effectiveness of investigators by providing them with software to store, search, share, match, and analyze large volumes of information. CETS has played a part in several international investigations, creating links that have helped apprehend offenders and, more important, that have led to the rescue of children in different countries. CETS helped law enforcement agencies follow hundreds of suspects simultaneously and eliminated duplicated work, making it easier for the agencies to follow up on leads, collect evidence, and apprehend online child sex offenders.[113]

---

[110] "Online Child Sexual Exploitation," Royal Canadian Mounted Police, http://www.rcmp-grc.gc.ca/ncecc-cncee/index-accueil-eng.htm .

[111] Personal communication with NCECC officials, July 22, 2008.

[112] Microsoft, "Microsoft Collaborates with Global Police to Develop Child Exploitation Tracking System for Law Enforcement Agencies," press release, April 7, 2005, http://news.microsoft.com/2005/04/07/microsoft-collaborates-with-global-police-to-develop-child-exploitation-tracking-system-for-law-enforcement-agencies/ .

[113] Microsoft, "Tool Thwarts Online Child Predators," press release, April 7, 2005, http://news.microsoft.com/2005/04/07/tool-thwarts-online-child-predators/ .

Cybertip.ca is the Canadian National Tipline for reporting online sexual exploitation of children. It was established in September 2002 as a pilot project and was officially launched in January 2005.[114] Private and public sectors joined with law enforcement agencies to create Cybertip.ca to combat the online victimization of children. It receives reports from the public regarding child pornography, luring (grooming), child sex tourism, child prostitution, and exploited children.[115] On average, Cybertip.ca receives more than 2,000 reports and 75,000 page views per month from the public. The child sexual exploitation reports are made up of the following:

- 90 percent relate to child sexual abuse images
- 8 percent relate to online luring
- 1 percent relate to traveling sex offenders
- 1 percent relate to children sexually exploited through prostitution[116]

In November 2006, Cybertip.ca started using the Cleanfeed system.[117] Access is blocked to websites that display images of prepubescent children being assaulted or children who were posed deliberately in a sexual manner. Canadian ISPs collaborate with the Canadian Centre for Child Protection (CCCP) as part of Project Cleanfeed Canada. This collaboration allows the CCCP to maintain a list of websites known to host images of child sexual abuse. Canadian ISPs have agreed to block the listed sites from their Canadian customers.[118] However, Cleanfeed is not capable of providing absolute blocking. Absolute blocking is a Sisyphean and relentless technological struggle that does not yield clear winners.

In 2008, Cybertip.ca designed a campaign to raise awareness about the techniques used by offenders to keep child sexual abuse a secret. Research has shown that children do not disclose 70 percent of sexual abuse that happens to them. Children avoid telling because they are often afraid of their parents' negative reactions or they think the abuser might harm them. The campaign aimed to educate children that some secrets should not be kept forever.[119] Emphasis was put on the length of time children were required to keep a

---

[114] See the website for the tipline at http://www.cybertip.ca/app/en/about . See also "Canadian Coalition against Internet Child Exploitation Releases National Action Plan," Cybertip.ca, press release, May 11, 2005, https://www.cybertip.ca/app/en/media_release_ccaice_action_plan .

[115] Akdeniz, *Internet Child Pornography*, 262.

[116] Personal communication with NCECC officials, July 22, 2008.

[117] See the Cleanfeed Canada website at https://www.cybertip.ca/app/en/projects-cleanfeed .

[118] Personal communication with NCECC officials, July 22, 2008; Akdeniz, *Internet Child Pornography*, 263.

[119] "Let's Keep This Our Little Secret," Cybertip.ca, 2008. For other Canadian educational campaigns, see https://www.cybertip.ca/app/en/projects-public_awareness .

secret. Some secrets – for example, those that do not have a precise time limit – should be disclosed. Indeed, we need to further educate children on how to protect themselves from strangers, how to protect themselves from people they do know who make sexual advances, and how to protect themselves from people who wish to extort or exploit them.

United States

In the United States, some argue that concern over child sexual abuse and child molesters has reached the status of a moral panic. Public sentiment generated by the threat attains a restless terrain of heightened emotions, keen concern, disconcerting fears and anxieties, hostility, and moral righteousness.[120] On April 16, 2008, I met with Rep. Rick Boucher, who was involved in passing the Digital Millennium Copyright Act of 1998.[121] At the very start of our meeting, Boucher told me that he was not concerned about the use of the Internet by terrorists, pedophiles, hatemongers, and criminals. He said there were other members of Congress who were concerned about those issues, and he was happy to refer me to them, but he was not. He did not believe that the Internet should be restricted, and he did not wish to see any government interference to that effect. When asked about use and abuse of the Internet in the future, Boucher was ready to relate only to the issue of use. His reluctance to speak of any boundaries to free expression was astounding.[122] Boucher saw his responsibility as a legislator to protect and expand Net freedom of expression, not to limit it in any way. Public safety considerations were of no significance to him.

Yet the facts are alarming: one in seven youths age 10 to 17 is solicited or approached by a sexual predator while online.[123] To address this concern among others, the United States has developed reporting, legislative, and law enforcement mechanisms. The National Center for Missing and Exploited Children (NCMEC) was established in 1984, two years after Congress passed the Missing Children's Act. Although the center is a private, nonprofit organization, it receives federal funding for certain core services and

---

[120] Pamela D. Schultz, "Naming, Blaming, and Framing: Moral Panic over Child Molesters and Its Implications for Public Policy," in *Moral Panics over Contemporary Children and Youth*, ed. Charles Krinsky (Farnham, UK: Ashgate, 2008), 95.

[121] Digital Millennium Copyright Act, Pub. L. No. 105-304, 112 Stat. 2860 (Oct. 28, 1998), codified at 17 U.S.C. sections 512, 1201-05, 1301-22; 28 U.S.C. section 4001, http://itlaw.wikia.com/wiki/Digital_Millennium_Copyright_Act .

[122] Interview with Rep. Rick Boucher, Rayburn House, Washington, DC, April 16, 2008.

[123] Daniela Deane, "Police Try MySpace to Deter Sexual Predators," *Washington Post*, May 2, 2008, B05.

private funding from corporations, foundations, and individuals. The center works in partnership with the US Department of Justice to help law enforcement officers find missing children and eliminate child sexual exploitation. NCMEC prepares reports and analyzes information received from different sources, mainly from the public and major ISPs (Yahoo!, Google, AOL). NCMEC serves as a clearinghouse of leads on crimes against children, identifying jurisdiction for law enforcement.[124] NCMEC processed 5,759 reports of online enticement of children by adults in 2009, 4,053 reports in 2010, and 3,638 reports in 2011.[125] In 2012 alone, at the request of law enforcement, NCMEC analysts reviewed more than 19 million child pornography images and videos.[126] Crime investigation remains in the hands of the police and the US Federal Bureau of Investigation (FBI).

As for legislative mechanisms, under federal law, offenders convicted of sexually abusing a child face fines and imprisonment. Offenders may face harsher penalties if the crime occurred under aggravated circumstances, which include using force or threats, inflicting serious bodily injury or death, or kidnapping a child in the process of committing child sexual abuse. A number of federal laws have been enacted to tackle child abuse offline and online.[127] In 2012 alone, the United States passed a number of important federal laws concerning the sexual exploitation of children. These include 18 U.S.C. section 1466A (2012), Obscene Visual Representations of the Sexual Abuse of Children; 18 U.S.C. section 1470 (2012), Transfer of Obscene Material to Minors; 18 U.S.C. section 2251 (2012), Sexual Exploitation of Children; 18 U.S.C section 2252B (2012), Misleading Domain Names on the Internet; and 18 U.S.C. section 2422 (2012), Coercion and Enticement.[128]

As for law enforcement, to catch sex offenders, law enforcement officers take advantage of the same Internet tools available to sex offenders. R. Stephanie Good describes her experience working with the FBI to tempt sex predators on the Internet, set a time and place, and seize the

---

[124] Interview with senior NCMEC officials, Alexandria, VA, April 2, 2008.

[125] Joseph Menn, "Social Networks Scan for Sexual Predators, with Uneven Results," Reuters, July 12, 2012, http://www.reuters.com/article/2012/07/12/us-usa-internet-predators-idUSBRE86B05 G20120712 .

[126] National Center for Missing and Exploited Children, "Every Child Deserves a Safe Childhood: 2012 Annual Report," http://www.missingkids.com/en_US/publications/NC171.pdf .

[127] 18 U.S.C. section 2241, Aggravated Sexual Abuse; 18 U.S.C. section 2242, Sexual Abuse; 18 U.S.C. section 2243, Sexual Abuse of a Minor or Ward; 18 U.S.C. section 2244, on Abusive Sexual Contact. See the Department of Justice's website at http://www.justice.gov/criminal/ceos/citizensguide/citizensguide_sexualabuse.html .

[128] See the list of federal laws concerning the sexual exploitation of children on the NCMEC website at http://www.missingkids.com/LegalResources/Exploitation/FederalLaw .

predators.[129] Just as a 50-year-old person can pose as a young boy to lure youth, so can a law enforcement officer assume a false persona to catch online child sex offenders. The FBI surfs online chat rooms and the like to identify and take action against sex offenders.[130]

One study showed that most investigators met their targets in chat rooms, through Internet Relay Chat (56 percent), or through instant messages (31 percent). Nearly half of all investigations began in sex-oriented chat rooms. Investigators communicated with suspected online child sex offenders mainly through chat rooms (55 percent), instant messaging (79 percent), and email (82 percent).[131] Experience shows that chats designed to gain the trust of sex offenders may lead offenders to disclose enough information to reveal their true identities.[132]

From time to time, the FBI or Interpol establishes a trap site, website, or bulletin board that either presents genuine child porn material or allows contributors to supply information about authentic sites and URLs.[133] In 2005, in the United States alone, active investigations represented a significant proportion (an estimated 25 percent) of all arrests for Internet crimes against minors.[134]

When one is fighting illegal activity, sage advice is to follow the trail of the money to deprive the operation of its oxygen – that is, the necessary funding. In 2006, NCMEC brought together the major credit card companies, including MasterCard, Visa, American Express, and Discover, along with banks, electronic payment networks, and Internet service companies, to cooperate with its activities. These companies do not wish to make money by helping child pornography. The companies joined NCMEC to build the Financial Coalition against Child Pornography, which bridges the gap between law enforcement and commercial companies to decrease monetary transactions online. The companies also can block such transactions and help track sellers and buyers. The major credit card issuers share information and take collective action against child pornography merchants. This effort has produced positive results in disrupting online child sex offenders' activities and in

---

[129] R. Stephanie Good, *Exposed: The Harrowing Story of a Mother's Undercover Work with the FBI to Save Children from Internet Sex Predators* (Nashville, TN: Thomas Nelson, 2007).

[130] Interview with a senior FBI official, Washington, DC, March 26, 2008.

[131] Kimberly J. Mitchell, Janis Wolak, and David Finkelhor, "Police Posing as Juveniles Online to Catch Sex Offenders: Is It Working?," *Sexual Abuse* 17, no. 3 (2005): 241–67, 252.

[132] James F. McLaughlin, "Characteristics of a Fictitious Child Victim: Turning a Sex Offender's Dreams into His Worst Nightmare," *International Journal of Communications Law and Policy* 9 (2004): 1–27, http://www.ijclp.net/files/ijclp_web-doc_6-cy-2004.pdf .

[133] Jenkins, *Beyond Tolerance*, 159.

[134] Mitchell, Wolak, and Finkelhor, "Police Posing as Juveniles," 241.

forcing offenders to seek less traditional payment mechanisms.[135] A growing number of child pornography sites accept bitcoin, offering buyers and sellers a way to cloak their identity.[136]

Since 1998, the Department of Justice has funded the CyberTipline.[137] As explained in chapter 7, the CyberTipline is operated by NCMEC to act as a national clearinghouse for reports of Internet-related child pornography and other Internet-related sex crimes committed against children.[138] The majority of concerns are with child pornography, child prostitution, and child sex tourism.[139] A 24-hour, toll-free telephone line, 1-800-THE-LOST (1-800-843-5678), is available in Canada, Mexico, and the United States for those who have information about missing and exploited children. I suggest establishing similar hotlines for alerts about violent messages that contain threats of murder. Readers of the Internet who might not wish to alert the police directly should be able to contact a similar cyber tipline and draw attention to violent threats and signals.

Finally, governments could require that all browser software and all mail and newsgroup reader software contain POWDER-sensitive technology. The Protocol for Web Description Resources (POWDER) provides a mechanism to describe and discover Web resources and helps Netusers to decide whether a given resource is of interest.[140] Once POWDER is mandated, its automated process can aid the discovery and regular review of the site's content and defend against the distribution of obscene material to minors. Each publisher of Internet material would be liable for the purposeful, knowing, or reckless distribution of such material. This approach would put the onus on the sender or publisher. It eliminates the massive, difficult task of the government or an independent board examining and providing ratings for an untold number of websites. It also helps regulate email and newsgroups.[141]

---

[135] Ernie Allen, president and CEO of NCMEC, testimony before the United States House of Representatives Committee on Energy and Commerce, Washington, DC, September 21, 2006, http://www.johnnygosch.com/images/TestimonyErnieAllenHouseEC-OISubcmte9-21-06.pdf ; discussion with senior NCMEC officials, Alexandria, VA, April 2, 2008.

[136] Kristen Schweizer, "Bitcoin Payments by Pedophiles Frustrate Child Porn Fight," *Bloomberg News*, October 10, 2014, http://www.bloomberg.com/news/2014-10-09/bitcoin-payments-by-pedophiles-frustrate-child-porn-fight.html .

[137] See the website for the CyberTipline at http://www.missingkids.com/CybertipLine .

[138] Janis Wolak, David Finkelhor, and Kimberly J. Mitchell, *Child-Pornography Possessors Arrested in Internet-Related Crimes: Findings from the National Juvenile Online Victimization Study* (NCMEC, 2005), ix, http://www.missingkids.com/en_US/publications/NC144.pdf .

[139] See also Stop It Now! at http://www.stopitnow.org/ .

[140] W3C Working Group, "Protocol for Web Description Resources (POWDER): Primer," W3C Working Group Note 1, http://www.w3.org/TR/powder-primer/ .

[141] Kevin W. Saunders, *Saving Our Children from the First Amendment* (New York: New York University Press), 176.

CYBERBULLYING

In the United States, 49 states have passed laws criminalizing bullying, 20 states have passed laws criminalizing cyberbullying, and 48 states have passed laws criminalizing electronic harassment.[142] New Hampshire and Pennsylvania require schools to implement anticyberbullying measures, including special training of teachers and education programs for pupils.[143] New Jersey instituted an antibullying bill of rights.[144] In the United Kingdom, several statutes may relate to cyberbullying, such as the Obscene Publications Act 1959 and the Computer Misuse Act 1990. In addition, the Criminal Justice and Public Order Act 1994 defines intentional harassment as a criminal offense, and police use the 1997 Protection from Harassment Act to prosecute those who post threatening statuses on Facebook and send offensive emails.[145] The Malicious Communications Act 1988 makes it an offense to send offensive or threatening electronic communications,[146] and the Communications Act 2003 includes the use of smartphones and Internet communications in the remit of the harassment laws.[147]

Schools have a crucial role to play. Amicable and supporting environments in schools have positive effects on students and reduce the likelihood of online

---

[142] Sameer Hinduja and Justin W. Patchin, "State Cyberbullying Laws," *Cyberbullying Research Center*, September 2014, http://www.cyberbullying.us/Bullying_and_Cyberbullying_Laws.pdf . See also National Conference of State Legislatures (NCSL), "Cyberbullying and the States," NCSL, July 9, 2010, http://www.ncsl.org/default.aspx?tabid=20753 .

[143] Independent Democratic Conference, "Cyberbullying: A Report on Bullying in a Digital Age," New York State Senate, Albany, September 2011, http://www.nysenate.gov/files/pdfs/final%20 cyberbullying_report_september_2011.pdf .

[144] "Anti-Bullying Bill of Rights (A-3466/S-2392): Provisions Impacting Students and Schools," Foundation for Educational Administration, Monroe, NJ, http://www.featrain ing.org/documents/bullying/Anti_Bullying_Bill_of_Rights_Bill_Provisions.pdf . See also Jeanette Rundquist and Ben Horowitz, "N.J. Schools Institute New Anti-Bullying Bill of Rights," *Star-Ledger*, September 14, 2011, http://www.nj.com/news/index.ssf/2011/09/ nj_schools_institute_new_anti-.html .

[145] In August 2009, Keeley Houghton became the first person to be convicted under the Protection from Harassment Act 1997. She bullied her victim for years and then posted the following status on Facebook: "Keeley is going to murder the bitch. She is an actress. What a f***ing liberty. Emily F***head Moore." Houghton pleaded guilty to this offense and was sentenced to three months detention in a young offenders institution. See Helen Carter, "Teenage Girl Is First to Be Jailed for Bullying on Facebook," *Guardian*, August 21, 2009, http://www.theguardian.com/ uk/2009/aug/21/facebook-bullying-sentence-teenage-girl ; Natasha Rigler, "Keeley Houghton: 'I Was Jailed for Cyber Bullying,'" *Reveal.com*, September 12, 2013, http://www.reveal.co.uk/real-life-stories/news/a514907/keeley-houghton-i-was-jailed-for-cyber-bullying.html .

[146] See the text of the Malicious Communications Act 1988 at http://www.legislation.gov.uk/ukpga/ 1988/27/section/1 .

[147] See the text of the Communications Act 2003 at http://www.legislation.gov.uk/ukpga/2003/21/ contents . See also "Cyberbullying: Bullying via the Internet or Mobile Phone," Kent Police, 2013, http://www.kent.police.uk/advice/personal/internet/cyberbullying.html .

and offline bullying. Research shows that the more youth are connected to their schools with a trusting, fair, pleasant, and positive climate, the lower their self-reported involvement in all forms of bullying – physical, verbal, and Internet – will be.[148] Schools need to explain what behavior is acceptable online if they are to promote an environment that does not tolerate any form of bullying (offline or online) and implement effective antibullying programs.[149] To varying degrees, cyberbullying supplements bullying that occurs in schools. Victims who are bullied at school are followed home through their computers and cell phones. Schools have the responsibility to fight bullying in all its manifestations first and foremost by (a) discussing the problem and bringing it into the open, (b) explaining the effects of bullying on the victims, and (c) working with students and parents to raise awareness and curb the problem. If educators know of a child who has been bullied and are made aware of threats that child has received, they should warn the child's parents. Schools should include cyberbullying as part of their bullying prevention strategies and include extensive discussions that address the importance of reporting bullying, of promoting responsible bystander behavior, and of confronting bullying by speaking about it, sharing the burden, and unifying efforts to overcome bullying.[150] The problem, however, is that often teachers do not know how to address the challenge. Their knowledge of technology often lags that of their students. They often do not know how to effectively handle new technologies and may not understand just how intrusive and menacing social networking sites and cell phones can be. Moreover, the core issue in addressing the challenge of cyberbullying is policy, and policy is directed by the school leadership. The policy should be (a) clear and unequivocal zero tolerance of bullying and cyberbullying both on and off campus, (b) tight collaboration with parents and the student council, (c) increased parent and student awareness of the potential or real problem, and (d) clear communication of measures that will be taken to stamp out any form of bullying.

[148] Kirk R. Williams and Nancy G. Guerra, "Prevalence and Predictors of Internet Bullying," *Journal of Adolescent Health* 41 (2007): S14–21, S19.

[149] Dan Olweus, *Bullying at School: What We Know and What We Can Do* (Cambridge, MA: Blackwell, 1993); Corinne David-Ferdon and Marci Feldman Hertz, "Electronic Media, Violence, and Adolescents: An Emerging Public Health Problem," *Journal of Adolescent Health* 41 (2007): S1–S5; Jonathan E. King, Carolyn E. Walpole, and Kristi Lamon, "Surf and Turf Wars Online: Growing Implications of Internet Gang Violence," *Journal of Adolescent Health* 41 (2007): S66–68; National Assessment Center, "At Risk Online: National Assessment of Youth on the Internet and the Effectiveness of i-SAFE Internet Safety Education, School Year 2005–06," i-SAFE, Carlsbad, CA, 2006.

[150] Patricia W. Agatston, Robin Kowalski, and Susan Limber, "Students' Perspectives on Cyber Bullying," *Journal of Adolescent Health* 41 (2007): S59–60.

The United Kingdom's Education Act 2011[151] grants teachers powers to tackle cyberbullying by providing them the ability to search for and, if necessary, delete inappropriate images (or files) on electronic devices, including mobile phones.[152] The Education and Inspections Act 2006[153] outlines some legal powers that relate to cyberbullying. Head teachers have the power "to such extent as is reasonable" to regulate pupils' conduct when they are off site or not under the control of a staff member.[154] This authority is of particular significance to cyberbullying, which is likely to take place outside school and can have a strong effect on school life. Section 3.4 of the "School Discipline and Pupil-Behaviour Policies" guidance provides advice on when schools might regulate off-site behavior.[155] The guidance emphasizes the unacceptability of pupils using mobile phones and other technological equipment to humiliate or bully others (e.g., sending abusive text messages, cyberbullying, and recording and transmitting abuse images such as "happy slapping").[156] In October 2005, David Morley, a 38-year-old bartender, was beaten to death by four young people. One of his killers approached him, saying, "We're filming a documentary about 'happy slapping.'" Morley was then punched and kicked to death. The attack had been filmed on a mobile phone.[157] The UK Department for Children, Schools, and Families is working with some of the most popular Internet sites among young people to help prevent and tackle cyberbullying.

In June 2012, Justice Secretary Ken Clarke announced the government's intention to strip away the cloak of anonymity that shields bullies, thereby ending the injustice of victims who are subjected to online abuse with little chance of finding out who is responsible. According to his proposal, ISPs and WHSs (Web-hosting sites) will have a defense against libel as long

---

[151] See the text of the Education Act 2011 at http://www.legislation.gov.uk/ukpga/2011/21/contents/ enacted .

[152] UK Department for Education, "Preventing and Tackling Bullying: Advice for Headteachers, Staff, and Governing Bodies," London, March 2014, https://www.gov.uk/government/uploads/ system/uploads/attachment_data/file/288444/preventing_and_tackling_bullying_march14. pdf .

[153] See the text of the Education and Inspections Act 2006 at http://www.legislation.gov.uk/ukpga/ 2006/40/contents .

[154] Education and Inspections Act 2006, section 89(5), http://www.education.gov.uk/publications/ eOrderingDownload/EducandInspectionsAct.pdf .

[155] UK Department for Children, Schools, and Families, "School Discipline and Pupil-Behaviour Policies: Guidance for Schools," DCSF Publications, Nottingham, UK, 2009, http://dera.ioe. ac.uk/11394/1/DCSF-00050-2010.pdf .

[156] Ibid.

[157] Chris Summers, "Violent Path of 'Happy Slapping,'" BBC News, January 23, 2006, http://news. bbc.co.uk/1/hi/uk/4478318.stm .

as they identify the authors of offensive material when asked to do so by the complainant.[158]

SUICIDE

There are documented cases of cybersuicides – both attempted and successful – influenced by the Internet.[159] In Britain alone, there have been at least 17 deaths since 2001 involving chat rooms or sites that give advice on suicide methods.[160] The Internet facilitates group suicides, providing a forum for like-minded people to meet to arrange their collective death.[161] Such behavior that encourages suicide constitutes akrasia – behavior that is stripped of any moral and social responsibility. It cannot be justified or legitimized.

Australia outlawed using the Internet to counsel or incite others to commit suicide or to promote and provide instruction on ways to do it. People who use the Internet to incite others to commit suicide or teach them how to kill themselves face fines of up to $A 550,000 under tough laws passed in 2005. Justice Minister Chris Ellison explained, "These offences are designed to protect the young and the vulnerable, those at greatest risk of suicide, from people who use the Internet with destructive intent to counsel or incite others to kill themselves."[162]

In September 2008, the British government considered rewriting the law on "suicide websites" to ensure that people know they are illegal. The government was concerned that people searching for information on suicide are more likely to find sites encouraging the act than sites offering support. Maria Eagle, who was then the justice minister, said, "Protecting vulnerable and

---

[158] James Slack, "The Unmasking of Internet Trolls: New Laws Will Make Websites Responsible for Vile Messages Unless They Reveal Identities of Bullies," *Mail Online*, June 12, 2012, http://www.dailymail.co.uk/news/article-2157937/New-laws-make-Facebook-Twitter-responsible-internet-trolls-unless-them.html . For discussion on the psychology of trolls, see Erin E. Buckels, Paul D. Trapnell, and Delroy L. Paulhus, "Trolls Just Want to Have Fun," *Personality and Individual Differences* 67 (September 2014): 97–102.

[159] Lucy Biddle, Jenny Donovan, Keith Hawton, Navneet Kapur, and David Gunnell, "Suicide and the Internet," *British Medical Journal* 336, no. 7648 (2008): 800–802; Ben Cubby, "Lost in a Tragic Web: Internet Death Pacts Increasing Worldwide," *Sydney Morning Herald*, April 24, 2007; S. Beatson, G. S. Hosty, and S. Smith, "Suicide and the Internet," *Psychiatric Bulletin* 24, no. 10 (2000): 434; Susan Thompson, "The Internet and Its Potential Influence on Suicide," *Psychiatric Bulletin* 23 (1999): 449–51.

[160] Mike Harvey, "Horror as Teenager Commits Suicide Live Online," *Times Online* (London), November 22, 2008, http://www.thetimes.co.uk/tto/news/world/americas/article1998615.ece .

[161] "Man, 34, Seeks Someone to Die With" [in Hebrew], *Walla!*, December 13, 2004, http://news.walla.co.il/?w=/402/639665 .

[162] "Australia Outlaws Using Internet to Incite Suicide," *New Zealand Herald*, June 25, 2005, http://www.nzherald.co.nz/world/news/article.cfm?c_id=2&objectid=10332573 .

young people must be a priority and a responsibility for us all. Suicide is a tragic phenomenon, especially for the families and friends left behind."[163] A year later, the director of public prosecutions issued guidance stating that a factor in making prosecution more likely in an assisted suicide case would be that "specific information" had been provided.[164] This phrase means that ISPs should not provide counseling on how to commit suicide, and they should not incite others to commit suicide. Note that the Suicide Act 1961 makes it illegal to promote suicide.[165] No website operator has been prosecuted to date.

TERROR AND CRIME

One of the prime responsibilities a state has to its citizens is to provide them with the ability to lead their lives in peace, free of existential threats and violence. Caution and reasonableness are prerequisites to address violent, antisocial challenges sensibly, without evoking moral panics. Evoking fear only plays into the hands of terrorists. Following the events of September 11, 2001, the United States has dedicated substantial resources to combat local and global terrorism. In an interview, Philip Mudd, associate executive assistant director of the FBI's National Security Branch, explained the mode of operation to counter terror on the Net. He said that the general approach is to look at people. The FBI receives information about individuals who might be a risk to security and then follows their actions, including on the Internet. The FBI examines the sites those individuals surf. According to Mudd, typically the FBI does not monitor the Net: "We go after people who we suspect, not websites. If the bad people go to sites, we follow them."[166] Mudd maintained that Internet providers should have integrity teams that instruct providers to take off dangerous content. He wanted to put the onus of responsibility on them. The FBI official further stressed the need to better understand the Internet's social networking sites, notably MySpace and Facebook. What are the implications of these tools for terrorism? Considerable effort is invested in tracking down terror-related Internet forums. Once the forums are tracked, the FBI is

---

[163] Ministry of Justice, "Suicide and the Internet: Updating the Law," news release, September 19, 2008; "Crackdown on 'Suicide Websites,'" BBC News, September 17, 2008. http://news.bbc.co.uk/2/hi/uk_news/7620676.stm .

[164] Kate Devlin, "Suicide Advice Websites Face Prosecution," *Telegraph*, September 24, 2009, http://www.telegraph.co.uk/health/healthnews/6223569/Suicide-advice-websites-face-prosecution.html .

[165] Suicide Act 1961, http://www.legislation.gov.uk/ukpga/Eliz2/9-10/60 . For further discussion, see parliamentary debate of February 6, 2013, *Daily Hansard*, http://www.publications.parliament.uk/pa/cm201213/cmhansrd/cm130206/debtext/130206-0003.htm .

[166] Interview with Philip Mudd, Washington, DC, March 25, 2008.

reluctant to shut them down. It prefers to keep an open eye on them. FBI officials are proud of their ability to provide security to American citizens after the shocking tragedy of September 11, 2001.

The United States, the United Kingdom, Canada, Australia, and New Zealand are using the Echelon program to intercept Internet transmissions.[167] The UK Terrorism Acts of 2000[168] and 2006[169] contain provisions criminalizing the encouragement of terrorism and the dissemination of terrorist publications. The acts prohibit (a) sharing of information that could be useful to terrorists; (b) videos of violence that contain praise for the attackers; (c) chat forums with postings inciting people to commit terrorist acts; and (d) instructions on to how to make weapons, poisons, and bombs. The acts also include notice and take-down provisions if the encouragement or dissemination takes place over the Internet.[170] Indeed, responsible governments should take down sensitive data about potential targets – for example, nuclear power plants, gas and oil stores, water pumps and supply, toxic-release inventories, lists of key factories, emergency services, medical computer systems, transportation systems, powerful generators, wastewater treatment plants, disaster control services, dams, bridges, and detailed maps of the nation's infrastructure. The UK acts provide a method for webmasters to be made aware of terrorist content, thus ensuring that they could not claim not to have known about the content if they were subsequently prosecuted.[171]

In March 2009, the British Home Office said it needed to monitor social networking sites such as Facebook to tackle terrorists and crime gangs who might use those sites. A spokesman said:

> The government has no interest in the content of people's social network sites, and this is not going to be part of our upcoming consultation. . . . We have been clear that the communications revolution has been rapid in this country and the way in which we collect communications data needs to change, so that law enforcement agencies can maintain their ability to tackle terrorism and gather evidence.[172]

---

[167] Tom Burghardt, "ECHELON Today: The Evolution of an NSA Black Program," Global Research, November 14, 2013, http://www.globalresearch.ca/echelon-today-the-evolution-of-an-nsa-black-program/5342646 .

[168] See the text of the Terrorism Act 2000 at http://www.legislation.gov.uk/ukpga/2000/11/contents .

[169] See the text of the Terrorism Act 2006 at http://www.legislation.gov.uk/ukpga/2006/11/pdfs/ukpga_20060011_en.pdf?timeline=true .

[170] "Reporting Extremism and Terrorism Online," Directgov, http://webarchive.nationalarchives.gov.uk/20120709035026/direct.gov.uk/en/crimejusticeandthelaw/counterterrorism/dg_183993 .

[171] Akdeniz, *Internet Child Pornography*, 239.

[172] "Social Network Sites 'Monitored,'" BBC News, March 25, 2009, http://news.bbc.co.uk/1/hi/uk_politics/7962631.stm .

The government emphasized that there was no intention to keep the content of conversations. This measure was not designed to invade privacy but rather to protect citizens from violence.

In 2010, the British government established the Counterterrorism Internet Referral Unit, which acts on tips from the public, the police, and the intelligence services. It has take-down power to instruct ISPs to modify or remove any unlawful content. During its first year of operation, the unit shut down 156 websites.[173]

Of Canada's 61,000 police officers, only a few hundred are dedicated to catching criminals online. The Royal Canadian Mounted Police (RCMP) National Security Criminal Investigations established a community outreach program to make possible regular exchange of information between the public and police. Under the program, law enforcement officers monitor child pornography and terrorist websites. However, they do not monitor sites that could potentially promote violence, and they are falling behind in flagging cyber clues.[174]

A shared Western problem is the lack of sufficient knowledge of the jihadist culture and history. Governments should invest in studying the radical narratives, in promoting awareness, and in educating young people. The online engagement should include the presentation of opposing viewpoints, thereby contributing to online debates and encouraging civic challenges to extremist narratives.[175] Often, however, Western analysts who fail to do their homework tend to perceive other cultures in their own image, thus failing to comprehend that radical Islam is fundamentally different from liberalism. The values of the two cultures are strikingly in opposition; the meaning of life is different, and concepts of good and evil are totally dissimilar. Whereas the concepts of not harming others and respecting others underpin liberal democracies, radical Islamists see no other way than to harm infidels and to show utmost disrespect to them by, for instance, decapitating them and exposing their enemy's last moments on the Internet. The concept of time is also manifestly different: Whereas Western people are able to think of themselves and the future of their children and grandchildren, radical Muslims think in terms of hundreds of

[173] Peter Neumann, "Countering Online Radicalization in America," Homeland Security Project, National Security Program, Bipartisan Policy Center, Washington, DC, December 2012, 25. For further discussion, see P. W. Singer and Allan Friedman, *Cybersecurity and Cyberwar: What Everyone Needs to Know* (New York: Oxford University Press, 2014).

[174] I thank Kevin Woodley for this information. See also Lloyd Robertson, "Web Links to Shooting," CTV, September 14, 2006.

[175] Neumann, "Countering Online Radicalization," 7, 35. For further discussion, see Singer and Friedman, *Cybersecurity and Cyberwar.*

years.[176] Thus, they are willing to concede significant sacrifices for winning the long-term battle, and they are unwilling to make compromises because they think that victory will eventually be theirs. The different concept of time strengthens the radicals' unyielding worldview, which is manifested by a resolute ability to make sacrifices and move on. The fight against terrorism also necessitates mastering foreign languages: Persian, Arabic, Urdu, Pashtu, Somali, Turkish, Dari, and Hindi, among others.

MONITORING

Government agencies may gain valuable information about terrorist and criminal activities by monitoring designated sites that are dangerous and antisocial. Furthermore, I call for monitoring websites that are likely to be used for creating social support groups for potential criminals. The idea is not to implement surveillance of the entire Internet, something that I oppose on principled, free speech grounds and that would be very costly and probably impractical, but to monitor the areas of the Internet that are potentially harmful in an effort to detect and forestall crimes. It is argued that monitoring certain sites where criminals voice their violent goals could potentially prevent unfortunate shooting sprees analyzed in chapter 5. The Internet is here to stay, but we must devise ways to deal with its less positive aspects.

In the United States, the intelligence agencies are not allowed to monitor US sites without predication. The FBI follows the US attorney general's regulations on how and when sites can be monitored.[177] Philip Mudd emphasized, "We monitor people suspected of criminal activity. We watch what they are doing, including their use of the Internet. We need predication to follow people. We follow information on links, but we do not surf the Internet."[178] I think there is a need to monitor sites where people may concoct, coordinate, or provide warnings of their aims to kill and maim others. We need to better understand the Internet's social networking sites, such as VampireFreaks, YouTube, MySpace, and Facebook. What are the implications of these tools on potential criminal activity, especially when this activity involves violence? There is a need to develop an elaborate psychological research area addressing the risky behaviors that people – especially youth – can display on social

---

[176] Joshua Alexander Geltzer, *U.S. Counter-terrorism Strategy and al-Qaeda* (New York: Routledge, 2009), 117; Benjamin Orbach, "Usama Bin Ladin and Al-Qaida: Origins and Doctrines," *Middle East Review of International Affairs* 5, no. 4 (2001): 54–68, 64. See generally Gerhard Böwering, "The Concept of Time in Islam," *Proceedings of the American Philosophical Society* 141, no. 1 (1997): 55–66.

[177] Interview with Philip Mudd, Washington, DC, March 25, 2008.

[178] Ibid.

networking sites.[179] Broadening research in this area can assist law enforcement agencies, as well as psychologists, psychiatrists, and educators, in identifying risky situations and preventing hideous crimes.

As mentioned in chapter 5, Liam Youens explicitly wrote on two separate websites that he intended to kill Amy Boyer. He even described how he would do it: "When she gets in I'll drive up to the car blocking her in, window to window I'll shoot her with my glock."[180] This sequence was exactly how he carried out the murder. The two websites he created were taken down after the crime. The companies that closed them down said that they had policies against such sites but no resources to monitor content.[181]

My intention is to draw the attention of governments, law enforcement agencies, and civil society groups to the urgent need to develop monitoring schemes for potentially problematic websites as a means of preventing homicide. The expectation for international cooperation by all segments of society is not based on any existing legal obligations but rather on the moral obligations that cross borders and cultures regarding the sanctity of life and the urgency to save lives and prevent crimes. The Internet business sector (ISPs, WHSs, and websites administrators and owners) bears an even heavier responsibility because the moral obligations imposed on it may become a legal obligation, as in the case of child pornography and cybercrime.[182] By "potentially problematic websites," I refer to websites that attract criminals who post their criminal ideas and intentions. Law enforcement agencies should devote more resources to the understanding of social networking on the Internet in an attempt to create flexible schemes for identifying those problematic websites and analyzing how criminals are using them.

Given the killing episodes described in chapter 5, studying social networking is a prudent and active step to forestall murder. If the police had monitored

---

[179] Kimberly J. Mitchell and Michele Ybarra, "Social Networking Sites: Finding a Balance between Their Risks and Benefits," *Archives of Pediatrics and Adolescent Medicine* 163, no. 1 (2009): 87–89; Megan A. Moreno, Malcolm R. Parks, Frederick J. Zimmerman, Tara A. Brito, and Dimitri A. Christakis, "Display of Health Risk Behaviors on MySpace by Adolescents," *Archives of Pediatrics and Adolescent Medicine* 163, no. 1 (2009): 27–34; Monica Barbovschi, "Meet the 'E-Strangers': Predictors of Teenagers' Online-Offline Encounters," *Cyberpsychology: Journal of Psychosocial Research on Cyberspace* 3, no. 1 (2009), article 4, http://www.cyberpsychology.eu/view.php?cisloclanku=2009061603&article=4 .

[180] See the reproduction of Youens's website at http://www.netcrimes.net/Amy%20Lynn%20Boyer_files/liamsite.htm .

[181] Alberto Moya, "An Online Tragedy," *48 Hours*, CBS News, March 23, 2000, http://www.cbsnews.com/stories/2000/03/23/48hours/main175556.shtml .

[182] See Council of Europe, Convention on Cybercrime, http://www.coe.int/t/DGHL/cooperation/economiccrime/cybercrime/default_en.asp . See also U.S Department of Justice, Computer Crimes and Intellectual Property Section: http://www.justice.gov/criminal/cybercrime/ .

such sites regularly, bloodshed could have been prevented. After the Saari shooting incident, the Internet monitoring unit of Finland's National Bureau of Investigation announced that it intended to upgrade its Internet monitoring for about a month, because after such a tragic event the cyber world usually sees a surge in activity. The Finnish Ministry of Interior also announced that it would upgrade its monitoring of online content.[183]

In the name of free speech, people object to any monitoring. As much as monitoring problematic websites can interfere with and violate the privacy of Netusers, the need to protect the lives of innocent people cannot be dismissed or ignored. At present, unless a person makes a specific threat, aimed at a specific person, there are no grounds to infringe on his or her freedom. Even explicit Net threats before an attack do not necessarily prompt action from police because officers do not patrol Web pages.

Alan Lipman, founder and executive director of the Center for the Study of Violence in Washington, D.C., said that often the people who committed school killings first told other people about their violent intent: "These individuals do not merely post onto the Internet. . . . They indicate their disturbance in clearly identifiable behaviors, time and again. What we need to do is to teach those who are in regular contact with adolescents and young adults how to identify these behaviors and how to act."[184] Writing statements that indicate rage, despair, and violent intent are a "leading indicator and precursor" of murderous episodes.[185]

On December 14, 2012, 20-year-old Adam Lanza killed 26 people, 20 of them children, inside Sandy Hook Elementary School in Newtown, Connecticut. This killing spree shocked the nation and may bring about a welcome policy change. In nearby New York City, top intelligence officials in the police department met to examine ways to search the Internet to identify potential "deranged" gunmen before they strike. They explored cybersearches of language that mass casualty shooters have used in emails and Internet postings in the past, aiming to identify a potential shooter in cyberspace; to engage that person in dialogue; and to intervene, "possibly using an undercover to get close, and take him into custody or otherwise disrupt his plans."[186]

---

[183] "NBI to Monitor Internet More Closely," *Helsinki Times*, September 25, 2008; "NBI Struggles to Monitor Internet for Potentially Dangerous Content," *Helsinki Times*, September 26, 2008.

[184] Perry Swanson and Kim Nguyen, "Web Rants Raise Red Flags for Violence: But Police Can Do Little to Prevent Attacks," *Gazette* (Colorado Springs), December 16, 2007.

[185] Ibid.

[186] Michael Wilson, "Police Dept. to Use Internet to Try to Stop Mass Shootings," *New York Times*, December 20, 2012, http://www.nytimes.com/2012/12/21/nyregion/police-dept-to-use-internet-to-try-to-stop-shootings.html .

Another move in the right direction is the report of the Bipartisan Policy Center's Homeland Security Project: it advises taking full advantage of violent extremists' and terrorists' presence in cyberspace and making utmost use of the information they are sharing with others. The report rightly notes that this information can be used "to gain strategic intelligence on terrorist groups' intentions and networks, on tactical intelligence on terrorist operations and the people who are involved in them, and on evidence that can be used in prosecutions."[187]

CONCLUSION

Aristotle's Golden Mean is a good guide for liberal democracies: For every polarity there is a mean that when practiced is a good benchmark for a life of moderation. The more we see the Golden Mean in each polarity, the better we find the true benchmarks of a life of wellness.[188] People have the freedom to express themselves, within reason. The two underpinning principles of liberal democracy are showing respect for others and not harming others. The first is derived from the Kantian deontological approach that perceives people as ends rather than means: "Act so that you use humanity, as much as in your own person as in the person of every other, always at the same time as an end and never merely as a means."[189] The second is derived from John Stuart Mill's liberal philosophy.[190] We should strive to uphold these principles on the Internet.

Furthermore, I reemphasize that one of the dangers in any political system is that the principles that underlie and characterize it may, through their application, bring about its destruction. Democracy, in its liberal form, is no exception. Liberal democracies need to find suitable answers to the challenges they are facing if they are to ensure that showing respect for others and not

---

[187] Neumann, "Countering Online Radicalization," 9. See also National Security Program, Homeland Security Project, "Jihadist Terrorism and Other Unconventional Threats," Bipartisan Policy Center, Washington, DC, September 2014, http://bipartisanpolicy.org/sites/default/files/BPC%20HSP%202014%20Jihadist%20Terrorism%20and%20Other%20Unconventional%20Threats%20September%202014.pdf .

[188] Aristotle, *Nicomachean Ethics*, ed. and trans. Martin Ostwald (Indianapolis, IN: Bobbs-Merrill, 1962). See also A. W. H. Adkins, "The Connection between Aristotle's *Ethics* and *Politics*," *Political Theory* 12, no. 1 (1984): 29–49; Richard Kraut, *Aristotle Political Philosophy* (New York: Oxford University Press, 2002).

[189] Immanuel Kant, *Groundwork for the Metaphysics of Morals*, trans. Allen W. Wood (New Haven, CT: Yale University Press, 2002), xviii.

[190] John S. Mill, *Utilitarianism, Liberty, and Representative Government* (London: J. M. Dent, 1948). For further discussion, see Raphael Cohen-Almagor, "Between Autonomy and State Regulation: J. S. Mill's Elastic Paternalism," *Philosophy* 87, no. 4 (October 2012): 557–82.

harming others will be sustained in the long run. Countries have moral and social responsibilities to their citizens.

To address the threat of hateful messages, countries may need to take legal action, as France did, inspired by a strong tradition of sovereign state regulation and with confidence that their values are of universal validity. In accordance with the promotional approach advocated in this book, regulation is warranted to protect vulnerable minorities. Liberal democracies regulate the economic market to positively affect the livelihood of certain segments of society, and they should also do the same in the marketplace of ideas. Countries can and should assert their jurisdictions by demanding that servers stop posting illegal, harmful content that is accessible in their jurisdictions. Moreover, to address cross-boundary, international rings of criminals and terrorists, international cooperation is needed. This subject is the concern of the next chapter.

9

International Responsibility

To see what is right and not to do it is want of courage.

–Confucius

On August 8, 2001, US Attorney General John Ashcroft and Chief Postal Inspector Kenneth Weaver announced Operation Avalanche. By this time, the two-year undercover operation had already made more than 100 arrests on charges related to child pornography, including the arrests of the two main criminals, Thomas and Janice Reedy.

The investigation centered on Landslide Productions, an Internet business engaged in advertising and conspiring to distribute child pornography. Landslide offered subscriptions to more than 250 websites, many of them devoted to child pornography. Landslide had 300,000 subscribers and earned as much as $1.4 million a month. Between 1996 and 1999, Landslide took in nearly $10 million, 85 percent of which came from child porn.[1] It was the largest commercial child pornography enterprise ever encountered up to that point. Thomas Reedy was sentenced to 1,335 years – 15 years for each of 89 charges – to run consecutively. His wife, Janice, was sentenced to 14 years in prison. More than 250 Americans were arrested, but the subscriber list contained names of many thousands of people from 60 countries, including 35,000 Americans.

The investigation required international cooperation to crack other rings. Police in Switzerland launched an investigation of 1,300 people in Operation Genesis. The German police launched Operation Pecunia to investigate more than 1,400 suspects. British authorities launched Operation Ore to investigate more than 7,200 names, Ireland conducted more than 100 raids

---

[1] Pip Clothier, "The World's Biggest Convicted Child Pornographer," *Independent* (London), May 13, 2003, http://www.crime-research.org/news/2003/05/Mess1306.html .

as part of Operation Amethyst, and Canada launched investigation in Operation Snowball.[2] This case is an example of how one child pornography investigation into the activities of individuals involved in a commercial website operation can lead to the apprehension of hundreds of offenders.

Illegal activity on the Internet is sophisticated. Criminals and terrorists strive to keep their identity, their modes of operation, and their vile plans secret. They operate in the dark Web – the alternate, covert side of the Internet that is crafted deliberately to lie beyond the reach of search engines. Criminals and terrorists use the technological tools described in chapter 2 to secure their privacy and anonymity and are quick to adapt innovations and exploit technological advantages as means to their ends. Clandestine modes of operation generate the necessary funding to keep them going. Terrorists, criminals, and online child sex offenders work in international cells and rings that contest geographic boundaries and that require substantial resources and close cooperation of law enforcement agencies to obstruct their activities.

In 1996, the International Law Commission adopted the Draft Code of Crimes against Peace and Security of Mankind and submitted it to the United Nations (UN) General Assembly. Acts described in the draft are crimes under international law and are punishable as such, whether or not they are punishable under national law. Article 2 declares, "A crime against the peace and security of mankind entails individual responsibility." Article 6 further postulates that responsibility for a crime against the peace and security of mankind also lies with the superiors of the individual who committed the crime if they knew or had reason to know, "in the circumstances at the time, that the subordinate was committing or was going to commit such a crime and if they did not take all necessary measures within their power to prevent or repress the crime."[3]

A significant step toward international cooperation was made in August 2000, when the Hoover Institution, the Consortium for Research on Information Security and Policy, the Center for International Security and Cooperation, and Stanford University drafted "A Proposal for an International Convention on Cyber Crime and Terrorism."[4] Cybercrime involves criminal activity in which computers or computer networks serve as the principal

2   "The U.S. Postal Inspection Service Teams with Internet Crimes against Children Task Forces in Operation Avalanche," US Postal Service, 2001, http://www.popcenter.org/problems/child_pornography/PDFs/USPIS_nd.pdf ; "Sting Operations," Adult Legal Services.com, http://adultlegalservices.com/sting.opperationa.asp .

3   The full text of the 1996 draft is available at http://legal.un.org/ilc/texts/instruments/english/draft%20articles/7_4_1996.pdf .

4   Abraham D. Sofaer, Seymour E. Goodman, Mariano-Florentino Cuéllar, Ekaterina A. Drozdova, David D. Elliott, Gregory D. Grove, Stephen J. Lukasik, Tonya L. Putnam,

means of committing criminal activity. Examples of criminal uses of Internet communications include cybertheft, phishing, online fraud, online money laundering, cyberextortions, cyberstalking, and cybertrespassing. The proposal recognizes that cybercrime is transnational and requires a transnational response. Cybercriminals exploit legal weaknesses and law enforcement deficiencies, thereby exposing states to dangers that are beyond their capacity to respond to unilaterally or bilaterally. The speed and technical complexity of cyberactivities require prearranged procedures for cooperation in investigating and responding to threats and attacks. The need was recognized for a multilateral convention that would espouse laws criminalizing dangerous cyberactivities, for enforcement of those laws or extradition of criminals for prosecution by other states, for cooperation in investigating criminal activities and in providing usable evidence for prosecutions, and for participation in formulating and implementing standards and practices that enhance safety and security.

On January 26, 2001, the European Commission issued a communication on "Creating a Safer Information Society by Improving the Security of Information Infrastructures and Combating Computer-Related Crime."[5] Following this communication, the commission established a European Union (EU) forum for deliberation in which law enforcement agencies, Internet service providers (ISPs), telecommunication operators, civil liberties organizations, consumer representatives, data protection authorities, and other interested parties were brought together to enhance cooperation at the EU level.

CONVENTION ON CYBERCRIME

In November 2001, the Council of Europe, consisting of 43 member states, approved the Convention on Cybercrime.[6] Its aim is to increase the efficiency of the fight against cybercrime by using the following means: (a) harmonizing national laws by creating a standardization between each state's domestic

and George D. Wilson, "A Proposal for an International Convention on Cyber Crime and Terrorism," Hoover Institution, the Consortium for Research on Information Security and Policy, the Center for International Security and Cooperation, and Stanford University, August 2000, http://www.iwar.org.uk/law/resources/cybercrime/stanford/cisac-draft.htm .

[5] Commission of the European Communities, "Communication from the Commission to the Council, the European Parliament, the Economic and Social Committee and the Committee of the Regions: Creating a Safer Information Society by Improving the Security of Information Infrastructures and Combating Computer-Related Crime," Brussels, January 26, 2001, http://eur-lex.europa.eu/legal-content/EN/TXT/?uri= CELEX:52000DC0890 .

[6] Council of Europe, Convention on Cybercrime, Budapest, November 23, 2001, http://conventions.coe.int/Treaty/en/Treaties/Html/185.htm .

criminal laws regarding cybercrime offenses, (b) improving investigative techniques by providing for domestic criminal procedural law essential powers for
the investigation and prosecution of such offenses as well as other offenses
committed by a computer system, and (c) creating an effective infrastructure
that will allow international cooperation. Forty-two countries had signed and
ratified the treaty as of September 22, 2014, including the United States. An
additional 11 states have only signed the convention, without ratifying it.[7] The
main offenses addressed by the convention are offenses against the confidentiality, integrity, and availability of computer data and systems; computer-
related offenses; content-related offenses, such as dissemination of child
pornography materials; and offenses related to infringements of copyright
and related rights.

Because nations are sensitive to their sovereignty, the convention does not
adequately address the problem of cross-border unauthorized disruption
crimes, and it fails to authorize remote cross-border searches, even in case of
emergency or hot pursuit. Instead, it requires a nation pursuing a cybercrim-
inal to consult with local officials before seizing, storing, and freezing data on
computers located in such countries. This step might give cybercriminals
precious time to cover their tracks.[8]

The convention provides for rapid enforcement assistance by, for example,
requiring the nation where a crime had originated to preserve and disclose
stored computer data at the request of the nation where the crime caused
damage. What is still needed is to harmonize each nation's cybercrime laws to
better facilitate extradition and information sharing and to enact appropriate
legislation that would allow expedited searches, seizures, and preservations of
information in the country.[9]

In 2002, the Council of Europe's decision-making body (the Committee of
Ministers) updated the convention by passing a provision that bans "any

[7] Council of Europe, Convention on Cybercrime, CETS No. 185, Status, http://conventions.
     coe.int/Treaty/Commun/ChercheSig.asp?NT=185&CM=8&DF=&CL=ENG .
[8] Jack L. Goldsmith, "The Internet and the Legitimacy of Remote Cross-Border Searches,"
     Public Law and Legal Theory Working Paper 16, Law School of the University of Chicago,
     2001, http://chicagounbound.uchicago.edu/cgi/viewcontent.cgi?article=1316&context=
     public_law_and_legal_theory . A remote cross-border search takes place when persons in
     one nation use computer networks to explore data on computers in another nation.
     Goldsmith argues that such searches are consistent with international law principles of
     enforcement jurisdiction.
[9] Jack Goldsmith and Tim Wu, Who Controls the Internet? Illusions of a Borderless World (New
     York: Oxford University Press, 2006), 166. For a useful overview of the EU policies regarding
     cybercrime, see Alain Megias, "European Union Policies Regarding Cybercrime," Information
     Policy (blog), April 25, 2011, http://www.i-policy.org/2011/04/european-union-policies-regarding-
     cybercrime.html .

written material, any image, or any other representation of ideas or theories, which advocates, promotes, or incites hatred, discrimination, or violence, against any individual or group of individuals, based on race, colour, descent, or national or ethnic origin, as well as religion if used as pretext for any of these factors."[10]

In January 2003, the Council of Europe issued an additional protocol to the Convention on Cybercrime.[11] This protocol deals with the criminalization of racist or xenophobic acts committed through computer networks. It aims to extend the scope of the Convention on Cybercrime and enhance cooperative efforts.

The first additional protocol stresses that "all human beings are born free and equal in dignity and rights."[12] *Racist and xenophobic materials* are defined in article 2 of the protocol as "any written material, any image or any other representation of ideas or theories, which advocates, promotes, or incites hatred, discrimination, or violence, against any individual or group of individuals, based on race, colour, descent, or national or ethnic origin, as well as religion if used as a pretext for any of these factors."[13] This definition refers to written material (e.g., texts, books, magazines, statements, and messages); images (e.g., pictures, photos, drawings); or any other representation of thoughts or theories of a racist and xenophobic nature in a format that can be stored, processed, and transmitted by a computer system.[14]

According to the protocol (articles 4 and 5), posting a message on a webpage that threatens a serious criminal offense against a person or group of people on the basis of race, color, or religion or making racist and xenophobic insults is considered to be an offense.[15] The protocol requires participating states to criminalize the dissemination of racist and xenophobic material through computer systems, racist and xenophobic-motivated threats and insults, and denial of the Holocaust and other genocides.[16] It criminalizes Internet hate speech, including hyperlinks to pages that contain offensive content.

[10] Julia Scheeres, "Europeans Outlaw Net Hate Speech," *Wired News*, November 9, 2002.

[11] Additional Protocol to the Convention on Cybercrime Concerning the Criminalisation of Acts of a Racist and Xenophobic Nature Committed through Computer Systems, Strasbourg, France, January 28, 2003, http://conventions.coe.int/Treaty/en/Treaties/Html/189.htm .

[12] Ibid.

[13] Ibid.

[14] Additional Protocol to the Convention on Cybercrime Concerning the Criminalisation of Acts of a Racist and Xenophobic Nature Committed through Computer Systems (ETS no. 189), *Explanatory Report*, November 7, 2002, section 12, http://www.inach.net/content/cctreatyad dexuk.html .

[15] Additional Protocol to the Convention on Cybercrime, January 28, 2003.

[16] Ibid.

The additional protocol is not part of the Convention on Cybercrime. The implementation and enforcement of this protocol depend on resources assigned to this cause by the respective governments. Furthermore, not all the countries that have signed the convention have also signed the additional protocol. The United States, for instance, has informed the Council of Europe that it will not become a party to the protocol because the protocol is inconsistent with American constitutional freedoms.[17] The dissemination of offensive material via the Internet is generally protected as free speech under the US Constitution.

The Council of Europe had operated on a few parallel paths to help states worldwide implement the Convention on Cybercrime and its additional protocol on xenophobia and racism. Having initiated the Project on Cybercrime, the council conducts an annual conference on cybercrime (Octopus Interface Conference) and runs ongoing activities in this matter.[18]

Because the Convention on Cybercrime is the only binding international instrument that deals with cybercrime, it received widespread international support. A growing need existed to create a platform for international consultation as provided in article 46 of the convention. To carry out these consultations, the council established the Cybercrime Convention Committee (T-CY), which first gathered in 2006.[19] The committee sends requests for information regarding the present legal situation in specific states, surveys proposals for international projects, and drafts questionnaires and reports. It also publishes recommendations and guidelines on various topics such as the following: a recommendation of the Committee of Ministers to member states on measures to promote the public service value of the Internet;[20] guidelines for the cooperation between law enforcement and ISPs against cybercrime;[21] and a recommendation of the Committee of

[17] A country that signed and ratified the main convention but not the protocol would not be bound by the terms of the protocol, and its authorities would not be required to assist other countries in investigating activity prohibited by the protocol.

[18] Council of Europe, "Action against Economic Crime," http://www.coe.int/t/DGHL/coopera tion/economiccrime/cybercrime/default_en.asp . For further discussion, see Indira Carr, ed., *Computer Crime* (Surrey, UK: Ashgate, 2009).

[19] Directorate General of Human Rights and Legal Affairs, "Information Document Concerning the T-CY," Secretariat memorandum, T-CY (2009) INF, Strasbourg, France, March 6, 2009, http://www.coe.int/t/dghl/cooperation/economiccrime/cybercrime/T-CY/Information_doc_ T-CY__2009_%2002%20-%20INF.pdf .

[20] Recommendation CM/Rec (2007)16 of the Committee of Ministers to Member States on Measures to Promote the Public Service Value of the Internet, adopted by the Committee of Ministers on November 7, 2007, http://www.coe.int/t/dghl/cooperation/economiccrime/cyber crime/T-CY/T-CY_2008_CMrec0711_en.PDF .

[21] Project on Cybercrime, "Guidelines for the Cooperation between Law Enforcement and Internet Service Providers against Cybercrime," adopted by the global Conference on Cooperation against Cybercrime, Council of Europe, Strasbourg, France, April 1–2, 2008,

Ministers to member states on measures to promote the respect for freedom of expression and information with regard to Internet filters.[22] The committee also promotes discussion of the legal issues surrounding the obligations of ISPs with respect to child pornography.[23]

In May 2007, the International Telecommunication Union launched the Global Cybercrime Agenda as a framework to coordinate the international response to violations of cybersecurity. The framework was designed to raise awareness and build partnerships with governments, nongovernmental organizations, industry, media, and other interested parties.[24] The 2009 conference, "Cooperation against Cybercrime,"[25] in Strasbourg, France, provided an opportunity for 300 experts from the public and private sectors as well as international and nongovernmental organizations in more than 70 countries to discuss the progress that had been made on the legislative level around the world as well as practical cooperation between law enforcement bodies and ISPs.[26] One of the issues discussed was the launching of the second phase of the Global Project on Cybercrime.[27] The project aims to help countries worldwide to implement the convention.

The goal of the first phase of the project (September 2006 to March 2009) was to establish the convention as the primary reference standard for cybercrime legislation globally. During that phase, a wide range of international and regional organizations adopted the convention. Among other things, the project assisted in creating global cooperation at all levels and provided legislative counseling and helped shape cybercrime legislation in many

http://www.coe.int/t/DGHL/cooperation/economiccrime/cybercrime/cy_activity_Interface2008/567_prov-d-guidelines_provisional2_3April2008_en.pdf .

[22] Recommendation CM/Rec (2008)6 of the Committee of Ministers to Member States on Measures to Promote the Respect for Freedom of Expression and Information with Regard to Internet Filters, adopted by the Committee of Ministers, March 26, 2008, http://www.coe.int/t/dghl/cooperation/economiccrime/cybercrime/T-CY/T-CY%20-%20Rec%20(2008)6_EN.pdf .

[23] Marco Gercke, "Obligations of Internet Service Providers with Regard to Child Pornography: Legal Issues," discussion paper (draft) for the Project on Cybercrime, Council of Europe, Strasbourg, France, March 4, 2009, http://www.coe.int/t/dghl/cooperation/economiccrime/cybercrime/T-CY/2079_out7_isp_liab_report1a%20_4%20March%202009.pdf .

[24] International Telecommunication Union, "Global Cybersecurity Agenda (GCA)," http://www.itu.int/osg/csd/cybersecurity/gca/global_strategic_report/chapt_5_iframe.htm .

[25] Octopus Interface 2009, "Cooperation against Cybercrime," Council of Europe, Strasbourg, France, March 10–11, 2009, http://www.coe.int/t/dghl/cooperation/economiccrime/cybercrime/cy%20activity%20interface%202009/Interface2009_en.asp .

[26] For a conference summary, see Octopus Interface Conference on Cooperation against Cybercrime, Council of Europe, Strasbourg, France, March 10–11, 2009, http://www.coe.int/t/dghl/cooperation/economiccrime/cybercrime/cy%20activity%20interface%202009/2079%20info9_SUMMARY1.pdf .

[27] Ibid.

countries in Asia, Africa, Latin America, and the Caribbean. In addition, modules for training judges were prepared.[28]

A few major issues were identified:

- A need for cooperation between the public and private sectors, especially between law enforcement agencies and ISPs[29]
- A need to protect personal data and privacy through security enhancement of cyberspace
- A need to create better measures for protecting children.[30]

During the second phase (March 2009 to December 2011), a broader implementation of the convention was promoted according to the needs identified during the first phase and in continuance of the operative steps mentioned. The funding came from the Council of Europe, with generous contributions by Microsoft, McAfee, and the Romanian government.[31]

The aim of the Council of Europe's actions against cybercrime through the T-CY, the annual conferences, and the Cybercrime Project is to issue practical recommendations that will assist states, law enforcement bodies, and business and private sectors in implementing the convention. At this point, the Convention on Cybercrime serves mainly as a legislative guideline or a model law for most countries. But through the Council of Europe initiatives, global cooperation is gathering momentum at all levels. This cooperation will advance as awareness of cybercrime concerns grow.

The United Nations has also mobilized to counter crime and terror on the Net. The UN Convention against Transnational Organized Crime addresses the most common forms of computer crime.[32] At the UN World Summit on the Information Society, held in Tunis in November 2005, the Internet Governance Forum was established. It is composed of

---

[28] Project on Cybercrime, "Global Project on Cybercrime (Phase 2)," February 20, 2009, http://www.coe.int/t/DGHL/cooperation/economiccrime/cybercrime/cy%20Project/2079%20adm%20pro%20summary1a%20_20%20Feb%202009.pdf .

[29] Guidelines were developed to help structure cooperative relations. See Project on Cybercrime, "Guidelines for the Cooperation between Law Enforcement and Internet Service Providers against Cybercrime."

[30] Project on Cybercrime, "Global Project on Cybercrime (Phase 2)."

[31] Council of Europe, Action against Economic Crime, Global Project on Cybercrime (Phase 2) website, http://www.coe.int/t/dghl/cooperation/economiccrime/cybercrime/cy%20Project%20global%20phase%202/projectcyber_en.asp .

[32] United Nations Office on Drugs and Crime, *United Nations Convention against Transnational Organized Crime and the Protocols Thereto* (New York: United Nations, 2004), http://www.unodc.org/documents/treaties/UNTOC/Publications/TOC%20Convention/TOCebook-e.pdf .

government representatives, public figures, business people, and industrialists.[33] The forum is a substantive body, a "form of international cooperation which is both inclusive and egalitarian ... with the opportunity to work together towards a sustainable, robust, secure, and stable Internet."[34] The forum cautioned against imposing unnecessary restrictions on Internet content, given the benefits of increased information flow. Simultaneously, it stressed that legitimate public policy objectives exist, "such as protecting the general public, and particularly children, from objectionable Internet content and prohibiting the use of the Internet for criminal activity."[35] In 2006, the UN General Assembly unanimously adopted the UN Global Terrorism Strategy (resolution 60/288), which resolved to work together to combat terrorism.

In November 2010, the European Union–United States Working Group on Cyber-Security and Cyber-Crime was established to tackle new threats to the global communication networks. The parties agreed to strengthen trans-Atlantic cooperation in cybersecurity by taking various measures, including expanding incident management response capabilities jointly and globally. They also agreed to engage the private sector; share good practices on collaboration with industry; and pursue specific engagement on key issue areas such as fighting botnets, securing industrial control systems (such as water treatment and power generation), and enhancing the resilience and stability of the Internet. Moreover, they agreed to continue to cooperate in efforts to remove child pornography from the Internet, including through work with domain-name registrars and registries.[36]

The same year, 2010, the European Commission approved and provided funding for a collaborative project involving academia, industry, and law enforcement, aimed at creating a network of Cybercrime Centres of Excellence for Training, Research, and Education in Europe. These centers develop training programs for use in the fight against cybercrime.[37]

---

[33] "Internet Governance," discussion with Professor William Dutton, Oxford Internet Institute, Oxford, UK, March 22, 2006; Internet Governance Forum website, http://www.intgovforum.org .

[34] Internet Governance Forum Secretariat, "The Internet Governance Forum (IGF) Second Meeting: Synthesis Paper," Rio de Janeiro, November 12–15, 2007, 1, http://www.intgovforum. org/Rio_Meeting/IGF.SynthesisPaper.24.09.2007.rtf .

[35] Ibid., 8–9.

[36] European Commission, "Cyber Security: EU and US Strengthen Transatlantic Cooperation in Face of Mounting Global Cyber-Security and Cyber-Crime Threats," press release, April 14, 2011, http://europa.eu/rapid/press-release_MEMO-11-246_en.htm .

[37] United Nations Office on Drugs and Crime, *The Use of the Internet for Terrorist Purposes* (New York: United Nations, 2012), 131. For further discussion, see Raphael F. Perl, "Terrorist Use of the Internet: Threat, Issues, and Options for International Co-operation," remarks before the Second International Forum on Information Security, Garmisch-Partenkirchen, Germany,

HOTLINES

Another important initiative is the voluntary establishment of Internet hotlines by ISPs from different countries. As explained in chapters 5 and 8, hotlines enable Netusers who may wish to remain unidentified to pinpoint disturbing content. Hotlines have to be transparent. Netcitizens should be aware – at the point of entry – of the persons or organizations responsible for running the hotline system and those persons and organizations on whose behalf hotlines are operated. Transparency also means that an explanation is provided as to which concerns will be processed, under what criteria, and by which public authorities. The reporting system should be explained in sufficient detail, Netcitizens should have the ability to track their concerns throughout the process, and they should be informed of the final outcome of the process.[38] To this end, organizations running hotline systems need to publish reports about their work.

Hotlines are associated in a global organization called the International Association of Internet Hotlines (INHOPE)[39] and enjoy the support of law enforcement agencies, local governments, and child welfare organizations. Because sites can be accessed from anywhere in the world, illegal content may be reported in a country other than the one in which the site is hosted. Once the source is traced, hotlines pass reports to the relevant host country. INHOPE has set processes for exchanging reports to ensure that a rapid response is made.

Hotlines deal mainly with illegal activity in chat rooms and hate speech. More specifically, they deal with child pornography and online grooming (mostly by pedophiles preying on children).[40] Through hotlines, Netcitizens can report something they suspect to be illegal on the Internet. The hotline investigates these reports to determine if those acts are illegal and, if so, traces the origin of the content. If the content is illegal, the hotline refers this information to local law enforcement agencies as well as to the ISP for removal.

In the United States, the National Center for Missing and Exploited Children's toll-free hotline, the CyberTipline, has received more than 2.5 million reports of suspected child sexual exploitation between 1998 and June

April 7–10, 2008; Wolfgang Benedek and Matthias C. Kettemann, *Freedom of Expression and the Internet* (Strasbourg, France: Council of Europe Publishing, 2014), 163–67.

[38] Jens Waltermann and Marcel Machill, eds., *Protecting Our Children on the Internet: Towards a New Culture of Responsibility* (Gütersloh, Germany: Bertelsmann Foundation, 2000), 48.

[39] See the INHOPE website at https://www.inhope.org/ .

[40] INHOPE, "INHOPE launches its 2013 statistics and infographics," press release, April 16, 2014, http://inhope.org/tns/news-and-events/news/14-04-16/INHOPE_launches_its_2013_stat istics_and_infographics.aspx .

2014.[41] In 2007 alone, the center received nearly 100,000 reports – more than 75 percent for online child pornography.[42] In the last quarter of 2010, the hotline handled an average of 262 service-related calls per day.[43]

In Australia, people can complain to three main charities: Crime Stoppers Australia,[44] the Alannah and Madeline Foundation,[45] and Kids Helpline.[46]

TERROR

Terrorism has global networks; therefore, tackling it requires international cooperation. Terrorists are being trained to be technologically savvy and to attack, disrupt, damage, and perhaps even destroy technology infrastructures and computer-based economic activities. Many modern terrorist groups share the pattern of the loosely knit network: decentralization, segmentation, and delegation of authority. These features make computer-mediated communication an ideal tool of coordination, information exchange, training, and recruitment.[47] The Internet has grown to be a key element in modern terrorist organizations' training, planning, and logistics.

In many respects, al-Qaeda has become a Web-directed guerrilla network.[48] Jihadi websites and chat rooms allow isolated young Muslims to engage with a worldwide network of like-minded people virtually united against their common enemies.[49] Younes Tsouli, a London terrorist known as "Terrorist 007," sent his postings, which included manuals on making suicide bombs and improvised explosive devices, over password-protected Internet forums called al-Ekhlaas and al-Ansar.[50] Many terrorists begin their journey to violent jihad

[41] See the website for the CyberTipline at http://www.missingkids.com/CyberTipline .

[42] Interview with senior officials, National Center for Missing and Exploited Children, Alexandria, VA, April 2, 2008; Jerry Markon, "Crackdown on Child Pornography," *Washington Post*, December 15, 2007, A1, A12.

[43] See the website for the CyberTipline at http://www.missingkids.com/CyberTipline .

[44] See the organization's website at http://www.crimestoppers.com.au/ .

[45] See the organization's website at http://www.amf.org.au/ .

[46] See the organization's website at http://www.kidshelp.com.au/ .

[47] Marc Sageman, *Leaderless Jihad: Terror Networks in the Twenty-First Century* (Philadelphia: University of Pennsylvania Press, 2008), 121; Gabriel Weimann, *Terror on the Internet: The New Arena, the New Challenges* (Washington, DC: US Institute of Peace Press, 2006), 116.

[48] Abdel Bari Atwan, *The Secret History of al Qaeda* (Berkeley: University of California Press, 2006), 122.

[49] Ibid., 144. Among the most well-known jihadi forums are al-Faloja, Al-Luyuth al-Islamiyyah, Alqimmah-Golaha Ansaarta Mujaahidiinta, al-Hanein, al-Jihad al-Alami, al-Leyoth, al-Ma'ark, al-Medad, al-Shamukh, al S'nam, Amanh, as-Ansar, at-Tahaddi, Jahad, Jahafal, Qimmah, Tawhed [Wal Jihad], and the Majahden Electronic Network.

[50] "Extremists Still Using Bomb Plotter's Website," *Mail on Sunday*, January 27, 2008, 18; Gordon Corera, "Al-Qaeda's 007," *Times* (London), January 16, 2008, 4. The *Al Qaeda Manual* is available at http://www.fas.org/irp/world/para/manualpart1_1.pdf .

on the Internet. Tens of thousands of Netusers frequent the password-protected jihadist message boards.[51] Radical forums create virtual communities, connect people, provide rich information on timely topics, reinforce beliefs, and normalize violent behavior. Diffusion in terrorist locations is made possible by Internet communications. Thousands of websites and bulletin boards offer videos, images, statements, and speeches that demonstrate the Internet's centrality to global terrorism.[52]

Security expert Rita Katz argued that only a handful of primary-source jihadist websites and password-protected online forums facilitate communication between al-Qaeda leaders and their followers.[53] The number of sites is not that significant. What is significant is the ability to communicate without being intercepted. Jihadi texts and videos are available for people who seek such guidance. Extreme religious ideologies are spread through websites and videotapes accessible throughout the world.[54] Great reverence is paid to the views of the militant leadership. Fatwas of religious sages legitimize and endorse violence. Anti-Western videos showing non-Muslims humiliating Muslims populate the Internet in an effort to win the hearts and minds of potential followers. Some of these followers travel to Iraq, Afghanistan, Pakistan, Somalia, Chechnya, Lebanon, Yemen, and other places to engage in fighting what they perceive as forces of evil. Police say that the Internet has taken on huge importance for militant groups, enabling them to share know-how (e.g., bomb making, suicide bombing, guerrilla operations), spread propaganda to a mass audience, plan operations, and recruit members.[55]

[51] Brian Michael Jenkins, RAND Corporation, "No Path to Glory: Deterring Homegrown Terrorism," statement before the House Committee on Homeland Security, Subcommittee on Intelligence, Information Sharing and Terrorism Risk Assessment, Washington, DC, May 26, 2010, 4. See, generally, Jarret M. Brachman, *Global Jihadism: Theory and Practice* (London: Routledge, 2009).

[52] James J. F. Forest, "Introduction," in *Teaching Terror: Strategic and Tactical Learning in the Terrorist World*, ed. James J. F. Forest (Lanham, MD: Rowman & Littlefield, 2006), 1–29, 9; "German Expert Sees 9,000 'Uncontrollable' Jihadist Websites," *BBC Monitoring International Reports*, September 13, 2007.

[53] Statement of Rita Katz, director of SITE Institute, in "Using the Web as a Weapon: The Internet as a Tool for Violent Radicalization and Homegrown Terrorism," Hearing before the Subcommittee on Intelligence, Information Sharing, and Terrorism Risk Assessment of the Committee on Homeland Security, House of Representatives, 110th Cong., 1st sess., November 6, 2007 (Washington, DC: US Government Printing Office, 2000), 14–34.

[54] Philip Bobbitt, *Terror and Consent: The Wars of the Twenty-First Century* (New York: Knopf, 2008), 57; Susan B. Glasser and Steve Coll, "The Web as Weapon," *Washington Post*, August 9, 2005, A1.

[55] Ingrid Melander, "EU States Share Monitoring of Militant Web Sites," Reuters, May 30, 2007. See also Jeffrey Stinson, "EU Discusses Anti-terror Efforts, Balance of Liberty and Security," *USA Today*, August 17, 2006, 7A.

Funding is essential for terrorist operations; thus, stifling sources of support is essential in the battle against terror. Terrorist organizations raise funds via the Internet by making email appeals or through their websites; by selling goods through their websites; through associated side businesses; through fraud, gambling, or online brokering;[56] and through online organizations that resemble humanitarian charity groups.

Many terrorist organizations have set up charities in the real world as well as the cyberworld. Multilateral bodies such as the FATF (Financial Action Task Force),[57] which was established to combat money laundering and terrorist financing, are instrumental in sharing information about the global charitable sector, improving oversight of national and international charities, devising methodologies for detecting terrorists masquerading as charities, and establishing international standards to combat such abuse.

On September 8, 2006, the UN adopted a Global Counter-terrorism Strategy. The strategy, in the form of a resolution and an annexed plan of action (A/RES/60/288), is a global instrument designed to enhance national, regional, and international efforts to counter terrorism. For the first time, all member states agreed to a common strategic approach to combating terrorism, thereby sending a clear message that terrorism is unacceptable in any form and resolving to take practical steps to fight it. Those practical steps range from strengthening state capacity to better coordinating UN counterterrorism activities.[58]

The European Convention on the Suppression of Terrorism was adopted in 1977.[59] Article 1 details the offenses covered by the convention, including attacks against life, kidnapping, and the use of firearms. The treaty was later supplemented by the Council of Europe Convention on the Prevention of Terrorism (2005). Article 5 of this convention explains that public provocation to commit a terrorist offense means "the distribution, or otherwise making available, of a message to the public, with the intent to incite the commission of a terrorist offence, where such conduct, whether or not directly advocating

---

[56] Madeleine Gruen, "White Ethnonationalist and Political Islamist Methods of Fund-Raising and Propaganda on the Internet," in *The Changing Face of Terrorism*, ed. Rohan Gunaratna (Singapore: Marshall Cavendish, 2004), 127–45, 139. See also Amos N. Guiora, *Homeland Security* (Boca Raton, FL: CRC Press, 2011), 111–35.

[57] See the organization's website at http://www.fatf-gafi.org/pages/aboutus/ .

[58] United Nations, "United Nations Action to Counter Terrorism," http://www.un.org/en/terrorism/strategy-counter-terrorism.shtml . For further discussion, see Paul J. Rabbat, "The Role of the United Nations in the Prevention and Repression of International Terrorism," in *A War on Terror? The European Stance on a New Threat, Changing Laws, and Human Rights Implications*, ed. Marianne Wade and Almir Mljevic (New York: Springer, 2011), 81–106.

[59] European Convention on the Suppression of Terrorism, Council of Europe, Strasbourg, France, January 27, 1977, http://conventions.coe.int/Treaty/en/Treaties/html/090.htm .

terrorist offences, causes a danger that one or more such offences may be committed."[60] Article 6 clarifies that recruitment for terrorism means "to solicit another person to commit or participate in the commission of a terrorist offence, or to join an association or group, for the purpose of contributing to the commission of one or more terrorist offences by the association or the group."[61] In turn, article 7 concerns training for terrorism, which means providing instruction in "the making or use of explosives, firearms, or other weapons or noxious or hazardous substances, or in other specific methods or techniques, for the purpose of carrying out or contributing to the commission of a terrorist offence, knowing that the skills provided are intended to be used for this purpose."[62]

In 2004, the Organization for Security and Co-operation in Europe (OSCE) decided that participating states "will exchange information on the use of the Internet for terrorist purposes and identify possible strategies to combat this threat, while ensuring respect for international human rights obligations and standards, including those concerning the rights to privacy and freedom of opinion and expression."[63] A year later, in 2005, the EU Council Framework Decision on attacks against information systems was enacted, requiring member states to ensure that illegally accessing information systems (article 2), illegally interfering with systems (article 3), and illegally interfering with data (article 4) are punishable as criminal offenses.[64]

In 2006, the European Union adopted a directive on the retention of data.[65] Article 5 specifies the categories of data to be retained: data necessary to trace and identify the source, destination, date, time, duration, and type of a communication; data necessary to identify Netusers' communication equipment or

[60] Council of Europe Convention on the Prevention of Terrorism, Warsaw, May 16, 2005, http://conventions.coe.int/Treaty/en/Treaties/Html/196.htm .

[61] Ibid.

[62] Ibid. For further discussion, see Thomas Wahl, "The European Union as an Actor in the Fight against Terrorism," *A War on Terror? The European Stance on a New Threat, Changing Laws, and Human Rights Implications*, ed. Marianne Wade and Almir Mljevic (New York: Springer, 2011), 107–70.

[63] OSCE, Ministerial Council Decision 3/04 on Combating the Use of the Internet for Terrorist Purposes, Sofia, Bulgaria, December 7, 2004, http://www.osce.org/mc/42647 .

[64] EU Council Framework Decision 2005/222/JHA of February 24, 2005, on attacks against information systems, http://eur-lex.europa.eu/LexUriServ/LexUriServ.do?uri=CELEX:32005 F0222:EN:NOT .

[65] "Directive 2006/24/EC of the European Parliament and the of the Council of 16 March 2006 on the retention of data generated or processed in connection with the provision of publicly available electronic communications services or of public communications networks and amending Directive 2002/58/EC," *Official Journal of the European Union* (April 13, 2006): L 105/54–63, http://eur-lex.europa.eu/LexUriServ/LexUriServ.do?uri=OJ:L:2006:105:0054:0063: EN:PDF .

what purports to be their equipment; and data required to identify the location of mobile communication equipment. Article 6 of the directive requires member states to retain these data for periods of not less than six months and not more than two years from the date of the communication.

In 2007, EU states began to share monitoring of militant websites, including sites linked to al-Qaeda. The challenge is enormous. Keeping tabs on the jihadi sites requires vigilance because statements and videos by individuals and groups appear for only a short period of time. Some member states under German lead responsibility are sharing the task of analyzing al-Qaeda's media department, as-Sahab ("the Clouds").[66] As-Sahab produced audiotapes and videotapes of Osama bin Laden, thereby aiding al-Qaeda in its international propaganda campaign. Another jihadi media organization, Al-Fajr Media Center, turned insurgency into a courageous journey and mayhem and violence into inspiring music videos.[67] In July 2007, the European Commission announced plans to frustrate terrorism by suppressing online guides on bomb making. European ISPs would face charges if they failed to block websites with bomb-making instructions.[68] The EU police agency, Europol, has built an information portal to allow exchange of information on monitoring of militant websites. The portal includes a list of links to monitored websites, statements by terrorists, and information designed to fight terrorism.[69] Whereas some countries (the United Kingdom, France, Italy, Spain) have a sense of urgency in stepping up combined efforts against the challenges because they fear terrorist operations on their soil, other countries that did not send troops to Afghanistan may think some of the proposed measures are excessive for their own needs.

In November 2007, the European Commission proposed that all 27 EU member states should make inciting terrorism over the Internet or using the Web for militant recruitment and training a criminal offense.[70] In 2008, Europe's

---

[66] "As-Sahab: Al Qaeda's Nebulous Media Branch," *Stratfor: Daily Terrorism Brief*, September 8, 2006; "As-Asahb, Al-Qaeda's Media Arm," Global Jihad, September 9, 2008, http://www.global jihad.net/view_page.asp?id=1132 ; "As-Ahab: Al-Qaeda Firing BM Rockets," LiveLeak, March 29, 2007, http://www.liveleak.com/view?i=dea_1175200207 ; "Assahab Media: Taliban Mujahideen Attack," Dailymotion, July 20, 2008, http://www.dailymotion.com/video/x66yjm_assahab-media taliban-mujahideen-att_news .

[67] "Al-Fajr Media Center: Video from Qaeda al-Jihad Islamic Maghreb – Part I," LiveLeak, April 7, 2008, http://www.liveleak.com/view?i=775_1207599624 ; "Al-Fajr Media Center: Video from Qaeda al-Jihad Islamic Maghreb – Part II," LiveLeak, April 7, 2008, http://www.live leak.com/view?i=2a9_1207598912 .

[68] Lewis Page, "EC Wants to Suppress Internet Bomb-Making Guides," *Register*, July 4, 2007.

[69] Ingrid Melander, "EU States Share Monitoring of Militant Web Sites," Reuters, May 30, 2007.

[70] Alexandra Zawadil, "Governments Struggle as Militants Refine Web Tactics," Reuters, November 16, 2007.

interior ministers announced a further crackdown on Internet terrorism, adding three new offenses – terrorist propaganda, recruitment, and training – to EU law. An amendment to the 2002 framework decision on combating terrorism was enacted. The amendment was intended to further harmonize the way terrorist offenses are tackled and punished in Europe and to deal with the Internet's "virtual training camps." Europe's amended framework decision makes it "easier for law enforcement authorities to get cooperation from Internet service providers," both in terms of identifying individuals and removing offending material.[71] It enables law enforcement authorities to investigate and track the diffusion of terrorist propaganda and know-how throughout the European Union.[72]

Terrorist organizations are the principal threat to the information infrastructure and the new digital economy.[73] To promote their ends, such organizations are resorting to clandestine methods that include passing encrypted messages, embedding codes using steganography (hiding large amounts of information within image and audio files), using the Internet to send violent threats, and hiring hackers to collect intelligence.

Views are divided regarding whether cyberterrorism – criminal acts perpetrated through computers that result in violence, death, or destruction and create terror for the purpose of coercing governments to change their policies – is a tangible threat. Whereas some argue that we need to see cyberterrorism as a concrete danger, others think that the threat is grossly exaggerated, creating hype and moral panics (see chapter 3). On the one hand, Dorothy Denning asserted that the possibility of cyberterrorism is a concern because al-Qaeda and other terrorist groups have become increasingly aware of the value of cyberspace to their objectives. They are adept at using the Internet to distribute propaganda and other information, collect data about potential targets and weapons, communicate with cohorts and supporters, recruit, raise money, and facilitate their operations. They have advocated conducting cyberattacks and have engaged in some hacking. New hacking groups have emerged with apparent ties to terrorists. Denning argued that the Internet has expanded damaging acts in support of terrorist objectives, regardless of whether those acts are characterized as cyberterrorism.[74]

[71] John Lettice, "Europe Moves against Internet's 'Virtual Training Camps,'" *Register*, April 21, 2008.

[72] "Council Framework Decision 2008/919/JHA of 28 November 2008, amending Framework Decision 2002/475/JHA on combating terrorism," *Official Journal of the European Union* (December 12, 2008).

[73] Athina Karatzogianni, *The Politics of Cyberconflict* (London: Routledge, 2006), 100.

[74] Dorothy E. Denning, "Terror's Web: How the Internet Is Transforming Terrorism," in *Handbook of Internet Crime*, ed. Yvonne Jewkes and Majid Yar, 194–213, 198. See also Dorothy E. Denning, "A View of Cyberterrorism Five Years Later," in *Internet Security:*

On the other hand, Joshua Green argued that following the terrorist attacks of September 11, 2001, the George W. Bush administration and security organizations repeatedly warned that terrorists might strike the nation's computer networks. However, those alarmists considerably overstated the risk of cyberterrorism. Private agendas as well as ignorance fueled the hysteria. The Bush administration accentuated the threat to generate more public anxiety about terrorism and thus garner more support for its budget on fighting terror. Technology companies, in turn, are eager to sell their security products and benefit from the fervor over cyberterrorism.[75]

The Bush administration might have exaggerated the cyberterror threat. It promoted the belief that a threat existed to basic order and underpinning values revered as sacred to our society. But the challenge of international webs of terror is real and requires international cooperation to counter its perils. In 2006, an extended cyberattack against the US Naval War College in Newport, Rhode Island, prompted officials to disconnect the entire campus from the Internet. A similar attack against the Pentagon in 2007 led officials to temporarily disconnect part of the unclassified network from the Internet. American defense officials acknowledged that the Global Information Grid, which is the main network for the US military, has to deal with more than 3 million daily intrusion attempts.[76] In December 2010, Internet Haganah, a global intelligence network dedicated to confronting global jihad online, confirmed that al-Qaeda has acquired the Wikileaked list of critical global infrastructure.[77] It would be unwarranted and irresponsible conduct on the part of the international security community to dismiss cyberterrorism as a futurist or insubstantial concern. The future, apparently, is already here.

Hacking, Counterhacking, and Society, ed. Kenneth E. Himma (Sudbury, MA: Jones & Bartlett, 2007), 123–40.

[75] Joshua Green, "The Threat of Cyberterrorism Is Greatly Exaggerated," in *Does the Internet Benefit Society?*, ed. Cindy Mur (Farmington Hills, MI: Greenhaven, 2005), 40–49, 42; Joshua Green, "Cyberterrorism Is Not a Major Threat," in *Does the Internet Increase the Risk of Crime?*, ed. Lisa Yount (Farmington Hills, MI: Greenhaven, 2006), 56–65. For further discussion on moral panics and terrorism, see David L. Altheide, *Terror Post 9/11 and the Media* (New York: Peter Lang, 2009), 99–116.

[76] Clay Wilson, "Botnets, Cybercrime, and Cyberterrorism: Vulnerabilities and Policy Issues for Congress," CRS Report for Congress, Congressional Research Service, Washington, DC, November 15, 2007. See also Philip B. Brunst, "Terrorism and the Internet: New Threats Posed by Cyberterrorism and Terrorist Use of the Internet," in *A War on Terror? The European Stance on a New Threat, Changing Laws, and Human Rights Implications*, ed. Marianne Wade and Almir Mljevic (New York: Springer, 2011), 81–106.

[77] Communications with Aaron Weisburd, director, Internet Haganah, December 8, 2010 and October 5, 2014.

CHILD PORNOGRAPHY

During the 1960s and 1970s, most Western nations did not regard sexual crimes against children as a high priority for law enforcement, partly because the general atmosphere of sexual liberalism promoted a much greater tolerance of most forms of sexual deviance. Denmark repealed its antipornography laws in 1969. Sweden followed suit in 1971. The result was a booming trade in child pornography, where such material appeared in the media and in ordinary shops.[78] At that time, expert opinion commonly held that sexual abuse or molestation was not a very widespread crime and that offenders were inadequate individuals in need of psychiatric help rather than violent predators.[79]

Since the 1980s, online child sex offenders and pedophiles have been connected to rings and networks, sexual violence, and murder, with the effect of becoming highly organized and violent figures.[80] In the early 1980s, child pornography magazines were still legally and publicly accessible in the Netherlands, posing severe difficulties for police in other European nations, who fought hard against importation. Matters changed with the rapid growth of awareness about the problem of child sexual abuse. From 1985 onward, a generalized American moral and social panic over sexual threats to children disseminated throughout Western Europe and Australia. A growing international consensus developed about the need to protect children. By 1986, virtually all the traditional avenues for obtaining child pornography in the United States had been firmly closed, raising the possibility of a thorough suppression of the whole child porn trade. Seeking out child porn almost automatically meant a confrontation with federal law enforcement.[81]

In 1989, the United Nations adopted the Convention on the Rights of the Child, which is a significant tool in the development of a coordinated approach to controlling and combating child pornography. Article 19.1 of the convention provides that

> States Parties shall take all appropriate legislative, administrative, social, and
> educational measures to protect the child from all forms of physical or mental
> violence, injury or abuse, neglect or negligent treatment, maltreatment or

---

[78] Ethel Quayle, "Child Pornography," in *Handbook of Internet Crime*, ed. Yvonne Jewkes and Majid Yar (Portland, OR: Willan), 343–68, 343.

[79] Philip Jenkins, *Beyond Tolerance: Child Pornography on the Internet* (New York: New York University Press, 2001), 188.

[80] Anneke Meyer, *The Child at Risk: Pedophiles, Media Responses, and Public Opinion* (Manchester, UK: Manchester University Press, 2007), 9.

[81] Jenkins, *Beyond Tolerance*, 41; Pamela D. Schultz, "Naming, Blaming, and Framing: Moral Panic over Child Molesters and Its Implications for Public Policy," in *Moral Panics over Contemporary Children and Youth*, ed. Charles Krinsky (Farnham, UK: Ashgate, 2008), 95–110.

exploitation, including sexual abuse, while in the care of parent(s), legal guardian(s), or any other person who has the care of the child.[82]

Under article 34, states parties undertake national, bilateral, and multilateral measures to prevent the following:

(a) The inducement or coercion of a child to engage in any unlawful sexual activity;

(b) The exploitative use of children in prostitution or other unlawful sexual practices;

(c) The exploitative use of children in pornographic performances and materials.[83]

Significant cooperative steps to fight global Internet child pornography were taken only during the first decade of the 21st century. In December 2003, the Virtual Global Taskforce (VGT) was established to fight child abuse online. The aims of the VGT are to build an effective, international partnership of law enforcement agencies that will help to protect children from online child abuse and to make the Internet a safer place; to identify, locate, and help children at risk; and to hold perpetrators appropriately to account.[84]

The VGT includes the Australian Federal Police, including the High Tech Crime Operations portfolio and Child Protection Operations; the Royal Canadian Mounted Police, National Child Exploitation Coordination Centre; the Dutch National Police; the Indonesian National Police; Interpol; Europol; the Italian Postal and Communication Police Service; the Korean National Police Agency's Cyber Bureau; the New Zealand Police; the Cybercrime Coordination Unit Switzerland; the Ministry of Interior for the United Arab Emirates; the UK National Crime Agency's Child Exploitation and Online Protection Command; and the US Department of Homeland Security, US Immigration and Customs Enforcement's Homeland Security Investigations and Operation Predator.[85] The VGT has established a comprehensive initiative that allows police from member countries to immediately review and respond to online reports of child sexual exploitation that have been submitted through the VGT tipline or website.

---

[82] Office of the High Commissioner for Human Rights, United Nations, Convention on the Rights of the Child, adopted and opened for signature November 20, 1989, entry into force September 2, 1990, http://www.ohchr.org/en/professionalinterest/pages/crc.aspx .

[83] Ibid.

[84] Virtual Global Taskforce, "Making the Internet Safer for Children," http://www.virtualglobal taskforce.com/what-we-do/ .

[85] Virtual Global Taskforce, "Who We Are," http://www.virtualglobaltaskforce.com/who-we-are/ member-countries/#australia .

In 2006, an international task force of 20 states, in all continents but Antarctica, was set up to combat pedophilia.[86] The task force is designed to exchange information between law enforcement agencies, conduct seminars, facilitate information, learn from one another's experiences, and cooperate to crack international rings of online child sex offenders who use the Internet to transfer and collect photos and videos of abused children.

In July 2007, the Council of Europe adopted the Convention on the Protection of Children against Sexual Exploitation and Sexual Abuse. Article 20 on child pornography accentuates article 9 of the Convention on Cybercrime.[87] It states that each party shall take the necessary legislative or other measures to ensure that the production, distribution, transmission, procurement, and possession of child pornography are criminalized.[88]

In turn, article 21 is about offenses concerning the participation of a child in pornographic performances. It provides:

1 Each Party shall take the necessary legislative or other measures to ensure that the following intentional conduct is criminalised:
   a recruiting a child into participating in pornographic performances or causing a child to participate in such performances;
   b coercing a child into participating in pornographic performances or profiting from or otherwise exploiting a child for such purposes;
   c knowingly attending pornographic performances involving the participation of children.[89]

Recently, the European Commission established the European Cybercrime Centre (EC3) at Europol as the focal point of the EU's fight against cybercrime. The EC3 is contributing to faster reactions to online crimes. It supports member states and the EU's institutions in building operational and analytical capacity for investigations and cooperation with international partners. Security officers gather information, exchange intelligence, and support multinational operations.[90]

---

[86] Interview with Shawn Henry, deputy assistant director, Cyber Division, Federal Bureau of Investigation, Washington, DC, March 26, 2008.

[87] Council of Europe, Convention on the Protection of Children against Sexual Abuse, CETS 201, Lanzarote, Spain, October 25, 2007, http://conventions.coe.int/Treaty/Commun/Que VoulezVous.asp?CL=ENG&NT=201 ; Council of Europe, Convention on Cybercrime, CETS 185, Budapest, Hungary, November 23, 2001, http://conventions.coe.int/Treaty/Com mun/QueVoulezVous.asp?NT=185&CL=ENG .

[88] Council of Europe, Convention on the Protection of Children against Sexual Exploitation and Sexual Abuse.

[89] Ibid.

[90] Europol, *The Internet Organised Crime Threat Assessment (iOCTA)* (Den Haag, Netherlands: European Police Office, 2014); Europol, https://www.europol.europa.eu/ec3 .

Protecting youth and fighting against illegal and harmful contents are vital. Proposals for solutions and initiatives are being discussed in the EU and other international bodies within their scope of responsibility (Organisation for European Co-operation and Development; Group of Eight; Council of Europe, United Nations Educational, Scientific, and Cultural Organization). Worth mentioning in this context are the guiding activities of the EU, in particular the Action Plan for a Safer Internet (1999–2004).[91] The plan served to fight illegal and harmful content, especially Internet content, throughout the European Community. The document addresses the EU member states as well as service providers and user organizations.

Online child sex offenders work in rings across many countries, and there have been dozens of international crackdowns of such rings. Here I mention only a few to illustrate the magnitude of the challenge.

Operation Starburst

Operation Starburst was the first major international operation against individuals using the Internet to trade child pornography. In July 1995, the British police conducted an international investigation of a ring that used the Internet to distribute graphic pictures of child pornography. The worldwide police operation resulted in the arrest of 15 offenders in Britain and a number of others in Canada, Germany, Hong Kong, Singapore, South Africa, and the United States.[92]

The Orchid Club

In 1996, members of the Orchid Club used the ring to share and exchange photos and videos of girls 5 to 10 years of age. The participants produced the majority of the material exchanged. While members of the ring were logged onto the videoconferencing system, a child was sexually abused, and the abuse was broadcast. At least 11 men in different countries watched the child being sexually abused in real time. Using videoconferencing software, the child abusers were able to communicate requests for different poses and abusive

---

[91] Europa, "Action Plan for a Safer Internet, 1999–2004." See also the EC3 webpage on Europol's website, http://europa.eu/legislation_summaries/information_society/internet/l24190_en.htm .

[92] Margaret A. Healy, "Child Pornography: An International Perspective," Computer Crime Research Center, August 2, 2004, http://www.crime-research.org/articles/536/6 ; Yaman Akdeniz, "Governance of Pornography and Child Pornography on the Global Internet: A Multi-layered Approach," in *Law and the Internet: Regulating Cyberspace*, ed. Lilian Edwards and Charlotte Waelde (Oxford: Hart, 1997), 223–41, http://www.cyber-rights.org/reports/governan.htm .

acts.[93] This was the first case of online broadcasting of live child abuse through a videoconferencing system.

The Wonderland Club

The Wonderland Club was a closed network of elite traffickers. Its investigation began in 1996 with a prosecution of 16 people in San Jose, California, who were charged with taking part in an online child porn network. Images were traded freely within the group, and some found their way into the wider child porn world.[94] One or more of the people charged cooperated with law enforcement, presumably in the hope of improving their own legal situation. That led to the identification of a British participant. US and British authorities together discovered the existence of the club and began an international investigation coordinated through Interpol.[95] The thorough investigation lasted two years. It revealed some 200 members in more than 40 countries, including Australia, Austria, Belgium, Finland, France, Germany, Italy, Norway, Portugal, Sweden, the United Kingdom, and the United States.[96] The investigation uncovered 750,000 photographs of children that were distributed by and to members throughout the world.[97] Allegedly, members of the Wonderland Club were required, as a condition of joining, to donate personal stockpiles of at least 10,000 child porn images, a figure quite in line with the collections commonly reported in arrests and seizures.[98]

Operation Web Sweep

The operation started in 2002 when US law enforcement agencies received a tip about a child pornography site. With the cooperation of the server operator, investigators said they determined that the site contained images of "clearly prepubescent" boys along with advertising describing the site's content and

---

[93] Marie Eneman, "The New Face of Child Pornography," in *Human Rights in the Digital Age*, ed. Mathias Klang and Andrew Murray (London: GlassHouse, 2005), 27–40, 31.

[94] Jenkins, *Beyond Tolerance*, 78.

[95] Ibid., 152; Rachel Downey, "Victims of Wonderland," *Community Care*, March 7, 2002, 30.

[96] Prepared statement of Philip Jenkins, professor of history and religious studies, Pennsylvania State University, before a hearing before the Subcommittee on Oversight and Investigations, Committee on Energy and Commerce, House of Representatives, 109th Cong., 2nd sess., September 26, 2006.

[97] Katherine S. Williams, "Child Pornography and Regulation of the Internet in the United Kingdom: The Impact of Fundamental Rights and International Relations," *Brandeis Law Journal* 41, no. 3 (2003): 463–505.

[98] Jenkins, *Beyond Tolerance*, 99–100. See also Waltermann and Machill, *Protecting Our Children on the Internet*, 93.

images. The site charged a membership fee of $19.99.[99] In February 2002, authorities disabled the site and then created a replacement at the same domain address. The website, which contained no illegal content, resembled the original. Previous subscribers were informed the site was rebuilding its collection of images. They could upload or transmit pictures to the site. The investigation identified nearly 200 potential suspects in 16 nations, including 29 states in the United States.[100]

Kids the Light of Our Lives

Kids the Light of Our Lives is said to be Britain's biggest-ever Internet pedophile ring. The victims, who were babies and teenagers, were often sexually assaulted live on the Web, viewed by hundreds of people who had booked paid appointments to see the abuse. At least 700 users from 35 countries participated. Some of them had published films and photographs showing themselves abusing their own children. The operation was masterminded by 28-year-old Timothy Cox (aka "Son of God"), who set up his site in December 2005 after a similar American chat room called Kiddypics was shut down. Cox sent out more than 11,000 images of the 76,000 images he had gathered. He also had 1,100 videos containing 316 hours of footage.[101]

The 10-month probe began in spring 2005 when the Child Exploitation and Online Protection Centre received a tipoff from its Canadian counterpart. Undercover officers infiltrated the chat room, posing as online child sex offenders, and lured Cox to provide them with vital evidence. After arresting Cox on September 28, 2006, police officers assumed Cox's identity and continued to operate his site for another 10 days before closing it down. Meanwhile, they gathered information about the dozens of abusers who flocked to the site. Agencies from all 35 countries, including Australia and the United States, were involved in the crackdown. Thirty-one children, some only a few months old and 15 from the United Kingdom, were rescued worldwide.[102]

A few months after Cox's arrest, police officers saw that the chat room was resurrected by Gordon MacIntosh, the 33-year-old manager of a

---

[99] "'Operation Web Sweep Targets Porn," Reuters, April 2, 2008, http://www.antionline.com/showthread.php?226986-Operation-Web-Sweep-targets-porn .

[100] "Sting Operations."

[101] "15 Children Saved from U.K.'s Biggest Paedophile Ring," *London Evening Standard*, June 19, 2007; D'Arcy Doran, "Global Pedophile Ring Busted: 31 Children Rescued," *Seattle Times*, June 19, 2007.

[102] "Paedophile Ring Smashed by Police," BBC News, June 18, 2007, http://news.bbc.co.uk/1/hi/uk/6763817.stm .

video-streaming company. After arresting MacIntosh in January 2007, police found more than 5,000 images on his computer and nearly 400 videos. Officers assumed his identity online and ran his chat room for three days while collecting information on offenders who traded images.[103]

The Swiss Operation

The level of sophistication of online child sex offenders is advancing all the time. In 2007, it was revealed that a hip-hop site in Switzerland was used to access videos of child pornography via encryption codes. Such sites are used to transfer large volumes of data: music files and videos. They are ideal for transferring child porn videos. Persons accessing the child porn portal on the hip-hop page site paid $10 to download one video and $500 for the full collection of 101 videos.[104] The site designer was unaware of how the site was abused. He had no idea that child porn files were implanted on his site and that the traffic included uploading of illegal videos. Sites can be protected against such abuse, but most ISPs do not invest in setting such protection.

The Swiss authorities became aware of this misuse and monitored the website. They discovered a ring spanning 78 countries and involving some 2,000 Internet Protocol addresses. Dozens of arrests and several convictions were made as a result of the investigation.[105]

Operation Rescue

In March 2011, it was revealed that 670 suspects were identified and 240 people were arrested for their involvement in an international pedophile ring that abused 230 children 2 to 14 years of age in more than 30 countries, including Australia, Belgium, Canada, Greece, Iceland, Italy, the Netherlands, New Zealand, Poland, Romania, Spain, Thailand, the United Kingdom, and the United States. Among the suspects were teachers and police officers.[106]

---

[103] "Global Pedophile Ring Investigation Nets 700 Suspects," Associated Press, June 18, 2007.

[104] "Child Porn Hidden in Swiss Hip-Hop Website," *Independent* (London), June 30, 2009, http://www.independent.co.uk/life-style/gadgets-and-tech/news/child-porn-hidden-in-swiss-hiphop-website-1724947.html .

[105] "Swiss Bust Child Pornography Ring," BBC News, June 28, 2009, http://news.bbc.co.uk/2/hi/europe/country_profiles/8123450.stm .

[106] CEOP Command, "Hundreds of Suspects Tracked in International Child Abuse Investigation," National Crime Agency, March 16, 2011, http://www.ceop.police.uk/Media-Centre/Press-releases/2011/hundreds-of-suspects-tracked-in-international-child-abuse-investiga tion/ ; "Police Arrest 184 in Worldwide Pedophile Ring," News.com.au, March 17, 2011, http://www.news.com.au/national/police-arrest-184-in-worldwide-paedophile-ring-europol/story-e6frfk p9-1226022892492 ; "Police Crack International Pedophile Ring," *Winnipeg Free Press*, March 17,

The ring was established through Amsterdam-based Boylover.net. This Internet forum that promoted sex between adults and young boys had some 70,000 members. The investigation spanned four years. It began when the British Child Exploitation and Online Protection Centre and the Australian Federal Police identified the website as a key online meeting place for abusers. After forum members made contact, they continued to communicate on private channels and to exchange and share pedophilic images using sophisticated technological means. Computers seized from those arrested contained huge quantities of images and films representing child abuse. The two law enforcement forces deployed officers to break into the private channels and to identify the members who were posing the most risk to children. They needed several months to infiltrate the encrypted security mechanisms. The ring was said to be the largest yet to be discovered.[107] The founder and owner of the website, Amir Ish-Hurwitz, 37, from the Netherlands, was jailed for three and one-half years on March 16, 2011, two years after the site was shut down in 2009. After his arrest, Ish-Hurwitz helped police crack the complex web of encryption measures, thus enabling police to begin covert investigations that included posing as children online. In late March 2011, Boylover.net was redirected to Varsityguys.com.

Operation Endeavour

Operation Endeavour began in 2012 and ended in early 2014. A joint investigation by the United Kingdom's National Crime Agency, the Australian Federal Police, and US Immigration and Customs Enforcement uncovered an organized crime group that facilitated the live-streaming of on-demand child sexual abuse in the Philippines. Twenty-nine international arrests were made, and 15 children in the Philippines between the ages of 6 and 15 were identified and safeguarded from further sexual abuse.[108]

Operation iGuardian

Operation iGuardian ran from May 28, 2013, to June 30, 2013. It was conducted by Homeland Security Investigations to identify and rescue victims of online

2011, http://www.winnipegfreepress.com/canada/police-crack-international-pedophile-ring-11815 4144.html .

[107] Dominic Casciani, "'World's Largest Paedophile Ring' Uncovered," BBC News, March 16, 2011, http://www.bbc.co.uk/news/uk-12762333 ; "Police Break Up the World's Largest Online Pedophile Ring," TNW, March 16, 2011, http://thenextweb.com/industry/2011/03/16/operation-rescue-breaks-up-worlds-largest-pedophile-ring/ ; "Operation Rescue Busts Online Pedophile Ring with 70,000 Members," iLookBothWays, March 16, 2011, http://ilookbothways.com/2011/03/16/operation-rescue-busts-online-pedophile-ring-with-70000-members/ .

[108] See the Virtual Global Taskforce's website at http://www.virtualglobaltaskforce.com/what-we-do/ .

sexual exploitation and to arrest sex abusers who owned, traded, and produced images of child pornography. In Brazil, Canada, Israel, Mexico, South Korea, the Philippines, Thailand, and the United States, 255 suspected child predators were arrested. The vast majority of them, 251, were men. Of the child predators, 17 were in positions of trust, including teachers and clergy members. Sixty-one children from the United States, Canada, Indonesia, and the Netherlands were rescued. Of these, 22 were nine years of age or younger, and 4 were less than three years of age. Forty-two of the victims were girls, and 19 were boys.[109]

These successful operations touch only the tip of the iceberg. The world community of online child sex offenders is estimated to include approximately three-quarters of a million predators searching for sites featuring desired images of children.[110] In the United Kingdom alone, there are 47,000 registered sex offenders and an estimated 150,000 to 200,000 sex offenders.[111] In the United States, during 2000 to 2008, there were 8,000 cases in which people were convicted for such criminal activity.[112] All crack-ring operations, international and national, resulted in the imprisonment of some thousands of people. Many thousands of websites and forums trade or sell child porn images.[113] A lot more needs to be done to combat child pornography on the Net successfully. According to Enough Is Enough, the online site dedicated to combating illegal pornography and sexual exploitation, child pornography is a $3 billion annual industry.[114]

Because child offenders operate in the global cyberarena, it is critical that law enforcement officers coordinate their activities. One officer told me that when he started to work on the investigation of one child abuse network, he did not know that other law enforcement agencies from other countries were working in the same online space. Consequently, UK law enforcement

---

[109] "Hundreds Arrested in U.S.-Led Operation against Sex Predators," *RT*, July 15, 2013, http://rt.com/usa/child-predators-arrested-guardian-131/ ; Jerry Seper, "ICE: 'Operation iGuardian' Arrests 255 Suspected Child Predators," *Washington Times*, July 15, 2013, http://www.washingtontimes.com/news/2013/jul/15/ice-operation-iguardian-arrests-255-suspected-chil/ .

[110] "Child Pornography Flourishes in a World with No Borders," United Nations Human Rights, November 26, 2009, http://www.ohchr.org/EN/NewsEvents/Pages/ChildPornography.aspx .

[111] Interview with Ruth Allen, Child Exploitation and Online Protection Centre, London, April 18, 2011.

[112] Interview with Shawn Henry, deputy assistant director, Cyber Division, Federal Bureau of Investigation, Washington, DC, March 26, 2008.

[113] Ron Scherer, "A Siege on the Child-Porn Market," *Christian Science Monitor*, March 16, 2006, http://www.csmonitor.com/2006/0316/p01s03-ussc.html .

[114] "Child Pornography," Enough Is Enough, http://www.enough.org/inside.php?tag=stat%20archives#3 ; see also Joseph Picard, "Four Sentenced in International Child Porn Ring," *International Business Times*, July 5, 2010.

officers may investigate US law enforcement officers who pose as child sex offenders, thus wasting time and resources. When it comes to arresting child offenders in different countries, coordination is absolutely essential; otherwise some offenders may be alerted by their friends and escape arrest.

The innovation of webcams provides greater opportunity for pedophiles to be part of sexual molestation of children without being nearby. Like Holly Hawkins and other Internet experts, I think that we can expect an immense increase in abuse of this technology. Indeed, the fight against child pornography is hard and frustrating. To quote an expert researcher of the electronic child porn world:

> Looking at the enormous amount of lolita-lovers out there, very, very few get arrested, the opposite of what most newbies [novices] seem to believe is the case, those that actually do get arrested, do not get arrested for downloading or uploading to abpep-t or visiting sites. Most people that get arrested do so for the following reasons: 1. They had to repair their PC when those repairing the PC discovered pics on the harddrive. 2. They have been trading thru email. 3. They have been using ICQ/IRC [chat-lines] for lolita business.[115]

HATE AND RACISM

In 1996, a governmental organization in Germany, Jugendschutz.net, and a nongovernmental organization in the Netherlands, Stichting Magenta, Meldpunt Discriminatie Internet, were the first organizations in the world to start a dedicated outfit to address the problems of racism, antisemitism, hate against Muslims, hate against gays, and other forms of discrimination or incitement to hatred, each in its own country. In 2002, they founded the International Network against Cyber Hate (INACH).

The exchange of information enhances the effectiveness of ISP-state cooperation and lobbying for international awareness about the harms and abuse of technology. INACH is "the international co-operation between complaints bureaus against discrimination, which allows the sharing of knowledge, the exchange of best practices, and coordinated measures against hate speech."[116] The vision of INACH is to act collectively against discrimination; to promote dignity, respect, citizenship, and responsibility; and to enable Netusers to exercise freedom of expression with respect for the rights and reputations of

---

[115] Prepared statement of Philip Jenkins, professor of history and religious studies, Pennsylvania State University, before a hearing before the Subcommittee on Oversight and Investigations, Committee on Energy and Commerce, House of Representatives, 109th Cong., 2nd sess., September 26, 2006. *abpep-t* is the acronym for Alt.Binaries.Pictures.Erotica.Pre-Teen.

[116] INACH Conference 2009, "Freedom of Speech versus Hate Speech," Amsterdam, November 9–10, 2009, http://www.inach.net/inach-conf-2009-program-public.pdf .

others. Netusers should be able to freely use the Internet without experiencing cyberhate. INACH's mission is to unite and empower organizations fighting cyberhate, to create awareness and promote attitude change about online discrimination, and to reinforce the rights of all Netusers.[117] INACH monitors the Internet and publishes overviews and reports about the situation in different countries. The network consists of 18 organizations in Europe and North America. INACH acts as an umbrella organization for hotlines specializing in racist and hateful content.

One obstacle to the international efforts to fight against hate is the varying definitions of *hate*. Having said that, I would like to highlight the universal international conventions that are pertinent to the fight against hatred. A former Canadian minister of justice, Professor Irwin Cotler, explained that international treaties are important because they state that hate speech does not enjoy the protection of free speech; hate speech is outside the ambit of protected speech.[118] Such declarations are important in denying legitimacy to hatemongers and in making their cyberlife more difficult because the business models of most ISPs seek to act within the consensus (see chapter 7).

Article 1 of the 1948 Universal Declaration of Human Rights declares, "All human beings are born free and equal in dignity and rights. They are endowed with reason and conscience and should act towards one another in a spirit of brotherhood." In article 2, it further accentuates:

> Everyone is entitled to all the rights and freedoms set forth in this Declaration, without distinction of any kind, such as race, colour, sex, language, religion, political or other opinion, national or social origin, property, birth, or other status. Furthermore, no distinction shall be made on the basis of the political, jurisdictional, or international status of the country or territory to which a person belongs, whether it be independent, trust, non-self-governing, or under any other limitation of sovereignty.[119]

Article 20(2) of the 1966 International Covenant on Civil and Political Rights states that "any advocacy of national, racial, or religious hatred that constitutes incitement to discrimination, hostility, or violence shall be prohibited by law."[120] Similarly, in article 4(a), the UN International Convention

---

[117] INACH, "Profile, Vision, and Mission," http://www.inach.net/mission.html .

[118] Interview with Irwin Cotler, Montreal, Canada, July 24, 2002.

[119] United Nations, "The Universal Declaration of Human Rights," http://www.un.org/en/docu ments/udhr/ . Vinton Cerf suggests that among the codified freedoms should be the right to expect freedom, or at least protection, from harm on the Internet. Vinton G. Cerf, "First, Do No Harm," *Philosophy and Technology* 24, no. 4 (2011): 463–65, 465.

[120] International Covenant on Civil and Political Rights, opened for signature on December 16, 1966, entry into force March 23, 1976, http://www.ohchr.org/en/professionalinterest/pages/ccpr. aspx .

on the Elimination of All Forms of Racial Discrimination requires its signatories to outlaw "all dissemination of ideas based on racial superiority or hatred, incitement to racial discrimination, as well as all acts of violence or incitement to such acts against any race or group of persons of another colour or ethnic origin, and also the provision of any assistance to racist activities, including the financing thereof."[121] States parties condemn all propaganda and all organizations that (a) are based on ideas or theories of superiority of one race or group of persons of one color or ethnic origin or (b) attempt to justify or promote racial hatred and discrimination in any form.[122]

In Europe, in addition to the Convention on Cybercrime, several important documents are noteworthy. Article 10 of the European Convention for the Protection of Human Rights and Fundamental Freedoms (1950) holds:

> The exercise of these freedoms . . . may be subject to such formalities, conditions, restrictions, or penalties as are prescribed by law and are necessary in a democratic society, in the interests of . . . public safety, for the prevention of disorder or crime, for the protection of health or morals, for the protection of the reputation or rights of others.[123]

Furthermore, article 3(c) of the Convention on the Prevention and Punishment of the Crime of Genocide (1948) requires contracting parties to punish direct and public "incitement to commit genocide."[124] I should also mention OSCE Ministerial Council Decision 9/09 of December 2, 2009, on Combating Hate Crimes, which, among other items, calls on the participating states

> to seek opportunities to co-operate and thereby address the increasing use of the Internet to advocate views constituting an incitement to bias-motivated violence including hate crimes and, in so doing, to reduce the harm caused by the dissemination of such material, while ensuring that any relevant measures taken are in line with OSCE commitments, in particular with regard to freedom of expression.[125]

[121] International Convention on the Elimination of All Forms of Racial Discrimination, adopted and opened for signature December 21, 1965, entry into force January 4, 1969, http://www.ohchr.org/EN/ProfessionalInterest/Pages/CERD.aspx .

[122] Ibid.

[123] Council of Europe, European Convention for the Protection of Human Rights and Fundamental Freedoms, Rome, November 5, 1950, entered into force September 3, 1953, http://conventions.coe.int/treaty/en/treaties/html/005.htm .

[124] United Nations, Convention on the Prevention and Punishment of the Crime of Genocide, December 9, 1948, 78 U.N.T.S. 277, http://www.hrweb.org/legal/genocide.html .

[125] Organization for Security and Co-operation in Europe, Ministerial Council Decision 9/09 on Combating Hate Crimes, Organization for Security and Co-operation in Europe, Athens, December 2, 2009, http://www.osce.org/cio/40695 .

In January 2009, in an answer to an oral question regarding Internet and hate crimes in Europe, Viviane Reding, who later became vice president of the European Commission, stated that the commission strongly rejects racism, xenophobia, and any type of hate speech.[126] The 2008 Council Framework Decision on combating certain forms and expressions of racism and xenophobia by means of criminal law sets out a common EU approach to racism and xenophobia.[127] Incitement to violence or hatred would also be punishable if committed by public dissemination and distribution of tracts, pictures, and other materials.

Apart from this legal approach, the European Commission is promoting trust and safer use of the Internet via its Safer Internet Plus Programme. With a budget of €55 million for the period from 2009 to 2013, the program is aimed at increasing public awareness and provides a weave of contact points for reporting illegal and harmful content and conduct, in particular on child sexual abuse material, grooming, and cyberbullying.[128]

The importance of such programs needs to be stressed, emphasizing that legal measures should not relieve the international community of its responsibilities to educate and gain awareness at all levels – primary and high schools as well as universities – of the values of liberal democracy: showing respect for others and not harming others. Education is vital in enshrining the values of liberty, tolerance, pluralism, and diversity in people's minds. At the same time, being cognizant of the democratic catch, people should be aware of the price we are required to pay for our adherence to these values. Liberty is not a prescription for anarchy, and tolerance has its limits. Otherwise, the foundations of democracy would be undermined. Education should alert and raise awareness to hate on the Internet in its various forms and attractions (music, video games, activities for kids). It should show why racism is logically incoherent, empirically unattainable, antidemocratic, and inhumane; why it is harmful; and who is targeted. It should teach the history of hate and the

---

[126] European Parliament, Parliamentary Questions, oral questions on the Internet and Hate Crimes, H-0048/09, January 20, 2009, http://www.europarl.europa.eu/sides/getDoc.do?type=QT&refer ence=H-2009-0048&language=EN ; http://www.europarl.europa.eu/sides/getDoc.do?type= CRE&reference=20090310&secondRef=ITEM-016&language=EN#2-448 .

[127] Council of the European Union, Council Framework Decision 2008/..../JHA on Combating Certain Forms and Expressions of Racism and Xenophobia, 16771/07, Brussels, February 26, 2008, http://register.consilium.europa.eu/pdf/en/07/st16/st16771.en07.pdf ; Council Framework Decision 2008/913/JHA of 28 November 2008 on Combating Certain Forms and Expressions of Racism and Xenophobia by means of Criminal Law, http://eur-lex.europa.eu/LexUriServ/ LexUriServ.do%3Furi=CELEX:32008F0913:EN:NOT .

[128] European Parliament, Debates, March 10, 2009, Question Time (Commission), Question No. 34: Subject: The Internet and Hate Crimes, http://www.europarl.europa.eu/sides/getDoc.do? type=CRE&reference=20090310&secondRef=ITEM-016&language=EN#2-448 .

connection between hate and some of the most horrific human catastrophes people have inflicted on other people. The fight against hate is difficult. Law alone will not suffice.

CONCLUSION

The Internet is ubiquitous. The ease of access to the Internet, its low cost and speed, its chaotic structure (or lack of structure), the anonymity that individuals and groups may enjoy, and the international character of the World Wide Web furnish all kinds of individuals and organizations with an easy and effective arena for their partisan interests.

The evolving technology increases the criminals' and terrorists' opportunities to participate in global and unlimited actions. The Internet became a most effective arena for criminals and terrorists to carry out their missions despite formidable odds and to sustain and expand their circles.[129] The cooperation is designed to facilitate one another's enterprises (drugs, for instance, facilitate terrorism) and is instrumental in devising ways to overcome national and international efforts to stifle their illegal activities.

Terrorists and online child sex offenders are using onion routers (see chapter 2) to preserve their anonymity and to protect their privacy. They use steganography and clandestine modes of communication as well as instant messaging, chat rooms, social networking sites, and file sharing. The international community has moral, social, and legal responsibilities to unite in combating such antisocial, violent activities. On some global issues, crosscountry cooperation is necessary to respond to global concerns. More and more countries understand the need to cooperate to tackle Net abuse. Given the magnitude of antisocial and violent phenomena on the Net, lack of such coordination would constitute utterly irresponsible, clear-eyed akrasia.

As the years have passed, people are more aware of the threats and of the need to provide security from them. Ignorance, whether circumstantial or normative, cannot serve as an excuse. The international community is

---

[129] Alex P. Schmid, "Links between Transnational Organized Crime and Terrorist Crimes," *Transnational Organized Crime* 2, no. 4 (1996): 40–82; Yvon Dandurand and Vivienne Chin, "Links between Terrorism and Other Forms of Crime," International Centre for Criminal Reform and Criminal Justice Policy, Vancouver, BC, 2004; Jeff Penrose, "The Nexus between Transnational Crime and Terrorism: Implications for Southeast Asia," presented at the annual meeting of the International Studies Association, Honolulu, HI, March 5, 2005; John Rollins, Liana Sun Wyler, and Seth Rosen, "International Terrorism and Transnational Crime: Security Threats, U.S. Policy, and Considerations for Congress," CRS Report for Congress, Washington, DC, March 18, 2010; Molly Land, "Toward an International Law of the Internet," *Harvard International Law Journal* 54, no. 2 (2013): 393–406.

expected to take the necessary responsible measures to promote security online. With the right cooperation, the international community has the capabilities to address the formidable challenges and provide appropriate answers. Failure to cooperate is inexcusable. Without responsible cooperation, Net abusers will prevail, and our children will suffer. Nations and responsible Netcitizens are obliged to ensure that future generations will be able to develop their autonomy, their individuality, and their capabilities in a free but also secure environment, both offline and online. Finding the right balance between freedom and security is tricky, yet it is doable and most necessary.

Conclusion

All that is necessary for the triumph of evil is that good men do nothing.
—Edmund Burke

It's not hard to make decisions when you know what your values are.
—Roy Disney

One of my students told me that he does not go to the movies anymore. I wondered why, and the student explained that going to the movies requires coordination (he does not like to go alone), money, and time, dictated by the movie theatre. It is too resource consuming. When I want to watch a movie, elucidated the student, I download it from the Internet, watch it at my leisure, and it does not cost me a penny. Because I do not like to sit for a long time without moving, I can stop the movie anytime I wish, go about my business, and resume watching the movie at will. Cinema is passé.

The Internet is a success story thanks to its open architecture and globally interoperable standards. It has affected virtually every aspect of society. The Internet has changed the way people study, conduct research, shop, travel, and promote their business. The Internet has changed industries, from culture and entertainment (books, music, movies) to commerce and banking. It is a quotidian network of interconnected multilayered networks. Unlike most communication media, Internet technology is based on global and nonproprietary principles. Its complex structure is the result of freedom and ability to engage in innovation.

The Internet encourages connectivity, communication, and creativity. Many people contribute to the Internet in various ways. We post ideas, blog, upload photos and video clips, write texts, organize petitions, and comment on others' contributions. The community of Netusers is vast and growing. What we need is to transform Netusers into Netcitizens – that is, Netusers with a sense of responsibility. And we need to reconcile Internet innovation, continued growth, and speed with society's best interests and security.

Throughout history, each major innovation in communications technology has caused distress and confusion similar to what society is experiencing today with the Internet. The introduction of writing, the printing press, the telegraph, the telephone, the radio, and the television – all raised similar issues. Key decisions determining the future of the Net cannot be made without insights from the past.[1]

The Internet is both a source of promise for our children and a source of concern. The promise is of Internet-based access to the information age, and the concern is that harm might befall people – especially vulnerable people such as our children – as they use the Internet.[2] The Internet may be regarded as simply a medium of distribution, but the Internet also has important societal and psychological roles to play.

In *Alice's Adventures in Wonderland* (1865), Lewis Carroll (Charles Dodgson) wrote:

> "Cheshire Puss," asked Alice. "Would you tell me, please, which way I ought to go from here?"
> "That depends a good deal on where you want to go," said the Cat.
> "I don't much care where," said Alice.
> "Then it doesn't matter which way you go," said the Cat.

This kind of attitude might be cherished by Internet experts, but it is not one I endorse. We need to set objectives for the new media. Creativity is wonderful, but it needs to have a sense of direction; it needs to develop with individual and social purpose. It needs to have a responsible vision.

In this book, I advocate for the need to weigh freedom of expression against social responsibility. Both are very important considerations, and we need to find the Golden Mean between them. We can reasonably expect people to know the difference between good and evil and then to act accordingly. Given their social context and basic capabilities, people should be expected to cooperate in the struggle against antisocial activities on the Internet.

Social networking sites have become a major communication platform in the past few years. As the platform grows in popularity, so does a body of research concerning social networking sites, stemming from diverse disciplines and using

---

[1] Jonathan Wallace and Mark Mangan, *Sex, Laws, and Cyberspace* (New York: Henry Holt, 1996), 194.

[2] Dick Thornburgh and Herbert S. Lin, eds., *Youth, Pornography, and the Internet* (Washington, DC: National Academies Press, 2002), 17.

various methodologies. We still have much to learn about the way people are communicating via such sites and what can be done to ensure that social networks will not become antisocial. We need to advance our knowledge of emerging social networking and the psychology of people who use the Internet for various purposes. Governments and Internet nongovernmental organizations should study the psychology of social networking.[3] More research is needed on the relationships between offline and online threats and risks, analyzing sociodemographic and psychological factors.[4] Clearly, we need to increase the moral and social responsibility of all parties concerned. In their work, danah boyd and Nicole Ellison mention four research pivots:[5] (a) impression management and friendship performance;[6] (b) networks and network structure;[7] (c) online versus offline connections;[8] and (d) privacy issues.[9]

---

[3] The British Home Office has put together a research team to study the psychology of social networking. Interview with Rosa Beer and Nisha Patel, senior research officers, Home Office, London, April 19, 2010. A good start is Clay Shirky, *Here Comes Everybody: The Power of Organizing with Organizations* (London: Penguin, 2008), and Yair Amichai-Hamburger, ed., *The Social Net: Understanding Human Behavior in Cyberspace* (Oxford: Oxford University Press, 2013).

[4] Leslie Haddon and Sonia Livingstone, "The Relationship between Offline and Online Risks," in *Young People, Media, and Health: Risks and Rights*, ed. Cecilia von Feilitzen and Johanna Stenersen (Göteborg, Germany: Nordicom, 2014), 21–32.

[5] danah m. boyd and Nicole B. Ellison, "Social Networks Sites: Definition, History, and Scholarship," *Journal of Computer-Mediated Communication* 13, no. 1 (2007): 210–30, http://onlinelibrary.wiley.com/doi/10.1111/j.1083-6101.2007.00393.x/abstract .

[6] See, for example, Joseph B. Walther, Brandon Van Der Heide, Sang-Yeon Kim, David Westerman, and Stephanie Tom Tong, "The Role of Friends' Appearance and Behavior on Evaluations of Individuals on Facebook: Are We Known by the Company We Keep?," *Human Communication Research* 34, no. 1 (2008): 28–49; danah m. boyd, "Taken Out of Context: American Teen Sociality in Networked Publics" (PhD diss., University of California, Berkeley, 2008), http://www.danah.org/papers/TakenOutOfContext.pdf ; Shanyang Zhao, Sherry Grasmuck, and Jason Martin, "Identity Construction on Facebook: Digital Empowerment in Anchored Relationships," *Computers in Human Behavior* 24, no. 5 (2008): 1816–36, http://astro.temple.edu/~bzhaooo1/Identity%20Construc tion%20on%20Facebook.pdf .

[7] See, for example, Bernie Hogan, "Analyzing Social Networks via the Internet," in *The Sage Handbook of Online Research Methods*, ed. Nigel Fielding, Raymond M. Lee, and Grant Blank (Thousand Oaks, CA: Sage, 2008), 141–60.

[8] See danah m. boyd, "Why Youth (Heart) Social Network Sites: The Role of Networked Publics in Teenage Social Life," in *Youth, Identity, and Digital Media*, ed. David Buckingham (Cambridge, MA: MIT Press, 2008), 119–42; Amanda Lenhart and Mary Madden, "Teens, Privacy, and Online Social Networks," Pew Internet Project, April 18, 2007, http://www.pew internet.org/Reports/2007/Teens-Privacy-and-Online-Social-Networks.aspx .

[9] Susan Barnes, "A Privacy Paradox: Social Networking in the United States," *First Monday* 11, no. 9 (2006); Matthew J. Hodge, "The Fourth Amendment and Privacy Issues on the 'New' Internet: Facebook.com and MySpace.com," *Southern Illinois University Law Journal* 31 (2006–2007): 95–123.

More coordination is needed on the national and international levels. For instance, the National Center for Missing and Exploited Children (NCMEC) should provide all Internet service providers (ISPs) with its image data bank. The main ISPs should share their bank of child pornography images with each other and with NCMEC. Consequently, whenever a person downloads an image, ISPs will automatically match it against the bank of images provided by NCMEC.

In the wake of September 11, 2001, the United States and the Organisation for Economic Co-operation and Development, in an effort to counter terror and other threats, issued guidelines for the security of information systems and networks. The guidelines were aimed at developing a culture of responsibility and security among governments, ISPs, and Netusers in an era of expanding technology and communication.[10]

At the 2011 conference of the Group of Eight, European leaders spoke of the need to bring "civility" to the Internet. French president Nicolas Sarkozy said that the Internet "is not a parallel universe which is free of rules of law or ethics or of any of the fundamental principles that must govern and do govern the social lives of our democratic states." He called for "collective responsibility" and "for everyone to be reasonable" in cyberspace.[11]

In April and May 2007, international cooperation was required to help Estonia overcome cyberattacks through botnets directed against the Estonian government's computer infrastructure. The crippling attacks flooded computers and servers and blocked legitimate users. Government websites that normally receive about 1,000 visits a day reportedly were receiving 2,000 visits every second, causing the repeated shutdown of some websites for several hours at a time or longer. The cyberattacks were attributed to Estonia's high dependence on information technology but limited resources for managing its infrastructure. Security experts said that the cyberattacks were unusual because the rate of the packet attack was very high and the series of attacks lasted weeks. Eventually, the North Atlantic Treaty Organization and the United States sent computer security experts to Estonia to help it recover from the attacks and to analyze the methods used and attempt to determine the source of the attacks.[12]

---

[10] Organisation for Economic Co-operation and Development, "OECD Guidelines for the Security of Information Systems and Networks," OECD Publications, Paris, http://www.oecd.org/dataoecd/16/22/15582260.pdf .

[11] Matthew Lasar, "Nazi Hunting: How France First 'Civilized' the Internet," Ars Technica, June 22, 2011, http://arstechnica.com/tech-policy/2011/06/how-france-proved-that-the-internet-is-not-global/ .

[12] Clay Wilson, "Botnets, Cybercrime, and Cyberterrorism: Vulnerabilities and Policy Issues for Congress," CRS Report for Congress, Congressional Research Service, Washington, DC, January 29, 2008.

UNIVERSALISM AND PARTICULARISM

The Internet is universal in nature, but societies do not adopt a universal common denominator to define the boundaries of freedom of expression. These boundaries vary from one society to another and are influenced by historical circumstances and cultural norms. Italy, Germany, and Spain are more sensitive to the evils of fascism than is the United States and rightly so. Whereas the United States and other liberal democracies may protect fascist speech, racism, and Holocaust denial, we would be most troubled if Italy, Germany, and Spain were *not* to adopt restrictive measures against Internet sites that promote fascist ideas.

Jan van Dijk considers four possible models of Internet governance: (a) *denationalized liberalism*, which is about the creation of new international governance institutions that are not based on states but on individual networkers; (b) *global governability*, which is about controlling the Internet through new institutions that transcend states; (c) *networked nationalism*, which accepts the power of networking but keeps it under the control of states through international treaties and networks; and (d) *cyber-conservatism*, which endorses subjecting the Internet to the authority of the state.[13] Although libertarians are likely to opt for the denationalized liberalism model,[14] I, together with van Dijk, support the networked nationalism model. International companies need to be cognizant of the plurality of interests and sensitivities of different national cultures. They cannot assume that the transcendent nature of the Internet puts their conduct above the law. National laws need to be respected both offline and online. Such responsible conduct needs to be supplemented by the responsible conduct of interest groups, of individual Netusers, and of international forums and associations. The creation of a responsible Internet requires the multifaceted efforts of all interested parties.

I stress Aristotle's Rule of the Golden Mean, that for every polarity there is a mean that, when practiced, provides good benchmarks for a life of moderation (see chapter 3). The more we see the golden mean in each polarity, the better we secure good benchmarks of a life of wellness. People have the freedom to express themselves – with responsibility and reason. Ethics is not only a question of dealing morally with a given world. It is also an issue of constructing the world, improving its nature, and shaping its development in the right

---

[13] Jan Van Dijk, *The Network Society* (London: Sage, 2012), 148–50.
[14] See, for example, Milton L. Mueller, *Networks and States: The Global Politics of Internet Governance* (Cambridge, MA: MIT Press, 2010), 81–106.

way.[15] There are no easy solutions to the problems that the Internet poses. An effective policy requires cooperation. This need is especially pressing on issues such as terrorism, child pornography, cyberbullying, hate, and bigotry. Given the agents' social context and basic capabilities, people are expected to cooperate in the struggle against antisocial activities on the Internet. Child pornography, racism, and terrorism – three of the most troubling phenomena on the Net – are international in character. Considerations of social responsibility dictate transnational cooperation to downsize their threats.

RECOMMENDATIONS

Governments could establish an international information agency or support private or quasi-public organizations to help providers understand each nation's regulatory standards and structures. They could update and extend to the networked world the mechanisms that currently exist for dealing with circumstances in which domestic laws conflict.[16] Governments could help spread the benefits of such mechanisms broadly by educating consumers about them and advocating their use.[17] Laws will continue to define the background set of rules within which private sector actors and mechanisms must operate.[18] While observing state laws, the international community should strive to promote awareness regarding social and moral standards shared at least by free democratic societies. Along with global opportunities come responsibilities. Those responsibilities define who we are and what kind of society we wish to inhabit.

Although awareness of child pornography and terrorism is on the rise among governments, civil society organizations, the business sector, law enforcement agencies, and individuals,[19] there is less awareness of murderers

---

[15] Luciano Floridi and J. W. Sanders, "Internet Ethics: The Constructionist Values of *Homo Poieticus*," in *The Impact of the Internet on Our Moral Lives*, ed. Robert J. Cavalier (Albany: State University of New York Press, 2005), 195–214, 195–96.

[16] National Research Council, *Global Networks and Local Values: A Comparative Look at Germany and the United States* (Washington, DC: National Academies Press, 2001), 227.

[17] Bradford L. Smith, "The Third Industrial Revolution: Policymaking for the Internet," *Columbia Science and Technology Law Review* 3 (2001–02): 1, http://www.columbia.edu/cu/stlr/html/Archive/.

[18] Ibid.

[19] Yaman Akdeniz, *Internet Child Pornography and the Law: National and International Responses* (Aldershot, UK: Ashgate, 2008); William R. Graham Jr., "Uncovering and Eliminating Child Pornography Rings on the Internet: Issues Regarding and Avenues Facilitating Law Enforcement's Access to 'Wonderland,'" *Law Review of Michigan State University, Detroit College of Law* 2 (2000): 457–84, 465; John Carr, "Theme Paper on Child Pornography for the 2nd World Congress on Commercial Sexual Exploitation of Children," Children and Technology Unit Nationwide Children's Hospital, London,

who use the Internet to publish their malicious plans. The good news is that we can use the international cooperation that was developed through many national and international Internet watch organizations to increase awareness among people and to operate a constant monitoring scheme for problematic websites such as VampireFreaks.com and other popular sites that provide social networking for criminal, antisocial ideas. Operating a monitoring scheme and educating people to alert and report whenever they encounter online threats of murder and other violent crimes could prevent murders, save lives, and cut down on crime. The monitoring scheme and surfers' alerts could help law enforcement agencies to track down the people who are planning crimes before they execute their plans. In addition, when public awareness regarding the subject increases, potential criminals will not receive praise from their readers but rather critical and opposing responses. This initiative, in turn, will fight the copycat phenomenon.

Some ISPs exhibit irresponsible akrasia in the face of Net hate. To address the challenge of hate on the Net, states and ISPs need to exchange information and enhance the effectiveness of their cooperation on human rights issues. They need to lobby for international awareness about the harms and abuse of technology, help support groups and institutions that want to set up tiplines alerting people about hate, and advance knowledge of emerging social networking and the psychology of people who use the Internet for various purposes. Hate poses a somber challenge that calls for serious consideration and redeeming answers. Responsible ISPs and Web-hosting companies should weigh freedom of expression against social responsibility and invest more effort into cleaning their servers of Net hate.

In "First, Do No Harm," Vinton G. Cerf suggests that those who make and operate the Internet and its applications have an ethical responsibility "to take steps to improve the ability of Internet-related technology to protect users from harm, to warn them when they are at risk, and to advocate domestic and international regimes to provide recourse when harms peculiar to the Internet environment occur."[20] Indeed, ISPs should continue to develop and embrace initiatives designed to protect Netusers – especially children. These initiatives include technological tools as well as educational campaigns. ISPs should carefully balance reasonable expectations of customer privacy with the need to ensure a safe and secure online environment. Industry should give due weight

---

2001, http://www.childcentre.info/robert/extensions/robert/doc/67ba32d30c03c842b7032932 f2e6ce74.pdf ; Yaman Akdeniz, "Regulation of Child Pornography on the Internet: Cases and Materials," Cyber-Rights & Cyber-Liberties, Leeds, UK, 2001, http://www.cyber-rights. org/reports/interdev.htm .

[20] Vinton G. Cerf, "First, Do No Harm," *Philosophy and Technology* 24, no. 4 (2011): 463–65, 464–65.

to societal considerations that may be essential to promote people's trust in it and to prevent abuse. A certain security level on the Internet needs to be ensured, as in any other industry. A senior security officer suggested that Internet providers should have integrity teams that instruct providers to take off inappropriate content.[21]

With continued development of technical solutions and innovation and with increased awareness of and adherence to basic Corporate Social Responsibility (CSR), a better structure is required. CSR should be part of any Web company's strategy vis-à-vis day-to-day operations. Indeed, CSR is a continuous, living process. Thus, social responsibility should influence ISPs and Web-hosting companies to scrutinize their servers, verifying that their servers do not become hubs for terror. Terrorist organizations will put their merchandise on identified servers that knowingly work with them. With good intelligence, security organizations will be aware of those sites and monitor their activity closely. The result will be fewer sites, and the sites that do exist will be identified and will be hosted in countries that sponsor terrorism. Businesses that oppose terrorism will stay away from such servers because associating themselves with terror will be perceived as bad for their business. The battle between the ideology of destruction and liberal ideologies of live-and-let-live will continue, and each camp will continue trying to attract adherents. However, when ideas translate into conduct, in the form of recruitment or raising funds for terror operations, concrete steps should be taken to apprehend the potential terrorists and to stifle the fundraising. Without new recruits and funding, terrorism withers away.

In the face of global threats, all relevant bodies must cooperate: voluntary and organized operations, civil society groups, governments, law enforcement agencies, and the business sector – especially ISPs, website administrators, and owners. When corporate activity causes harm that transgresses national laws, nations can and should assert their regulatory authority. The threat of multiple regulatory exposures will not destroy the Internet. A formal strategic planning effort is needed that is positively linked to CSR.[22] Firms will have to filter content geographically to comply with local law for only a small fraction of their communications. True, this effort will impose costs on multinational Internet firms, which will have to adjust to this cost of doing business just as real-space multinationals do. But in light of the Internet's many efficiencies,

---

[21] Interview with a senior security officer, Washington, DC, March 25, 2008.
[22] Jeremy Galbreath, "Drivers of Corporate Social Responsibility: The Role of Formal Strategic Planning and Firm Culture," *British Journal of Management* 21, no. 2 (2010): 511–25.

this cost will be trivial in the long run.[23] Corporations have ethical and social responsibilities to avoid social and ecological harm and to promote the well-being of their communities.

Responsible people in whatever capacity have a moral obligation to alert the designated victim and the legal authorities about harmful threats. When human life is at stake, we all have a moral obligation to assist others and to prevent harm. Responsible Netcitizens do not stand idly by when they witness targeting and victimizing of others and when they read threats to commit murder.

VISION

We have a universally interconnected electronic communication system that is based on a variety of linkable electronic carriers, using radio, cable, microwave, optical fiber, and satellites and delivering to every home and office a vast variety of different kinds of mail, print, sound, and video through an electronic network of networks.[24] Communication systems and networks are likely to continue their growth and to develop new applications that will affect our lives in different ways. Two of the founding fathers of the Internet, J. C. R. Licklider and Robert W. Taylor, wrote an article in 1968 that attempted to foresee the future: they predicted that people would not send a letter or a telegram, would seldom make a telephone call, and would seldom make a purely business trip, because linking consoles would be so much more efficient.[25] They were wrong on all counts. Netusers send more emails than letters but still resort to letters. The nature of email is such that it is most useful when one wishes to send a laconic message or to convey en masse impersonal information. For more personal and complex communications, many people resort to other, more private channels, with more cues about the recipient's reaction to the message. It is one thing to ask your spouse to buy milk on the way home or to remind your business partner to submit an application today; it is quite another thing to coordinate a trip to the Bahamas or to explain what you did last night with "that stranger." On delicate matters, people, Netusers included, prefer the phone or a private meeting.

With the advancement of mobile technology, people are making far more phone calls than before. People are hooked on their cell phones and take them

---

[23] Jack Goldsmith and Tim Wu, *Who Controls the Internet? Illusions of a Borderless World* (New York: Oxford University Press, 2006), 160–61.

[24] Ithiel de Sola Pool, *Technologies of Freedom: Our Free Speech in an Electronic Age* (Cambridge, MA: Harvard University Press, 1983), 233–34.

[25] J. C. R. Licklider and Robert W. Taylor, "The Computer as a Communication Device," *Science and Technology* 76, no. 2 (1968): 20–41, 38.

wherever they go. They also text, something that Licklider and Taylor did not envisage. People travel far more than before, including long distances. The Internet certainly has made our lives easier in many respects, but it did not replace mail, telephone, and travel.

Leonard Kleinrock's vision is that Internet technology will be everywhere, always accessible, and always on; anyone will be able to plug in from any location with any device at any time, and the device will be invisible.[26] The Internet has achieved the first three elements but not the last two. Kleinrock sees intelligent software agents deployed across the network, whose function it is to mine data, act on that data, observe trends, carry out tasks, and adapt to their environment. He sees considerably more network traffic generated, "not so much by humans, but by these embedded devices and these intelligent software agents."[27] He sees huge amounts of information "flashing across networks instantaneously, with this information undergoing enormous processing and informing the sophisticated decision support and control systems of our society."[28] Kleinrock foresees that the Internet will essentially be an invisible infrastructure "serving as a global nervous system for the peoples and processes of this planet."[29]

With time, we can assume that more people will be able to connect to computers and to the Internet wherever they are. Both will be portable and easily accessible through superior cell phones that will become sophisticated media centers, enabling traditional (phone, radio, and television) and new media (Internet, Kindle books, recorded books, games, music, camera, video, and new forms of entertainment). People will be able to listen to and appear on the radio; watch television and appear on its programs while on the move in land, sea, and air; connect to programs via the Internet; and contribute to but also manipulate the content of those programs. Advanced voice-recognition capabilities will make both the keyboard and the mouse obsolete. A team at Dartmouth College has already created an eye-tracking system that lets a user operate a smartphone with eye movement. Solutions will be sought to overcome the growing challenges of maintaining privacy, reducing noise levels, and isolating the cacophonies of sound in public places. The establishment of sophisticated validation mechanisms of data and identity, along with the establishment of integrity teams, will be unavoidable.

[26] Leonard Kleinrock, "History of the Internet and Its Flexible Future," *IEEE Wireless Communications* 15, no. 1 (2008): 8–18, 15.

[27] Ibid., 17.

[28] Ibid.

[29] Ibid.

Technology will develop to enhance connectivity between media and our senses, engaging our senses more fully with tinier and more powerful speakers deep inside human ears; 3-D innovations will enable our bodies to feel sensations and to taste edible products we see on our portable screens. Obviously, social responsibility codes as to how to behave in public will require some readjustment. These wonderful innovations should be accompanied with fine awareness of their consequences on individuals and society at large – because every innovation is open to use, but also abuse.

Futurist Jim Dator, in a speech to the World Futures Studies Federation World Conference in 1993, said:

> As the electronic revolution merges with the biological evolution, we will have – if we don't have it already – artificial intelligence, and artificial life, and will be struggling even more than now with issues such as the legal rights of robots, and whether you should allow your son to marry one, and who has custody of the offspring of such a union.[30]

Dator is correct about the expansion of robot technology in all spheres of life. Humans invent robots that eventually replace them. Fewer people will be required to do the same amount of work more efficiently. Developed robots will be more and more efficient and less prone to suffer from human errors.

Visionary scientist Vannevar Bush wrote that when we create or absorb material, we use one or more of our senses. The impulses that flow in the arm nerves when we type convey to our fingers the translated information that reaches our eyes or ears for our fingers to strike the keyboard. Bush asks whether these currents might be intercepted either in the original form in which information is conveyed to the brain or in the metamorphosed form in which they then proceed to the hand. To be transmitted, sounds and sights are reduced to currents in electronic circuits. Bush further asks whether we must always transform to mechanical movements to proceed from one electrical phenomenon to another.[31] What Bush is suggesting is to connect the computer directly to our human cognition, thus skipping the mediation of senses: instead of using our eyes to read data on the computer screen and transmit the information to our brain, the brain will connect directly to the database and comprehend the information. There is still a way to go until humanity reaches that stage.

Today, the growth of cloud computing is providing powerful new ways to easily build and support new software. Because companies and individuals

---

[30] Quoted in "Imagining the Internet: A History and Forecast," School of Communications, Elon University, Elon, NC, http://www.elon.edu/e-web/predictions/150/1960.xhtml .

[31] Vannevar Bush, "As We May Think," *Atlantic Monthly*, 176, no. 1 (1945): 101–8.

can "rent" computing power and storage from services like the Amazon Elastic Compute Cloud, someone with a good idea can much more easily and rapidly turn it into an online service. This development is leading to an explosion in new uses for the Internet and a corresponding explosion in the traffic flowing across the Internet.[32] However, as the promise is great, so are the challenges. The new frontiers offer the prospect of exciting ventures, but at the same time, they open new avenues for abuse. Technology needs to advance in a responsible way. We must not put the carriage before the horse, pushing initiatives forward without thinking of their consequences. Challenges should not be left to an unidentified "someone" to fix. Instead, that "someone" is you: you the Netuser, you the reader, you the businessperson, you the government official, and you the decision maker at the international level. Responsibility is crucial to ensure a safer, tranquil world for our children.

We can predict that further incidents will occur in which states will assert their sovereignty on the Net. More pressure will be put on ISPs to take proactive steps to regulate content, and more innovation will enable human connectedness whenever people are awake, notwithstanding their physical location.

One book cannot address all the problematic issues that may be encountered on the Internet. But the issues I have addressed are important and instructive, providing us with the tools for addressing other concerns. The actors may change. Today it is GoDaddy and tomorrow it might be ByeMommy. Names are not important. The guiding principles for analysis are important, and they may serve us for many years to come. The structure of Net intermediaries may change as well. As said, in the future more information will likely pass via cell phones. Thus, responsibility may shift from ISPs to mobile phone carriers. Human creativity and innovation exceed present human imagination. The underpinning principles of social responsibility, accountability, and answerability; of CSR; and of respect for others and not harming others should continue to serve as a beacon for the conduct of all people concerned. The promotional approach (as explained in chapter 6), with its emphasis on these shared liberal norms, should guide the policies of ISPs, Web-hosting services, and liberal democracies.

[32] Michael R. Nelson, "A Response to Responsibility of and Trust in ISPs by Raphael Cohen-Almagor," *Knowledge, Technology, and Policy* 23, no. 3 (2010): 403–7. See also Christopher Millard, "Data Privacy in the Clouds," in *Society and the Internet: How Networks of Information and Communication Are Changing Our Lives*, ed. Mark Graham and William. H. Dutton (Oxford: Oxford University Press, 2014), 333–47; Yorick Wilks, "Beyond the Internet and the Web," in Graham and Dutton, *Society and the Internet*, 360–73.

CLEANET

The time has come to consider the introduction of a new browser funded by an affluent person with a sense of social responsibility, a nongovernmental organization, or a group of organizations (such as the Deliberative Democracy Consortium) wishing to establish a better Internet future for our children.[33] The new browser will be called CleaNet and will have no connections with any government. Because of the potential governmental tendency to restrict out-of-favor political speech under the pretense of "dangerous" and "terrorist" speech, no government will be involved in this delicate, deliberative process.

Deliberative democracy directly involves citizens in decision-making processes on matters of public concern. It requires the setting up of institutions for public reason through which knowledge is exchanged and ideas crystallized via mechanisms of deliberation and critical reflection. Democratic procedures establish a network of pragmatic considerations and a constant flow of relevant information. People present their cases in persuasive ways, trying to bring others to accept their proposals. Processes of deliberation take place through an exchange of information among parties who introduce and critically test proposals. Deliberations are free of any coercion, and all parties are substantially and formally equal, enjoying equal standing, equal ability, and equal opportunity to table proposals, offer compromises, suggest solutions, support some motions, and criticize others. Each participant has an equal voice in the process and tries to find reasons that are persuasive to all to promote the common good.[34] The technology at hand enables direct participation of people, eliminates geographic distances, and re-creates direct Athenian-style democracy. It empowers good citizenship and public partnership in promoting shared social values and norms. Because the Internet affects each of us, we have a vested interest in attempting to make it a social tool that enables the promotion of social good. Following Jürgen Habermas's ideas on deliberative democracy and the importance of having access to different publics and organizations in the international civil society, some argue that

---

[33] The Deliberative Democracy Consortium is a network of practitioners and researchers representing more than 50 organizations and universities, collaborating to strengthen the field of deliberative democracy. The consortium supports research activities and aims to advance practice at all levels of government around the world. See the consortium's website at http://www.deliberative-democracy.net/ .

[34] Joshua Cohen, "Deliberation and Democratic Legitimacy," in *The Good Polity: Normative Analysis of the State*, ed. Alan Hamlin and Philip Petit (Oxford: Blackwell, 1989), 17–34, 22–23, http://philosophyfaculty.ucsd.edu/faculty/rarneson/JCOHENDELIBERATIVE%20DEM.pdf ; Jürgen Habermas, *Between Facts and Norms: Contributions to a Discourse Theory of Law and Democracy* (Cambridge: Polity, 1996), 304–8.

the Internet will be stable in the long run only if Netusers generally perceive it as a legitimate instrument – that is, only if they perceive the Internet as right and good and based on shared values and norms.[35]

Mutual recognition, respect, and equal protection are essential. Habermas explained that democracies are associations of free and equal persons. Such an association is structured by relations of mutual recognition in which each individual is respected as free and equal. According to Habermas, every person should receive a threefold recognition: "they should receive equal protection and equal respect in their integrity as irreplaceable individuals, as members of ethnic or cultural groups, and as citizens, that is, as members of the political community."[36]

The first step will be to convene a Netcitizens Committee to decide what should be excluded from the new browser. The committee should determine the agreed-on problematic topics that are regarded as unprotected speech.

A public open call for Netcitizens Committee members will be issued, and the process will be conducted with transparency, full disclosure, and open deliberation and debate. Clear deadlines for each step of the process will be outlined to ensure that the process does not drag on for many months. The Netcitizens Committee will be determined by a special select committee, nominated by the owners of the new browser. Nongovernmental organizations in the fields of new media, human rights, freedom of expression societies, and institutions that promote social responsibility will be invited to serve on the CleaNet select committee.

The Netcitizens Committee will include representatives of ISPs and web-hosting companies, Internet experts, media professionals, Internet scholars, government officials, representatives of human rights and minority rights organizations[37] and freedom of speech organizations, computer engineers, judges, lawyers, and other interested parties.

The Netcitizens Committee will include no less than 100 people and no more than 400 people, depending on the number of applicants willing to commit themselves for the responsible work at hand. This committee needs to be a working committee. It cannot be too big.

---

[35] Jürgen Habermas, *Moral Consciousness and Communicative Action* (Cambridge, MA: MIT Press, 1990); Habermas, *Between Facts and Norms.* See also James S. Fishkin, *Democracy and Deliberation: New Directions for Democratic Reform* (New Haven, CT: Yale University Press, 1993); John S. Dryzek, *Foundations and Frontiers of Deliberative Governance* (Oxford: Oxford University Press, 2012); Zsuzsanna Chappell, *Deliberative Democracy: A Critical Introduction* (Houndmills, UK: Palgrave-Macmillan, 2012).

[36] Habermas, *Between Facts and Norms,* 496.

[37] This representation is crucial because minorities frequently face difficulty in having an equal voice and equal standing in decision-making processes.

Members will commit to work for one year, renewable for two more years at most. After one year, the least active members will be asked to leave, and they will be replaced by others. One-third of the committee is expected to change each year. Such a reshuffle is advisable and productive. It keeps the committee energetic, engaged, viable, and fresh with ideas.

Because the work is hard and demanding, with considerable societal implications, members of the Netcitizens Committee will be paid for their work. The payment should not be too meager nor should it be very substantial. It should be enough to provide an incentive, denoting the responsible work at stake, but it should not be the main job of the Netcitizen. A sum between €1,000 and €2,000 per month is recommended.

The first issue on the agenda is to detail what should be ousted from the Net and parameters for identifying problematic, antisocial speech. The committee will consider the wide needs and interests of the public in an open, transparent, and critical way. All committee members will have the opportunity to participate and voice an opinion, to present arguments, to submit criticisms and reservations, and to respond to counterarguments. No one will ever be excluded from the deliberative process. The committee will try to reach a consensus in delineating the scope of legitimate and acceptable Net speech. In the absence of a consensus, decisions may be made by vote, but the committee needs to make every effort to reach a consensual decision that reflects the widest possible public needs and interests. Members of the committee need to recognize that the widest possible consensus will ensure the legitimacy of their decisions.

Because the committee represents a Western liberal tradition, the scope of what is legitimate and acceptable should be as wide as possible. Whenever the committee members consider restricting speech, the onus for limiting free expression is always on those who wish to limit expression. Committee members should therefore bring concrete evidence to justify restriction. The speech must be dangerous or harmful. The danger or harm cannot be implicit or implied. If speech can be prohibited only because its danger might be implied from an unclear purpose that is open to interpretation, then the scope for curtailing fundamental democratic rights is too broad, and the slippery-slope syndrome becomes tangible. The implicit way is not the path that liberals should tread when pondering restrictions on freedom of expression. Nevertheless, committee members should be vigilant in protecting democracy. But mere suspicions ("bad tendencies") will not be enough to override basic freedoms.

Throughout the process, participants will exercise their *communicative freedoms*, a term Habermas applied to activities that seek to achieve mutual understanding through reasoned discourse. The open, deliberative discourse allows everyone to participate in the processes of opinion and will formation in which citizens exercise their autonomy.[38] When the list of requirements is concluded, it should be handed to software engineers to design the algorithm for excluding material.

In a sense, CleaNet will be an enhanced, citizen-based form of server filtering. Detailed terms of fair conduct will be drafted. Only material that is deemed problematic by at least 80 percent of the votes will be listed for exclusion. A separate list, "under review," should include debatable speech to be considered and debated periodically until a resolution is reached: either to permit it or to filter it from CleaNet. The "under review" list will also include problematic material with restricted access for which Netusers will have to sign up. It will be the responsibility of the ISPs and web-hosting companies to retain the list and to cooperate with law enforcement whenever required.

CleaNet will be launched in a special press conference, notifying the public of its availability, rationale, and significance. CleaNet should be freely downloadable with open access for all. Netusers will have a choice: retaining their present browsers, adding CleaNet as an alternative (primary or secondary) browser, or replacing their present browsers with CleaNet.

CleaNet will be attentive to societal and cultural norms. For instance, although Holocaust denial is not problematic in the United States, it is quite problematic in Germany, Israel, and other democracies. CleaNet should pay special attention to such sensitive matters.

Although one may assume that international consensus will exist about excluding certain antisocial material – child pornography, cyberbullying, and the promotion of violent crime and terrorism – from CleaNet, such a consensus will likely not exist regarding hate and bigotry. The notable exception will probably be the United States. However, such tolerant norms should not bind other countries that believe their Net should be free of bigotry and hatred. They may opt to filter that material. That is, the browser will be the same. Its configuration may vary from one country to another, in accordance with the country's basic norms. At the same time, because the entire idea is based on converging freedom with social responsibility, people may opt to use CleaNet in any of its versions or opt not to use it at all.

---

[38] Habermas, *Between Facts and Norms*, 118–27. For further discussion, see Karen Mossberger, Caroline J. Tolbert, and Ramona S. McNeal, *Digital Citizenship: The Internet, Society, and Participation* (Cambridge, MA: MIT Press, 2007); John Parkinson and Jane Mansbridge, eds., *Deliberative Systems: Deliberative Democracy at the Large Scale* (Cambridge: Cambridge University Press, 2012).

Once CleaNet has been implemented and marketed, the government of each country will push its adoption in the public sector. Only government agencies that have a specific interest in studying antisocial material should be granted permission to use other browsers. Otherwise, we can assume that the public sector has no need to have access to, for instance, child pornography, criminal speech, terrorism, and bigotry.

On CleaNet, search engines will not keep their ranking algorithms secret. Quite the opposite. They will proudly announce, in a transparent and explicit way, that the ordering of search results is influenced by standards of moral and social responsibility, a commitment to preserving and promoting security online and offline, and adherence to the liberal principles we hold dear: promoting liberty, tolerance, and human dignity; showing respect for others; and not harming others.

The assumption is that once people become aware of the advantages of CleaNet, they will prefer it to their present browsers. Open discussions about the merits and flaws of the new browser will grow. Attempts will be made to remedy the flaws.

The entire process of debating, implementing, and browsing with CleaNet will be transparent and open for critique and feedback. Netcitizens will be welcomed to provide criticism on the CleaNet hotline and will receive an answer within 24 hours. Netcitizens will have the option to make their feedback public or private, with or without attribution.

Paid officers will screen the hotline and pass thought-provoking complaints to a Complaints Committee. The Complaints Committee will be a subcommittee of the Netcitizens Committee and will include 20 to 40 members. They will receive an additional monthly payment of €500 to €1,000 for their work. It is assumed that sitting on this committee will be a great honor and privilege, thus giving no reason for a higher salary despite the hard work involved. The Complaints Committee will study the complaints and will issue a reasoned response within a month. It is assumed that some Netusers will seek to admit into CleaNet unauthorized sites. Companies or organizations that were excluded from CleaNet will protest. The paid officers or the Complaints Committee will study every complaint and respond in a timely fashion.

By the end of each year, both the Netcitizens Committee and the Complaints Committee will issue an annual report about their work, which will be freely available to all interested parties on the CleaNet website. The report will be as detailed as possible, including the terms of practice, how the terms were implemented, reflections on the past year's work, lessons learned, reasoning for specific decisions, and recommendations for the future.

An International Steering Committee of national representatives will be formed to learn from one another's experiences, to cooperate in case of need, to exchange views, and to deliberate on sensitive issues. As Habermas explained, such public discourse filters reasons and information, topics, and contributions in such a way that the discourse outcome enjoys a presumption of rational acceptability. At the same time, public discourse establishes relations of mutual understanding that are free of violence in the sense that participants seek uncoerced agreement rather than dominating or manipulating others. Habermas described the forms of communication that constitute political discourse as structures of mutual recognition.[39]

The hotline will be operated by a team of paid professionals who will provide effective and speedy response to all questions and criticisms. The hotline will provide easy accessibility, high availability, and an assured response. Both queries and answers will be transparent. They will be posted on the hotline's website. Transparency also means that the rules and procedures according to which concerns are processed will be explained at the point of entry. The system should be explained in detail, and additional help made available if needed. Netusers should have the ability to track their concern throughout the process, and they should be informed of the outcome of the process.[40] The Netcitizens Committee will make publicly available annual reports on the basic statistics and experiences of the Complaints Committee and the hotline.

One may ask, "How is CleaNet different from any of the multiple commercial products that offer filtering of Internet and Web-based content?" Well, to start with, CleaNet will be the result of democratic and open deliberation involving citizens. The decision-making process will involve concerned citizens who will decide together what the future Internet should look like. They will be involved in an ongoing process, offering reasoning and counter-reasoning where everything is put on the table for discussion. Furthermore, CleaNet will be more comprehensive than any existing filter. Whereas some filters are designed to help parents ensure that their children will not encounter pornography on the Net (e.g., Net Nanny[41]), and others are designed to filter hate (e.g., HateFilter[42]), CleaNet will be a transparent browser that provides Netusers with the ability to surf the Internet in a friendly

---

[39] Habermas, *Between Facts and Norms*, 147–57.

[40] Jens Waltermann, and Marcel Machill, eds., *Protecting Our Children on the Internet: Towards a New Culture of Responsibility* (Gütersloh, Germany: Bertelsmann Foundation, 2000), 48.

[41] See the features of this parental control software at http://www.netnanny.com/ .

[42] See the features of this software developed by the Anti-Defamation League at http://www.internet-filters.net/hatefilter.html .

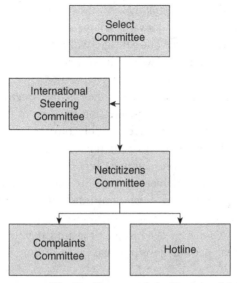

FIGURE 9.1. CleaNet Framework for Decision Making

environment, free of the antisocial, evil material that is now so prevalent and accessible via existing browsers. In addition, CleaNet will be a pragmatic, fluid tool, sensitive to cultural norms and open to contestation. It is designed by the people, for the people, answering people's needs and concerns. CleaNet is suggested precisely because no existing filter can achieve the desired outcome of a clean Internet, with full transparency in regard to the relevant considerations and the citizens' ability to deliberate, exchange ideas, and influence cybersurfing.

See figure 9.1 for the organizational framework of CleaNet. This is a rough proposal. I hope it will attract deliberations and challenges, evoke attention, and gather momentum. With the participation of many concerned citizens in the deliberative process, CleaNet may come to the world as a more refined tool, the result of collective minds aiming to construct a better Internet for this and future generations.

CONCLUSION

This is the first comprehensive book that pits freedom of expression and social responsibility on the Internet against each other. Freedom of expression is important, and so is social responsibility. Norms of social behavior on the Net should be developed to ascertain that the Internet fulfills its vast potential in the best possible way. This book makes a modest attempt to highlight some concerns and to find a way forward. I hope it will provoke attention and discussion. Although not everyone will agree with all of my recommendations, I hope many will work in one way or another to ensure a constructive Web environment for our children.

A major bone of contention relates to the role that Internet intermediaries and gatekeepers should play. I am pressing for a more widely shared and consistent commitment by ISPs to responsible, proactive, and not merely law-abiding conduct. I have shown that the tide of proactive CSR is rising, and now is a good time for more corporate officers to join this flow. As more companies adopt and implement responsible codes of conduct, the risk of incompatible responses across different jurisdictions can be lowered. Hence, the role that governments take upon themselves can become narrower and more focused.[43] We always need to guard against potential political abuse aimed at advancing partisan governmental (to be distinguished from public) interests.

Netusers, readers, ISPs, liberal democracies, and the international community all should share the basic liberal principles of showing respect for others and not harming others. We need, of course, to take into account the temper of the time. Experience is significant. The level of tolerance is in flux. We need to evoke awareness about abuse of the Internet for promoting antisocial, criminal activities and the appropriate ways to counter those activities. The discussion, no doubt, will continue for many years to come.

---

[43] Robert Madelin, "The Evolving Social Responsibilities of Internet Corporate Actors: Pointers Past and Present," *Philosophy and Technology* 24, no. 4 (2011): 455–61, 460.

Glossary

Access providers: Service providers that connect content producers and Netusers to each other through the Internet. Once the connection to the network has been made, access providers have no influence on either what material moves through the wires or where it goes. They are similar to the postal system in that they do not know the contents of the messages they deliver.

Advanced Research Projects Agency (ARPA): An agency established in 1958 by the U.S. Department of Defense to mobilize American universities and research laboratories to develop strategic information technology and computer network research and development.

Akrasia: The state of acting against one's better judgment. Akrasia denotes recklessness and weakness of will.

Al-Qaeda: An international terrorist organization evolved from the Maktab al-Khidamat, a mujahideen resistance organization that fought against the Soviet invasion of Afghanistan during the 1980s. Al-Qaeda aims to (a) drive Americans and American influence out of all Muslim nations, especially Saudi Arabia; (b) destroy Israel; (c) topple pro-Western dictatorships around the Middle East; (d) unite all Muslims; and (e) establish by whatever means necessary an Islamic nation that adheres to the rule of the first caliphs.

Anonymity: The ability to engage in communication while concealing one's identity.

ARPANET (ARPA Network): The world's first advanced computer network. ARPANET was a network of computers designed to transmit information by packet switching.

Autonomous system (AS): A unit of router policy, either a single network or a group of networks, controlled by a common network administrator on behalf of an institution, business company, or Internet service provider. Each AS is assigned a globally unique number called an Autonomous System Number (ASN).

Authentication: The process whereby a degree of confidence is established, usually about the identity of a person with whom a party is dealing. Authentication ensures that the individual is who he or she claims to be.

Bitcoin: A software-based online payment system. It is the first fully implemented decentralized cryptocurrency that provides a simple way to exchange money at very low cost without a central repository.

BITNET: Because It's Time (originally, "Because It's There") Network, established in 1981 between the City University of New York and Yale University. BITNET expanded to more than 500 US and 1,400 international universities and research institutions but became obsolete by the mid 1990s.

Bit Torrent: A peer-to-peer protocol and free, open-source file-sharing program especially for large files.

Blog: Short for weblog, an online journal. The activity of updating a blog is called blogging, and someone who keeps a blog is a blogger.

Blogosphere: The world of blogs.

Bot: A computer program that runs remotely on another machine, usually without the computer owner's awareness.

Botnet: A collection of software robots, or bots, that run automatically and autonomously. Botnets infect PCs so that they forward transmissions (including spam or viruses) to other computers on the Internet, usually without the awareness of those computers' owners. A botnet is also known as a *zombie army*.

Browser: A software program that is used to access the Internet and to locate and display Web pages.

Bulletin Board System (BBS): A form of asynchronous communication that allows participants to create topical groups in which a series of messages can be listed, one after another.

Carnivore: A computer surveillance device that allowed security agents to examine everything that a suspect does on the Internet. The Internet equivalent of a wiretap, Carnivore allowed agents to examine and keep copies of every bit of information sent to and from an individual. In 2005, it was reported that the FBI has effectively abandoned Carnivore.

Chat room: An area of the Internet where Netusers can exchange text-only messages in real time with other people all over the world.

Child Exploitation and Online Protection (CEOP): A UK law enforcement agency established in 2005 that is dedicated to eradicating the sexual abuse of children online and offline.

Child Exploitation Tracking System (CETS): A software solution that supports effective, intelligence-based child exploitation policing by enabling collaboration and information sharing across police services.

Child pornography: Audiovisual material that uses children in a sexual context.

Cipher: Any encryption algorithm. Ciphers can be classified according to whether they are symmetric or asymmetric (i.e., public key algorithms) and by whether they operate on their data as a stream or divided into blocks.

Ciphertext: Text encrypted by a cipher.

Cleanfeed: A Web content–filtering system launched by British Telecom in the summer of 2004. Cleanfeed was created for blocking online child sexual abuse.

CleaNet: An idea for a future browser that would be clean of terror, crime, child pornography, cyberbullying, and hate speech. It would be operated by Netcitizens via tools of deliberate democracy, where ideas are aired and priorities are set with the declared aim of establishing a better Internet for children via proactive cooperation with socially responsible Internet service providers.

Cloud computing: A new generation of computing that uses distant servers for data storage and management, thereby allowing the device to use smaller and more efficient chips that consume less energy than standard computers. Netusers do not physically store their data but leave that responsibility to the data server.

Computer-mediated communication (CMC): Human communication that occurs through the use of two or more computers.

Collaboratory: A system that allows the development of tools to link computers, thereby creating a rich environment for computer-based collaboration. The term merges the words *collaboration* and *laboratory*.

Columbine High School shooting: An event that occurred on April 20, 1999, when two high school seniors, Eric Harris and Dylan Klebold, enacted an all-out, well-planned assault on Columbine High School in Littleton, Colorado. Harris and Klebold killed 12 people and wounded 24 others before committing suicide. They became role models for some people who entertain murderous thoughts, a few of whom have imitated the boys' shooting spree.

Content net neutrality: A principle that accentuates freedom of expression and endorses nondiscrimination of content on the Internet.

Cookies: Bits of data put on a hard disk when a Netuser visits certain websites. Cookies can contain virtually any kind of information, such as the last time a person visited the site; the person's favorite sites; and other similar, customizable information. The most common use of this data is to make it easier for people to use websites that require a username and password. The cookie on the hard disk has the username and password in it, so people do not have to log in to every page that requires that information. Cookies can be used to track people as they go through a website and to help gain statistics about what types of pages people like to visit.

Corporate Social Responsibility (CSR): The continuing commitment by businesses to behave ethically and contribute to economic development while improving the quality of life of workers and their families as well as of the local community and society at large.

Cracker: Common term used to describe a malicious hacker. Crackers get into all kinds of mischief, including breaking, or "cracking," copy protection on software programs, breaking into systems and causing harm, changing data, or stealing. Some hackers regard crackers as less educated individuals who cannot truly create their own work and simply steal other people's work to cause mischief or for personal gain.

Cracking: The modification of software to remove or disable features that are considered undesirable by the person cracking computer software, usually related to protection methods (see also *hacking*).

craigslist: A network of online communities that features online classified advertisements with sections devoted to jobs, housing, goods, services, romance, local activities, personals, gigs, and discussion forums.

Crawler: An Internet bot, also called a *spider*, that systematically browses the World Wide Web. See *also search engines*.

Crime-facilitating speech: Speech that is devised and used to promote crime.

Cryptography: The science of data encryption or coding. Cryptography involves taking an intelligible message and translating it into unintelligible gibberish that could become meaningful only by means of a key (see also *encryption, steganography*).

Cryptology: The science that deals with hidden, disguised, or encrypted communications.

CTIRU: The British Counterterrorism Internet Referral Unit, established by the government in 2010.

Cyberbullying: Using digital technologies to harass, humiliate, intimidate, embarrass, or torment others. Cyberbullying often occurs among minors.

Cybercrime: Illegal activities that involve the exploitation of access to and use of network or system processes. Often cybercrime combines traditional criminal activity with information technology–related crime; this combination is sometimes called a *blended attack*. Examples of criminal uses of Internet communications include cybertheft, online fraud, online money laundering, cyberextortion, cyberstalking, and cybertrespass.

Cyber revenge: Use of the Internet to settle accounts or take revenge for the disliked conduct of another. Cyber revenge commonly involves invasion of privacy.

Cyberspace: A term coined by William Gibson in his 1984 novel *Neuromancer*, referring to a whole range of information resources in the digital world.

Cyberstalking: Using the Internet to place a victim in reasonable fear of death or bodily injury.

Cyberterrorism: The convergence of cyberspace and terrorism. Cyberterrorism involves premeditated activity that uses electronic communication networks to further unlawful and violent ends, and it usually refers to computer-based attacks intended to intimidate or coerce governments or societies in pursuit of goals that are political, religious, or ideological.

Cybertheft: Criminal activities mediated by computer systems for the stealing of money, goods, or intellectual property.

Cyberthreats: Direct threats or distressing material that raises concerns that a Netuser may be considering committing a violent act against himself or herself or others.

Cyberwarfare: Attacks on governmental, financial, and other important information and information systems in an effort to disable official websites and networks, disrupt or disable essential services, steal or alter classified data, and cripple financial systems, among other things.

Dark Web: The alternate, clandestine side of the Internet that is used by criminals, terrorists, and extremist groups to foster cooperation; spread their ideas; and promote their antisocial, criminal, and violent causes. The Dark Web contains data crafted deliberately to lie beyond the reach of search engines. It is also known as the *Deep Web* or *Deep Net*.

Data Encryption Standard (DES): A popular commercial encryption algorithm. Originally developed at IBM, DES was chosen by the US National Bureau of Standards as the government-standard encryption algorithm in 1976. Since then, it has become a domestic and international encryption standard and has been used in thousands of applications.

Data retention: The storage of data for backup and historical purposes. Data retention refers to the storage of Internet traffic and transaction data, especially of emails, telecommunications, and visited websites.

Decryption: The activity of decoding encryption.

Deep Web: See *Dark Web*.

Democratic catch: Theory that holds that the foundations of democracy might undermine it and bring about its end.

Denial of service (DoS) attack: An attack that takes place when access to a computer or to network resources is intentionally blocked or degraded. This blockage or degradation occurs as a result of malicious action taken by one or more Netusers who use coordinating computers to send repeated requests to a Web page, ultimately overwhelming the server for that page. The attacker destroys files, degrades processes or storage capability, blocks resources, or causes the shutdown of a process or of the system.

Digital signature: A digital code that can be attached to an electronically transmitted message that uniquely identifies the sender. Its purpose is to guarantee that the individual sending the message really is who he or she claims to be. Digital signatures are especially important for e-commerce and are a key component of most authentication schemes. To be effective, digital signatures must be unforgeable. A number of encryption techniques guarantee this level of security.

Domain Name System (DNS): The service that translates Internet names. DNS helps Netusers to find their way around the Internet.

eBay: The largest online auction and shopping website. Founded in 1995, eBay allows people and businesses to buy and sell a broad variety of goods and services worldwide. The website offers an online global trading platform where millions of items are traded every day and where nearly everyone can trade practically anything.

Echelon: An automated global interception and relay system operated by the intelligence agencies in five nations: the United States, the United Kingdom, Canada, Australia, and New Zealand. Satellite receiver stations and spy satellites allegedly give Echelon the ability to intercept any telephone, fax, email, or Internet message sent by any individual and to inspect its contents.

E-jihad: The way information technology is applied by al-Qaeda and other organizations to organize logistics for their campaigns and to develop their strategic intelligence.

Encryption: Technology designed to hide certain information from the eyes of others by creating a private language between the information source and those trying to gain access to the information. Encryption is a means of securing critical infrastructure and protecting sensitive information. The power of encryption systems resides in the power of the algorithm and the size of the key: the longer the key, the larger the equation (see *cryptography*, *steganography*).

Engine for content analysis (ECA): A search engine that provides medium-range security by searching for the specific content that can be an indicator for the types of harmful behavior that an engine for relationship analysis uncovers (see *engine for relationship analysis*).

Engine for relationship analysis (ERA): A search engine that monitors, analyzes, and assesses online relationships as they develop by examining live chat and instant messenger conversations.

EuroISPA: The pan-European association of European Internet service providers associations.

Exploit: A piece of software, chunk of data, or sequence of commands that causes irregular behavior to occur in computer software or hardware by taking advantage of a bug, glitch, or vulnerability.

Facebook: A social network website founded in February 2004 by Mark Zuckerberg, Eduardo Saverin, Dustin Moskovitz, and Chris Hughes. Originally a network for American universities, Facebook was later extended to anyone with a registered email address. The site is free to join and makes a profit through advertising revenue. It is the largest social network on the Internet, with more than 1 billion users.

File sharing: A process in which devices controlled by Netusers interact directly with each other to transfer files between them. The practice is generally associated with the sharing of video clips, movies, music, images, and software through dedicated websites.

File Transfer Protocol (FTP): A program that enables Netusers to transfer files between computers on a network.

Filtering: Using software to restrict the information that one may find on the Web. Filtering can be used in various ways to disrupt or prevent spam, hoaxes, viruses, and other malware from reaching Netusers.

Firewall: A computer network's security system, composed of hardware and software, that separates a network into two or more parts for security purposes. It controls access between two networks, such as a local area network and the Internet or the Internet and a computer.

First Amendment: Part of the Bill of Rights to the US Constitution. The First Amendment states: "Congress shall make no law . . . abridging the freedom of speech, or of the press."

Flickr: An online photo management and sharing application.

Foursquare: A social networking site that allows users to share their location with friends through GPS-enabled mobile devices. Established in 2009 by Dennis Crowley and Naveen Selvadurai, the site is also a game in which users earn points and badges for checking in at the bars and restaurants that they visit.

Freenet: A free software that enables anonymous sharing of files, browsing and publishing of "freesites" (websites accessible only through Freenet), and chatting on forums without fear of censorship. Freenet is decentralized and difficult to detect. Communications are encrypted and routed through other nodes to make it extremely difficult to determine who is requesting the information and what its content is.

Google: An American public corporation that earns revenue from advertising related to its Internet search, email, online mapping, office productivity, social networking, and video-sharing services as well as through sales of advertisement-free versions of the same technologies.

Goth: A culture that first arose in the United Kingdom during the 1980s as an offshoot of the post–punk rock era. Members of the culture (known as *goths*) favor black clothing; dyed black, bleach blond, red, or purple hair; body piercing; and extreme pallor. Some goths have a morbid fascination with death and darkness. Popular cultural influences include Mary Shelley's *Frankenstein*, Anne Rice's *Vampire Chronicles*, movies such as *Interview with the Vampire* and *Dracula*, and television series such as *Buffy the Vampire Slayer*. Vampires are a clique among the goth culture and take part in rituals that are throwbacks to their Eastern European roots.

Grooming: The process of reducing a child's resistance to sexual abuse and molestation with the aim of securing the victim's cooperation.

Hacker: An unauthorized person accessing a computer or information system.

Hacking: A term associated with two different meanings: (a) creative problem solving when faced with complex technical problems, or (b) illicit and often illegal activity involving unauthorized access to information technology systems (see *cracking*).

Haganah: A global intelligence network dedicated to confronting global jihad online. *Haganah* means defense in Hebrew.

Happy slapping: The recording and transmitting of images of abuse, usually among youth.

Hate crime: Harm inflicted on a victim who is chosen because of his or her actual or perceived race, color, religion, national origin, ethnicity, gender, disability, or sexual orientation.

Hate speech: A bias-motivated, hostile, malicious speech aimed at a person or a group of people because of some of their actual or perceived innate characteristics.

Hoax: A fabricated story or image that alters facts. Often spread through email, chat rooms, forums, and blogs, hoaxes can be merely innocuous tales or can be malicious efforts to crash servers and spread viruses.

Honeynet: A network of hosts that is specifically designed to be compromised. The main purpose of such networks is to study hackers' behaviors, tools, motives, and possible contra measures.

Honeypot: An Internet site designed to lure criminals in a scheme to study their behavior and eventually catch them.

Hosting: See *Internet hosting service.*

Hotline: A line of communication that enables Netusers to respond to problematic Internet content by drawing attention to the content. Netusers can communicate their concerns by phone, fax, or email. Hotlines require easy accessibility, high availability, and assured responses from their workers to be effective. The hotline receives a report and, if necessary, sets in motion a process of response. The response involves processing the report, providing the Netuser with feedback, and deciding whether to forward the report to law enforcement or a self-regulatory authority.

HyperText Markup Language (HTML): Coding language used to create Internet documents. HTML facilitates finding information and moving between resources. The codes (called *tags*) are embedded in the text and define the page layout, fonts, graphic elements, and hypertext links to other documents on the Web.

Hypertext Transfer Protocol (HTTP): The communication protocol used on the Internet to transfer data and communicate between computers.

I2P: An effort, also called the *Invisible Internet Project*, to build, deploy, and maintain a network to support secure and anonymous communication. There is no central point in the I2P network on which pressure can be exerted to compromise the integrity, security, or anonymity of the system. The network supports dynamic reconfiguration and has been designed to make use of additional resources as they become available. All aspects of the network are open and freely available.

ICQ: A popular instant messaging software program developed by the Israeli company Mirabilis. ICQ is pronounced "I seek you."

International Data Encryption Algorithm (IDEA): A symmetrical encryption algorithm developed by ETH Zürich. The encoder and decoder use the same key, a sophisticated block cipher with a 128-bit key. IDEA is considered to be among the best publicly known algorithms.

Incitement: Speech that directly provokes illegal action.

Instant Message (IM): A two-way, real-time, private dialogue between two Netusers. The user initiating an IM sends an invitation to talk to another (specific) user who is online at the same time.

Interface Message Processor (IMP): The original message-switching node on ARPANET. IMP was a system-independent interface that could be used by any computer system to access the ARPANET.

International Association of Internet Hotlines (INHOPE): An association founded in 1999 under the European Commission's Safer Internet Action Plan. INHOPE represents Internet hotlines all over the world, supporting them in their aim to respond to reports of illegal content and to make the Internet safer.

Internet: The vast global system of interconnected networks. The Internet is the global data communication capability realized by the interconnection of public and private telecommunication networks using Internet Protocol (IP), Transmission Control Protocol (TCP), and the other protocols required to implement IP internetworking on a global scale, such as Domain Name System (DNS) and packet-routing protocols. The Internet Engineering Task Force defines the Internet as "a loosely organized international collaboration of autonomous, interconnected networks" that "supports host-to-host communication through voluntary adherence to open protocols and procedures defined by Internet Standards."

Internet Architecture Board (IAB): A committee of the Internet Engineering Task Force. One of the IAB's responsibilities is to provide oversight of, and occasional commentary on, aspects of the architecture for the protocols and procedures used by the Internet. It also oversees the process used to create Internet standards.

Internet Assigned Numbers Authority (IANA): The body responsible for coordinating some of the key elements that keep the Internet running smoothly. Specifically, IANA allocates and maintains unique codes and numbering systems that are used in the technical standards ("protocols") that drive the Internet. IANA's activities can be broadly grouped into three categories: management of domain names, coordination of the global pool of IP addresses, and management of protocol assignments.

Internet Corporation for Assigned Names and Numbers (ICANN): A corporation established in 1998 to coordinate, at the broadest level, the global Internet's systems of unique identifiers and in particular to ensure the stable and secure operation of the Internet's unique identifier systems. ICANN coordinates the Domain Name System (DNS), Internet Protocol (IP) addresses, space allocation, protocol identifier assignment, generic (gTLD) and country code (ccTLD) top-level domain name system management, and root server system management functions. ICANN is also responsible for accrediting the domain name registrars. *Accredit* means to identify and set minimum standards for the performance of registration functions, to recognize persons or entities meeting those standards, and to enter into an accreditation agreement that sets forth the rules and procedures applicable to the provision of registrar services. These services were originally performed under US government contract by the Internet Assigned Numbers Authority (IANA) and other entities. ICANN took to performing the IANA functions.

Internet Engineering Task Force (IETF): An organized activity of the Internet Society. The IETF's mission is to make the Internet work better from an engineering point of view. This committee of more than 100 working groups is designed to ensure the working of the Internet protocols.

Internet governance: The development and application by governments, the private sector, and civil society, in their respective roles, of shared principles, norms, rules, decision-making procedures, and programs that shape the evolution and use of the Internet.

Internet Haganah: See *Haganah*.

Internet Hosting Service: A service that runs Internet servers, thereby allowing organizations and individuals to serve content to the Internet.

Internet Protocol (IP): A standard procedure for regulating data transmission between computers. It is part of the TCP (Transmission Control Protocol)/IP suite of protocols that allows the various machines that make up the Internet to communicate with each other.

Internet Protocol (IP) address: An identifier for a computer or device on a TCP (Transmission Control Protocol)/IP network. It is the numeric Internet protocol address that uniquely identifies a connection to a specific Internet location. The format of an IP address is a 32-bit numeric address written as four numbers separated by periods. Each number can be between zero and 255.

Internet Relay Chat (IRC): An Internet conferencing system where users do not communicate directly with each other. Instead, the server broadcasts all messages to all users of a particular channel. IRC enables people all over the world to hold conversations on topics of interest.

Internet service provider (ISP): A company or other organization that provides access to the Internet, usually for a fee, thereby enabling Netusers to establish contact with the public network.

Internet Society (ISOC): A global cause-driven organization, founded in 1992, that is dedicated to ensuring that the Internet stays open, transparent, and defined by Netusers. It provides a platform for sharing ideas and for technological innovations.

Internet telephony: Two-way transmission of audio over the Internet, allowing users to employ the Internet as the transmission medium for telephone calls. It is also commonly referred to as voice over Internet Protocol (VoIP).

Internet Watch Foundation (IWF): A foundation established in 1996 by the UK Internet industry to provide hotlines for the public and for information technology professionals to report criminal online content in a secure and confidential way. IWF works in partnership with the online industry, law enforcement, the government, and international partners to minimize the availability of child sexual abuse images hosted anywhere in the world, criminally obscene adult content hosted in the United Kingdom, incitement to racial hatred content hosted in the United Kingdom, and nonphotographic child sexual abuse images hosted in the United Kingdom.

Jihad: Struggle, strife, or effort. Muslims use the term to describe four different kinds of struggle: (a) *jihad nafsii* (self-improvement), (b) *jihad shaitani* (struggle with the devil), (c) *jihad al-kufar* (struggle against the infidel), and (d) *jihad al-munafikin* (struggle against hypocrites).

Joint Academic Network (Janet): A network established by and connecting UK universities. Janet is the backbone network for the UK university system of academic and research computers.

JuicyCampus.com: A website, now defunct, that aimed at becoming the "world's most authentic college website," with content generated by college students for college students.

Justin.tv: A website, now shut down, that allowed users to produce and watch live-streaming video. As in the case of YouTube, Justin.tv user accounts were called *channels*, and users were encouraged to broadcast a wide variety of user-generated live video content, called *broadcasts*.

Keystroke Logger: A program used by both the hackers' community and the security community. Keystroke Logger is usually stealth-type software injected between the keyboard and operating system to record every keystroke.

Keystroke monitoring: Audit trail software or a specifically designed device that records every key struck by a user and every character of the response that the information system returns to the user.

Magic Lantern: An information-gathering technique used by the US Federal Bureau of Investigation for the handling of encrypted information on the computers of criminals and terrorists.

Malicious software: A variety of computer codes designed to disrupt or corrupt the computer's normal operation. Malicious software includes viruses and Trojan horses.

Mashable: Website and blog founded in 2005 by Pete Cashmore. Mashable is an extensive resource for news in social and digital media, technology, and web culture. This social media guide is popular with bloggers and Twitter and Facebook users. It is said to be one of the top 10 most profitable blogs in the world.

Moral panic: An unwarranted or excessive reaction to a perceived social problem of deviance or crime often produced by media (mis)representation.

Mujahideen Secrets: A jihadi encryption program that provides users with five encryption algorithms, advanced encryption keys, and data compression tools.

MySpace: Social networking website founded in July 2003 by Tom Anderson and Chris DeWolfe. MySpace was the most popular social networking site in the United States, with more than 100 million users, until it was overtaken internationally by its main competitor, Facebook.

Netcitizen: A responsible user of the Internet. Netcitizens are also called *Netizens*.

Netiquette: Internet etiquette, or the set of commonly accepted rules that represent proper Net behavior.

Netizen: See *Netcitizen*.

Net Nanny: A brand of content control software that offers simple, easy-to-use setup assistance to help parents determine what online activities are appropriate for their children. Net Nanny enables a parent or guardian to monitor everything that passes through the computer.

Net neutrality: The guiding principle for Internet use in the United States and the result of the First Amendment to the U.S. Constitution, which allows wide scope for freedom of expression. According to this principle, Internet service providers should not be allowed to discriminate between the content that users post and access online.

Netuser: A user of the Internet.

Network backbone: Infrastructure that includes the long-distance lines and supporting technology that transport large amounts of data between major network nodes.

Newsgroups: Electronic bulletin boards that enrolled members can access whenever they choose so that they can read messages posted by other members or post their own, thereby contributing to collective discussion around the newsgroup's specific topic. Newsgroups are varied, changeable, and chaotic. Today, there are hundreds of thousands of newsgroups. Netusers can post messages, pictures, sound files, and video files. The people who run the newsgroups are usually anonymous.

Onion routing: An Internet-based system to prevent eavesdropping and traffic analysis attacks. The system uses an *onion router*, a router that builds a new chain of servers and encryption keys for each Netuser at frequent intervals (see also *Tor*).

Open-source software: A software system whose code is directly accessible to the user. The software is free. No fees or royalties are charged.

Packet: A unit of data sent over the Internet. It includes headers and a sequential number, which allow the packet to be sent to its destination, using one of many possible routes, and then assembled with other packets to form a complete data file.

Pedophilia: A clinical term used to refer to a psychosexual disorder involving a sexual preference for children.

Peer-to-peer (P2P) communication: A communication model in which each party has the same capabilities and either party can initiate a communication session. In some cases, P2P communications are implemented by giving each communication node both server and client capabilities. In recent usage, P2P has come to describe applications in which Netusers can use the Internet to exchange software, music files, photos, video clips, and movies with each other directly or through a mediating server.

Pharming: An attack used by hackers to redirect a website's traffic to another (usually bogus) website.

Phishing: An attempt to fraudulently acquire sensitive information, such as usernames, passwords, or credit card details, from Netusers. The attempt is often made through an email that is crafted to appear as if it was sent from a legitimate company. Typically, the phishing email will purport to have come from a bank, possibly warning of suspicious activity on one's account. The message may include legitimate bank logos and might link to a site that looks identical to a familiar online banking site but is purely designed to capture sensitive information.

Platform for Internet Content Selection (PICS): Technology intended to give Netusers control over the kinds of material to which they and their children have access. PICS enables "self-rating" by content providers of material that they create and distribute. It allows individuals to use ratings and labels from a diversity of sources to control the information that they and those under their supervision receive. PICS has been superseded by the Protocol for Web Description Resources (POWDER).

Podcast: A form of audio broadcasting that uses the Internet.

Portal: A website or service that offers a broad array of resources and services, such as email, search engines, subject directories, and forums. Yahoo! is one of the largest Internet portals today.

Pretty Good Privacy (PGP): A widely available software package that permits Netusers to use encryption when exchanging messages. Developed by Philip R. Zimmermann in 1991, PGP has become a de facto standard for email security. PGP can also be used to encrypt files being stored so that they are unreadable by other users or intruders.

Protocol: See *Internet Protocol*.

Protocol for Web Description Resources (POWDER): Protocol that provides a mechanism to describe and discover Web resources and helps Netusers make decisions regarding whether a given resource is of interest.

Proxy server: A server that sits between a client application, such as a Web browser, and a real server. It intercepts all requests to the real server and filters them to see if it can fulfill the requests itself. A proxy server may optionally alter the client's request or the server's response, and sometimes it may serve the request without contacting the specified server. Proxies are used to keep the machines that are behind them anonymous, to block undesired sites, to bypass security controls, to scan transmitted content for malware before delivery, and to circumvent regional restrictions. When used in conjunction with a firewall, a proxy server's identity (as well as its connected personal computers) is completely masked or hidden from other users. Secure sites operate through the use of proxy servers.

Racism: An ideology or set of beliefs that assumes that races are distinct human groups that have specific characteristics. Those characteristics define a racial group's cultures, beliefs, and moralities, usually in ways that devalue and render members of other racial groups inferior and of less value. Racism involves persecution, humiliation, degradation, a display of enmity and hostility, or incitement of violence against a target group because of its color, racial affiliation, or national ethnic origin.

Radicalization: The process whereby individuals and groups become political extremists in views and, potentially, in deeds.

Real-time message filter (RMF): A filtering system that offers short-range protection by either screening out or blocking completely any inappropriate content before it reaches its intended recipient.

Revenge porn: Homemade pornography uploaded to the Internet to humiliate a former girlfriend or boyfriend after a particularly vicious breakup.

Rogue software: Software that uses malware or malicious tools to advertise or install itself or to force computer users to pay for removal of nonexistent spyware. Rogue software often installs a Trojan horse to download a trial version or executes other unwanted actions.

Rootkit: Stealth software tools used by a third party who gains access to another person's computer system. A rootkit can be used to alter files or execute processes without the consent and knowledge of the owner of the computer system.

Router: A server on the Internet that keeps track of Internet addresses.

RSA: A public-key encryption system that employs a powerful algorithm. RSA is named after its originators, Ron Rivest, Adi Shamir, and Leonard Adelman.

Search engines: Internet tools that identify information by the uniform resource locator (URL), or Web address, as well as by hidden identifiers. Each search engine uses a crawler or spider with its own set of rules that determine how documents are gathered.

Second Life: A free 3D virtual world where Netusers can socialize, connect, and create using free voice and text chat. Members assume an identity and take up residence in Second Life, creating a customized avatar or personage to represent themselves.

Server: A computer that provides a particular service. Web servers display Web pages, and email servers handle mail messages.

Sexting: The sharing of explicit texts and nude photos, usually by teenagers through cellphones. However, sexting is also happening on the Internet and in other age groups. As technology has advanced and cell phones have developed the capability to record and send photos and video, the practice of sending suggestive and explicit pictures has increased, especially among teens.

Sextortion: The use of sexually explicit or suggestive photos to blackmail the victim shown in the photos.

Skype: A popular, easy-to-use Internet program that enables people to call, to talk, and to video chat with other Skype users.

Sniffers: Programs secretly installed on the communication channels to eavesdrop on network traffic, examining all traffic on selected network segments.

Social network–hosting service: A web-hosting service that specifically hosts the creation of web-based social networking services alongside related applications.

Social networking: The practice of interacting and forming social relationships with others by using Internet services designed for that purpose.

Social networking site (SNS): A platform for communicating and sharing information between Netusers ("friends") on an online forum.

Social responsibility: An ethical framework that relates to the societal implications of a given conduct for which an agent is responsible.

Spam: Unsolicited email that pitches products, updates, and other messages to thousands of people who perceive it as junk mail. Spam abuses electronic messaging systems to indiscriminately send unwanted bulk messages, usually so that the sender can make a profit. Spammers use software applications that run automated tasks over the Internet (see *bots*).

Spider: See *crawler.*

Spyware: Software that is secretly installed on a Netuser's computer that monitors use of the computer in some way without the user's knowledge or consent. Some spyware also sends information about the user to another user over the Internet.

Steganography: Literally, "hidden writing." Unlike cryptography, which creates an unreadable version of a message for anyone without the key, steganography involves spread-spectrum radio transmissions, in which parts of a message (or even parts of individual bits) are sent on pseudo-randomly varying radio frequencies. Without the right equipment, the signal is merely electronic white noise. Similarly, it is possible to use steganography to conceal information in digital files (see *cryptography, encryption*).

Stormfront: Website established by Don Black on January 11, 1995, that is considered the first hate site on the World Wide Web. Since its creation, Stormfront has served as a veritable supermarket of online hate, distributing antisemitism and racism. Black's site became home to the webpages of other extremists, such as Aryan Nations and Ed Fields, racist publisher of *The Truth at Last*, a hate-filled newspaper.

Super Columbine Massacre RPG: A violent video game released in 2005. *Super Columbine Massacre RPG* is based on the Columbine High School shootings by Eric Harris and Dylan Klebold in Littleton, Colorado. The game delves into the morning of April 20, 1999, and allows players to relive that day through the eyes of Harris and Klebold.

SurfWatch: Software package that lets parents or educators limit access to sexually explicit material on the Internet without restricting the access rights of other Netusers.

Telnet: "Telecommunications network," a protocol that provides a way for Netusers to connect to multiuser computers. Telnet was the original Internet when the Net first launched in 1969. Telnet was built to be a kind of remote control that could be used to manage mainframe computers from distant terminals.

Terrorism: The threat or employment of violence against noncombatant targets for political, religious, or ideological purposes by subnational groups or clandestine individuals who are willing to justify all means to achieve their goals.

Tor: The most popular onion router. Tor allows Netusers to connect anonymously to any TCP/IP (Transmission Control Protocol/Internet Protocol) service on the Internet. Using Tor protects Netusers from traffic analysis. Activists in different parts of the less democratic world use Tor to maintain anonymity when communicating about socially sensitive issues and accessing websites that are blocked by their governments (see also *Onion router*).

Transmission Control Protocol/Internet Protocol (TCP/IP): Set of rules that computers on a network use to organize data into packages, put them into the right order on arrival at their destination, and check them for errors.

TCP/IP allows different hosts on the Internet to establish a connection with each other and exchange streams of data.

Trojan horse: Software program that performs like any real program that a user might wish to run but also performs unauthorized actions behind the scenes. The term was coined by a hacker, Dan Edwards, after the mythical Trojan horse that was able, under the guise of a harmless gift, to put behind enemy lines the group of Greek warriors concealed in its interior. A Trojan horse might be a malicious, security-breaking program that is disguised as something benign, such as a directory lister, archiver, game, or virus-destroying program. Unlike worms, Trojan horses are not capable of replication.

Trolling: A troll joins a listserv or other online community with the intention of taking over the communicative bandwidth of the group by provoking others and drawing attention to his or her own agenda, ideas, and theories. Trolls may not always intend to be destructive, but they are infamously so.

Twitter: A free social networking site, founded in 2006 by Jack Dorsey, that combines Short Code Messaging—Short Message Service—with social grouping.

Uniform resource locator (URL): A webpage address. The URL is the unique identifier for a page and functions like a postal or email address. It includes information about the location of the host computer, the location of the website on the host computer, the name of the web page, and the file type of each document. For instance, the URL http://www.internet-guide.co.uk/index.html breaks down as follows: *http://*, which is short for Hypertext Transfer Protocol; *www.*, which is the page on the World Wide Web; *internet-guide.co.uk*, which is the domain name; and *index.html*, which is the webpage inside the folder.

Usenet: A worldwide, decentralized, distributed Internet discussion system consisting of a set of newsgroups with names that are classified hierarchically by subject. The system evolved from general-purpose architecture and was run by volunteers with common interests. Usenet resembles bulletin board systems in most respects and is the predecessor to the various Internet forums that are widely used today. The original Usenet groups did not have a central server. Instead, a system of servers, usually on individual computers, served to hold the network together.

VampireFreaks: A social networking site dedicated to goth culture.

Video-hosting service: A service that allows individuals to upload video clips to an Internet website. The video host will then store the video on its server and show the individual different types of code that will allow others to view this video. The website used to host the videos is usually called the *video-sharing website*. Among the most popular video-hosting services are MySpace, YouTube, Flickr, Big Think, Google Videos, and Yahoo! Screen.

Virtual hosting: The ability of Internet service providers to host websites or other services for different entities on one computer while giving the appearance that they exist on separate servers. For instance, with virtual hosting, one might have websites from two separate, distinct organizations residing on and being served from one particular server (with one particular Internet Protocol address).

Virus: A segment of a computer code or a program that will copy its code into one or more "host" programs when it is activated. When the embedded virus is executed, it might corrupt a computer's hard disk and destroy its data. This process tends to be invisible to the user.

Voice over Internet Protocol (VoIP): See *Internet telephony*.

Web-hosting service: A service that allows individuals and organizations to provide their own website accessible through the World Wide Web. Web-hosting companies provide space on a server that they own for use by their clients as well as providing Internet connectivity. Web hosts can also provide data center space and connectivity to the Internet to servers they do not own, which are then located in their data center—a practice called *colocation*.

Wikipedia: A free, multilingual, open-content online encyclopedia created through the collaborative effort of a community of Netusers known as *Wikipedians*. Anyone registered on the site can create an article for publication; registration is not required to edit articles. The site's name comes from the Hawaiian word *wiki*, meaning quick, which refers to a technology used for creating collaborative websites, and the word *encyclopedia*. Wikipedia's articles provide links to guide the user to related pages with additional information.

WiredSafety: The largest organization in the world that provides help and safety online.

World Wide Web (WWW): A subsection of the Internet that consists of documents residing on different computers in the world. The WWW is a globally connected network that contains at present millions of webpages incorporating text, graphics, audio, video, animation, and other multimedia elements.

Worm: A software program that propagates by itself, across a network, into the systems of vulnerable computers. A worm enters a system by exploiting bugs or overlooked features in commonly used network software already running on the targeted system. It is not a virus because it does not attach itself to other programs or interfere with their operation. It is a bit of extra computer code that could harmlessly sit on someone's machine or could corrupt files or do other damage that its author commands. Worms often exist purely in memory; hence, they avoid the file system and, more importantly, are invisible to the file-scanning antivirus software.

Yahoo!: A World Wide Web directory started by David Filo and Jerry Yang at Stanford University in 1994. The two began compiling and categorizing webpages and in less than two years made Yahoo! into one of the most popular search engines and websites in the world. The company also provides multiple other Web services, including a directory (Yahoo! Directory), email, news, maps, advertising, auctions, and video sharing (Yahoo! Screen). The name Yahoo! is short for Yet Another Hierarchical Officious Oracle.

YouTube: Founded in 2005 by Chad Hurley, Steve Chen, and Jawed Karim, who were all early employees of PayPal. YouTube is the leader in online video, allowing Netusers to share original videos worldwide through a Web experience. YouTube allows people to easily upload and share video clips across the Internet through websites, mobile devices, blogs, and email. In November 2006, YouTube was bought by Google for $1.65 billion and is now operated as a subsidiary of Google.

Zombie: A computer that a remote attacker has accessed and set up, using a program known as a *bot*, to forward transmissions (including spam and viruses) to other computers on the Internet. The purpose is usually either malice or financial gain.

Selected Bibliography

GENERAL

Abbate, Janet. *Inventing the Internet.* Cambridge, MA: MIT Press, 2000.

Abel, Richard L. *Speaking Respect, Respecting Speech.* Chicago: University of Chicago Press, 1998.

Abend, Gabriel. *The Moral Background: An Inquiry into the History of Business Ethics.* Princeton, NJ: Princeton University Press, 2014.

Ackerman, Bruce A. *Social Justice in the Liberal State.* New Haven, CT: Yale University Press, 1980.

Allport, Gordon W. *The Nature of Prejudice.* Cambridge, MA: Addison-Wesley, 1954.

Amichai-Hamburger, Yair, ed. *The Social Net: Understanding Human Behavior in Cyberspace.* Oxford: Oxford University Press, 2005.

Aristotle. *Nicomachean Ethics.* Edited and translated by Martin Ostwald. Indianapolis, IN: Bobbs-Merrill, 1962.

Bader, Eleanor J. "The Patriot Act Has Led to Increased Censorship in the United States." In *Censorship: Opposing Viewpoints*, edited by Andrea C. Nakaya, 103–7. Farmington Hill, MI: Greenhaven, 2005.

Balkin, Jack M., Beth S. Noveck, and Kermit Roosevelt. "Filtering the Internet: A Best Practices Model." In *Protecting Our Children on the Internet: Towards a New Culture of Responsibility*, edited by Jens Waltermann and Marcel Machill, 199–261. Gütersloh, Germany: Bertelsmann Foundation, 2000.

Barak, Aharon. "Freedom of Expression and Its Limitations." In *Challenges to Democracy: Essays in Honour and Memory of Professor Sir Isaiah Berlin*, edited by Raphael Cohen-Almagor, 167–88. Aldershot, UK: Ashgate, 2000.

Baran, Paul. "On Distributed Communications." Memorandum RM-3420-PR, RAND, Santa Monica, CA, 1964, http://www.rand.org/content/dam/rand/pubs/research_memoranda/2006/RM3420.pdf .

Barendt, Eric. *Freedom of Speech.* New York: Oxford University Press, 2007.

Beckett, David. "Internet Technology." In *Internet Ethics*, edited by Duncan Langford, 13–46. New York: St. Martin's Press, 2000.

Belkin, Danny. "The Internet Will Speed Up Human Evolution." In *The Future of the Internet*, edited by Tom Head, 33–42. Farmington Hills, MI: Greenhaven Press, 2005.

Benedek, Wolfgang, and Matthias C. Kettemann. *Freedom of Expression and the Internet*. Strasbourg, France: Council of Europe Publishing, 2014.

Benkler, Yochai. *The Wealth of Networks: How Social Production Transforms Markets and Freedom*. New Haven, CT: Yale University Press, 2006.

Bessette, Joseph. *The Mild Voice of Reason: Deliberative Democracy and American National Government*. Chicago: University of Chicago Press, 1994.

Biddle, Lucy, Jenny Donovan, Keith Hawton, Navneet Kapur, and David Gunnell. "Suicide and the Internet," *British Medical Journal* 336, no. 7648 (2008): 800–802.

Black, Hugo L. "The Bill of Rights." *New York University Law Review* 35 (1960): 865–81.

Boatright, John R. *Ethics and the Conduct of Business*. Upper Saddle River, NJ: Pearson Prentice Hall, 2009.

boyd, danah m., and Nicole B. Ellison. "Social Networks Sites: Definition, History, and Scholarship." *Journal of Computer-Mediated Communication* 13, no. 1 (2007): 210–30. http://onlinelibrary.wiley.com/doi/10.1111/j.1083-6101.2007.00393.x/abstract.

Brenkert, George G., and Tom L. Beauchamp, eds. *The Oxford Handbook of Business Ethics*. New York: Oxford University Press, 2010.

Bryce, James. *Modern Democracies*. London: Macmillan, 1924.

Buckingham, David, ed. *Youth, Identity, and Digital Media*. Cambridge, MA: MIT Press, 2008.

Bunt, Gary R. *Islam in the Digital Age: E-Jihad, Online Fatwas, and Cyber Islamic Environments*. London: Pluto Press, 2003.

Bush, Vannevar. "As We May Think." *Atlantic Monthly* 176, no. 1 (1945): 101–8.

Bynum, Terrell Ward. "Norbert Wiener's Vision: The Impact of 'the Automatic Age' on Our Moral Lives." In *The Impact of the Internet on Our Moral Lives*, edited by Robert J. Cavalier, 11–25. Albany: State University of New York Press, 2005.

Campbell, Tom, and Seumas Miller, eds. *Human Rights and the Moral Responsibilities of Corporate and Public Sector Organisations*. Dordrecht, Netherlands: Kluwer, 2004.

Carroll, Archie B. *Business and Society: Managing Corporate Social Performance*. Boston: Little, Brown, 1981.

——— "Corporate Social Responsibility." *Business and Society* 38, no. 3 (1999): 268–95.

——— "The Pyramid of Corporate Social Responsibility: Toward the Moral Management of Organizational Stakeholders." *Business Horizons* 34, no. 4 (1991): 39–48.

——— "A Three-Dimensional Conceptual Model of Corporate Social Performance." *Academy of Management Review* 4, no. 4 (1979): 497–505.

Carroll, Archie B., and Ann K. Buchholtz. *Business and Society: Ethics, Sustainability, and Stakeholder Management*. 8th ed. New York: South-Western College, 2011.

Castells, Manuel. *Communication Power*. Oxford: Oxford University Press, 2009.

——— *The Internet Galaxy: Reflections on the Internet, Business, and Society*. Oxford: Oxford University Press, 2001.

Cavalier, Robert J., ed. *The Impact of the Internet on Our Moral Lives*. Albany: State University of New York Press, 2005.

Cerf, Vinton G. "Computer Networking: Global Infrastructure for the 21st Century." Computing Research Association, Washington, DC, 1995. http://www.cs.washington.edu/homes/lazowska/cra/networks.html.

——— "First, Do No Harm." *Philosophy and Technology* 24, no. 4 (2011): 463–65.

"The Scope of Internet Governance." In *Internet Governance Forum (IGF): The First Two Years*, edited by Avri Doria and Wolfgang Kleinwächter, 51–56. Geneva: IGF Office, 2008.

Cerf, Vinton G., and Robert Kahn. "A Protocol for Packet Network Interconnection." *IEEE Transactions on Communications* 22, no. 5 (1974): 637–48.

Ceruzzi, Paul E. *A History of Modern Computing*. 2nd ed. Cambridge, MA: MIT Press, 2003.

Chappell, Zsuzsanna. *Deliberative Democracy: A Critical Introduction*. Houndmills, UK: Palgrave-Macmillan, 2012.

Christians, Clifford G. "Self-Regulation: A Critical Role for Codes of Ethics." In *Media Freedom and Accountability*, edited by Everette E. Dennis, Donald M. Gillmor, and Theodore L. Glasser, 35–53. Westport, CT: Greenwood, 1989.

Christians, Clifford G., John P. Ferré, and P. Mark Fackler. *Good News: Social Ethics and the Press*. New York: Oxford University Press, 1993.

Christians, Clifford G., and Kaarle Nordenstreng. "Social Responsibility Worldwide." *Journal of Mass Media Ethics* 19, no. 1 (2004): 3–28.

Christians, Clifford G., and Michael Traber, eds. *Communication Ethics and Universal Values*. Thousand Oaks, CA: Sage, 1997.

Citron, Danielle Keats. "Cyber Civil Rights." *Boston University Law Review* 89 (2009): 61–125.

Clark, David, Frank Field, and Matt Richards. "Computer Networks and the Internet: A Brief History of Predicting Their Future." Advanced Network Architecture, Massachusetts Institute of Technology, Cambridge, MA, January 2010. http://groups.csail.mit.edu/ana/People/DDC/Predicting-the-future-3-0.pdf .

Coates, Joseph H. "The Internet Will Enhance Human Life." In *The Future of the Internet*, edited by Tom Head, 10–18. Farmington Hills, MI: Greenhaven Press, 2005.

Cohen, Joshua. "Deliberation and Democratic Legitimacy." In *The Good Polity: Normative Analysis of the State*, edited by Alan Hamlin and Philip Petit, 17–34. Oxford: Blackwell, 1989. http://philosophyfaculty.ucsd.edu/faculty/rarneson/JCO HENDELIBERATIVE%20DEM.pdf .

Cohen, Stanley. *Folk Devils and Moral Panics*. London: Routledge, 1987.

Cohen-Almagor, Raphael. "Between Autonomy and State Regulation: J. S. Mill's Elastic Paternalism." *Philosophy* 87, no. 4 (2012): 557–82.

The Boundaries of Liberty and Tolerance: The Struggle against Kahanism in Israel. Gainesville: University Press of Florida, 1994.

ed. *Challenges to Democracy: Essays in Honour and Memory of Professor Sir Isaiah Berlin*. Aldershot, UK: Ashgate, 2000.

"Content Net Neutrality: A Critique." In *Luciano Floridi's Philosophy of Technology: Critical Reflections*, edited by Hilmi Demir, 151–67. Dordrecht, Netherlands: Springer, 2012.

The Democratic Catch: Free Speech and Its Limits [in Hebrew]. Tel Aviv: Maariv, 2007.

"In Internet's Way." In *Ethics and Evil in the Public Sphere: Media, Universal Values, and Global Development*, edited by Robert S. Fortner and P. Mark Fackler, 93–115. Cresskill, NJ: Hampton Press, 2010.

"Internet Responsibility and Business Ethics: The Yahoo! Saga and Its Aftermath." *Journal of Business Ethics* 106, no. 3 (2012): 353–65.

"Responsibility of and Trust in ISPs." *Knowledge, Technology, and Policy* 23, no. 3 (2010): 381–96.

"Responsibility of Net Users." In *The Handbook of Global Communication and Media Ethics*, vol. 1, edited by Robert S. Fortner and P. Mark Fackler, 415–33. Oxford: Wiley-Blackwell, 2011.

The Scope of Tolerance: Studies on the Costs of Free Expression and Freedom of the Press. London: Routledge, 2006.

Speech, Media, and Ethics: The Limits of Free Expression. Houndmills, UK: Palgrave, 2005.

Cohen-Almagor, Raphael, Ori Arbel-Ganz, and Asa Kasher, eds. *Public Responsibility in Israel* [in Hebrew]. Tel Aviv and Jerusalem: Hakibbutz Hameuchad and Mishkanot Shaananim, 2012.

Conn, Kathleen. *The Internet and the Law: What Educators Need to Know*. Alexandria, VA: Association for Supervision and Curriculum Development, 2002.

Connolly, Dan. "A Little History of the World Wide Web," World Wide Web Consortium, 2000. http://www.w3.org/History.html .

Consalvo, Mia, and Charles Ess, eds. *The Handbook of Internet Studies*. Chichester, UK: Wiley, 2011.

Coppel, Philip. *Information Rights: Law and Practice*. Oxford: Hart, 2010.

Crane, Andrew, Dirk Matten, Abagail McWilliams, Jeremy Moon, and Donald S. Siegel, eds. *The Oxford Handbook of Corporate Social Responsibility*. Oxford: Oxford University Press, 2009.

Curran, James, and Jean Seaton. *Power without Responsibility: How Congress Abuses the People through Delegation*. London: Routledge, 2009.

Daniels, Peggy, ed. *Policing the Internet*. Farmington Hills, MI: Greenhaven, 2007.

Dann, G. Elijah, and Neil Haddow. "Just Doing Business or Doing Just Business: Google, Microsoft, Yahoo!, and the Business of Censoring China's Internet." *Journal of Business Ethics* 79, no. 3 (2008): 219–34.

Darwall, Stephen L. "Two Kinds of Respect." *Ethics* 88, no. 1 (1977): 36–49.

Deibert, Ronald J., and Nart Villeneuve. "Firewalls and Power: An Overview of Global State Censorship of the Internet." In *Human Rights in the Digital Age*, edited by Mathias Klang and Andrew Murray, 111–24. London: GlassHouse, 2005.

Deibert, Ronald, John Palfrey, Rafal Rohozinski, and Jonathan Zittrain, eds. *Access Denied: The Practice and Policy of Global Internet Filtering*. Cambridge, MA: MIT Press, 2008.

De Laat, Paul B. "Trusting Virtual Trust." *Ethics and Information Technology* 7, no. 3 (2005): 167–80.

Demir, Hilmi, ed. *Luciano Floridi's Philosophy of Technology: Critical Reflections*. Dordrecht, Netherlands: Springer, 2012.

DeMott, Andrea B., ed. *The Internet*. San Diego, CA: Greenhaven Press, 2007.

Denning, Dorothy E. "Comments on Responsibility of and Trust in ISPs." *Knowledge, Technology, and Policy* 23, no. 3 (2010): 399–401.

de Sola Pool, Ithiel. *Technologies of Freedom: Our Free Speech in an Electronic Age*. Cambridge, MA: Harvard University Press, 1983.

Doria, Avri, and Wolfgang Kleinwächter, eds. *Internet Governance Forum (IGF): The First Two Years*. Geneva: IGF Office, 2008. http://www.intgovforum.org/cms/documents/publications/172-internet-governance-forum-igf-the-first-two-years/file .

Dovidio, John F., Peter Glick, and Laurie A. Rudman, eds. *On the Nature of Prejudice: Fifty Years after Allport.* Oxford: Blackwell, 2005.

Dryzek, John S. *Deliberative Democracy and Beyond: Liberals, Critics, Contestations.* Oxford: Oxford University Press, 2002.

———. *Foundations and Frontiers of Deliberative Governance.* Oxford: Oxford University Press, 2012.

Dubow, Charles. "The Internet: An Overview." In *Does the Internet Benefit Society?*, edited by Cindy Mur, 7–13. Detroit, MI: Greenhaven Press, 2005.

Duff, Alistair S. *A Normative Theory of the Information Society.* New York: Routledge, 2012.

Dutton, William H., and Adrian Shepherd. "Trust in the Internet as an Experience Technology," *Information, Communication, and Society* 9, no. 4 (2006): 433–51.

Düwell, Marcus, Jens Braarvig, Roger Brownsword, and Dietmar Mieth, eds. *The Cambridge Handbook of Human Dignity: Interdisciplinary Perspectives.* Cambridge: Cambridge University Press, 2014.

Dworkin, Ronald. *Justice for Hedgehogs.* Cambridge, MA: Harvard University Press, 2011.

———. *Sovereign Virtue: The Theory and Practice of Equality.* Boston: Harvard University Press, 2002.

———. *Taking Rights Seriously.* London: Duckworth, 1977.

Dwyer, Susan. "Enter Here – At Your Own Risk: The Moral Dangers of Cyberporn." In *The Impact of the Internet on Our Moral Lives*, edited by Robert J. Cavalier, 69–94. Albany: State University of New York Press, 2005.

Edwards, Lilian, and Charlotte Waelde, eds. *Law and the Internet: Regulating Cyberspace.* Oxford: Hart, 1997.

Ehrlich, Howard J. *Campus Ethnoviolence and the Policy Options.* Baltimore: National Institute against Prejudice and Violence, 1990.

Elster, Jon, ed. *Deliberative Democracy.* Cambridge: Cambridge University Press, 1998.

Emerson, Thomas I. *The System of Freedom of Expression.* New York: Random House, 1970.

Esler, Brian W. "Filtering, Blocking, and Rating: Chaperones or Censorship?" In *Human Rights in the Digital Age*, edited by Mathias Klang and Andrew Murray, 99–110. London: GlassHouse, 2005.

Ess, Charles. "Moral Imperatives for Life in an Intercultural Global Village." In *The Impact of the Internet on Our Moral Lives*, edited by Robert J. Cavalier, 161–93. Albany: State University of New York Press, 2005.

Feinberg, Joel. "Legal Paternalism." *Canadian Journal of Philosophy* 1, no. 1 (1971): 105–24.

Fenton, Natalie, "The Internet and Social Networking." In *Misunderstanding the Internet*, edited by James Curran, Natalie Fenton, and Des Freedman, 123–48. Abingdon, UK: Routledge, 2012.

Fielding, Nigel, Raymond M. Lee, and Grant Blank, eds. *The Sage Handbook of Online Research Methods.* Thousand Oaks, CA: Sage, 2008.

Fishkin, James S., and Peter Laslett. *Debating Deliberative Democracy.* Oxford: Wiley-Blackwell, 2003.

FitzPatrick, William J. "Moral Responsibility and Normative Ignorance: Answering a New Skeptical Challenge." *Ethics* 118, no. 4 (2008): 589–613.

Floridi, Luciano. *The Ethics of Information.* Oxford: Oxford University Press, 2013.
    The Fourth Revolution: How the Infosphere Is Reshaping Human Reality. Oxford:
    Oxford University Press, 2014.
    ed. *The Cambridge Handbook of Information and Computer Ethics.* Cambridge:
    Cambridge University Press, 2010.
    Information: A Very Short Introduction. Oxford: Oxford University Press, 2010.
Floridi, Luciano, and J. W. Sanders. "Internet Ethics: The Constructionist Values of
    Homo Poieticus." In *The Impact of the Internet on Our Moral Lives,* edited by
    Robert J. Cavalier, 195–214. Albany: State University of New York Press, 2005.
Fortner, Robert S., and P. Mark Fackler, eds. *Ethics and Evil in the Public Sphere:
    Media, Universal Values, and Global Development.* Cresskill, NJ: Hampton
    Press, 2010.
    eds. *The Handbook of Global Communication and Media Ethics.* Oxford: Wiley-
    Blackwell, 2011.
Friedman, Lauri S., ed. *The Internet: Introducing Issues with Opposing Viewpoints.*
    Farmington Hills, MI: Greenhaven, 2008.
Fukuyama, Francis. *Trust: The Social Virtues of and the Creation of Prosperity.* New
    York: Free Press, 1995.
Galbreath, Jeremy. "Drivers of Corporate Social Responsibility: The Role of Formal
    Strategic Planning and Firm Culture." *British Journal of Management* 21, no. 2
    (2010): 511–25.
Gane, Nicholas, and David Beer. *New Media: The Key Concepts.* Oxford: Berg, 2008.
Garfinkel, Simson L. *PGP: Pretty Good Privacy.* Sebastopol, CA: O'Reilly, 1995.
Gentile, Douglas A., ed. *Media Violence and Children: A Complete Guide for Parents
    and Professionals.* Westport, CT: Praeger, 2003.
Goldsmith, Jack L. "Against Cyberanarchy." *University of Chicago Law Review* 65,
    no. 4 (1998): 1199–1250.
Goldsmith, Jack L., and Tim Wu. *Who Controls the Internet? Illusions of a Borderless
    World.* New York: Oxford University Press, 2006.
Graham, Mark, and William. H. Dutton, eds. *Society and the Internet: How Networks of
    Information and Communication Are Changing Our Lives.* Oxford: Oxford
    University Press, 2014.
Gralla, Preston. *How the Internet Works.* 8th ed. Indianapolis, IN: Que, 2007.
Gruen, Madeleine. "Innovative Recruitment and Indoctrination Tactics by Extremists:
    Video Games, Hip-Hop, and the World Wide Web." In *The Making of a Terrorist:
    Recruitment, Training, and Root Causes,* edited by James J. F. Forest, 11–22.
    Westport, CT: Praeger, 2006.
    "White Ethnonationalist and Political Islamist Methods of Fund-Raising and
    Propaganda on the Internet." In *The Changing Face of Terrorism,* edited by
    Rohan Gunaratna, 127–45. Singapore: Marshall Cavendish, 2004.
Gutmann, Amy, and Dennis F. Thompson. *Why Deliberative Democracy?* Princeton,
    NJ: Princeton University Press, 2004.
Haas, Larry. *Safeguarding Your Child Online: How to Deal with Stranger Contact and
    Pornography on the Internet.* Philadelphia: Xlibris, 2000.
Habermas, Jürgen. *Between Facts and Norms: Contributions to a Discourse Theory of
    Law and Democracy.* Cambridge: Polity, 1996.
    Moral Consciousness and Communicative Action. Cambridge, MA: MIT Press, 1990.

Hafner, Katie, and Matthew Lyon. *Where Wizards Stay Up Late: The Origins of the Internet*. New York: Simon & Schuster, 1996.

Hamdani, Assaf. "Who's Liable for Cyberwrongs?" *Cornell Law Review* 87, no. 4 (2002): 901–57.

Hamlin, Alan, and Philip Petit, eds. *The Good Polity: Normative Analysis of the State*. Oxford: Blackwell, 1989.

Haney, Craig, Curtis Banks, and Philip Zimbardo. "Interpersonal Dynamics in a Simulated Prison." *International Journal of Criminology and Penology* 1 (1973): 69–97.

Harding, Matthew. "Manifesting Trust." *Oxford Journal of Legal Studies* 29, no. 2 (2009): 245–65.

"Responding to Trust." *Ratio Juris* 24, no. 1 (2011): 75–87.

Hauben, Michael, and Ronda Hauben. *Netizens: On the History and Impact of Usenet and the Internet*. Los Alamitos, CA: IEEE Computer Society Press, 1997.

Henderson, Harry. *Internet Predators*. New York: Facts on File, 2005.

Henthoff, Nat. *Free Speech for Me – But Not for Thee: How the American Left and Right Relentlessly Censor Each Other*. New York: Harper Collins, 1992.

Heyman, Steven J. *Free Speech and Human Dignity*. New Haven, CT: Yale University Press, 2008.

Hier, Sean P., ed. *Moral Panic and the Politics of Anxiety*. London: Routledge, 2011.

Horrigan, Bryan. *Corporate Social Responsibility in the 21st Century: Debates, Models, and Practices across Government, Law, and Business*. Northampton, MA: Edward Elgar, 2010.

Ihlen, Øyvind, Jennifer L. Bartlett, and Steve May, eds. *The Handbook of Communication and Corporate Social Responsibility*. Chichester, UK: Wiley-Blackwell, 2011.

Jacquette, Dale. *Journalistic Ethics: Moral Responsibility in the Media*. Upper Saddle River, NJ: Pearson, 2007.

Jaffe, J. Michael. "Riding the Electronic Tiger: Censorship in Global, Distributed Networks." In *Liberal Democracy and the Limits of Tolerance: Essays in Honor and Memory of Yitzhak Rabin*, edited by Raphael Cohen-Almagor, 275–94. Ann Arbor: University of Michigan Press, 2000.

Jensen, Michael C. "Value Maximization, Stakeholder Theory, and the Corporate Objective Function." *Business Ethics Quarterly* 12, no. 2 (2002): 235–56.

Johnson, David R., and David G. Post. "Law and Borders: The Rise of Law in Cyberspace." *Stanford Law Review* 48, no. 5 (1996): 1367–1402.

Johnson, Deborah G., and Keith W. Miller. *Computer Ethics*. Upper Saddle River, NJ: Pearson, 2009.

Jonas, Hans. *The Imperative of Responsibility: In Search of an Ethics for the Technological Age*. Chicago: University of Chicago Press, 1984.

Kant, Immanuel. *Foundations of the Metaphysics of Morals*, edited by Robert Paul Wolff. Indianapolis, IN: Bobbs-Merrill, 1969.

Karatzogianni, Athina. *The Politics of Cyberconflict*. London: Routledge, 2006.

Karst, Kenneth L. "Threats and Meanings: How the Facts Govern First Amendment Doctrine." *Stanford Law Review* 58, no. 5 (2006): 1337–1412.

Kellerman, Aharon. *The Internet on Earth: A Geography of Information*. Oxford: Wiley, 2002.

Kerr, Michael, Richard Janda, and Chip Pitts. *Corporate Social Responsibility: A Legal Analysis*. Markham, ON: LexisNexis, 2009.

Kiesler, Sara, ed. *Culture of the Internet*. Mahwah, NJ: Lawrence Erlbaum Associates, 1997.

Kizza, Joseph Migga. *Ethical and Social Issues in the Information Age*. London: Springer, 2007.

Klang, Mathias, and Andrew Murray, eds. *Human Rights in the Digital Age*. London: GlassHouse, 2005.

Kleinrock, Leonard. "Creating a Mathematical Theory of Computer Networks." *Operations Research* 50, no. 1 (2002): 125–31.

——— "An Early History of the Internet." *IEEE Communications Magazine* 48, no. 8 (2010): 26–36.

——— "History of the Internet and Its Flexible Future." *IEEE Wireless Communications* 15, no. 1 (2008): 8–18.

——— "Information Flow in Large Communication Nets." PhD thesis proposal, Massachusetts Institute of Technology, Cambridge, MA, May 31, 1961, http://www. lk.cs.ucla.edu/data/files/Kleinrock/Information%20Flow%20in%20Large%20Com munication%20Nets.pdf .

Köcher, Renate. "Representative Survey on Internet Content Concerns in Australia, Germany, and the United States of America." In *Protecting Our Children on the Internet: Towards a New Culture of Responsibility*, edited by Jens Waltermann and Marcel Machill, 401–56. Gütersloh, Germany: Bertelsmann Foundation, 2000.

Kohl, Uta. *Jurisdiction and the Internet: A Study of Regulatory Competence over Online Activity*. Cambridge: Cambridge University Press, 2007.

Kohlberg, Lawrence. *The Philosophy of Moral Development: Moral Stages and the Idea of Justice*. Vol. 1 of *Essays on Moral Development*. San Francisco: Harper & Row, 1981.

Kohn, Marek. *Trust: Self-Interest and the Common Good*. New York: Oxford University Press, 2008.

Kotler, Philip, and Nancy Lee. *Corporate Social Responsibility: Doing the Most Good for Your Company and Your Cause*. Hoboken, NJ: Wiley, 2005.

Kouzes, Richard T., James D. Myers, and William Wulf. "Collaboratories: Doing Science on the Internet." *IEEE Computer* 29, no. 8 (1996): 40–46.

Kueng, Lucy, Robert G. Picard, and Ruth Towse, eds. *The Internet and the Mass Media*. Thousand Oaks, CA: Sage, 2008.

Langford, Duncan, ed. *Internet Ethics*. New York: St. Martin's Press, 2000.

Larmore, Charles. *Patterns of Moral Complexity*. Cambridge: Cambridge University Press, 1987.

Lee, Ki-Hoon, and Dongyoung Shin. "Consumers' Responses to CSR Activities: The Linkage between Increased Awareness and Purchase Intention." *Public Relations Review* 36, no. 2 (2010): 193–95.

Leiner, Barry M., Vinton G. Cerf, David D. Clark, Robert E. Kahn, Leonard Kleinrock, Daniel C. Lynch, Jon Postel, Lawrence G. Roberts, and Stephen S. Wolff. "A Brief History of the Internet," Internet Society, 2003. http://www.isoc.org/internet/history/ brief.shtml .

——— "The Past and Future History of the Internet." *Communications of the ACM* 40, no. 2 (1997): 102–8.

Lessig, Lawrence. *Code and Other Laws of Cyberspace.* New York: Basic Books, 1999.
Free Culture: How Big Media Uses Technology and the Law to Lock Down Culture and Control Creativity. New York: Penguin, 2004.
The Future of Ideas: The Fate of the Commons in a Connected World. New York: Vintage, 2002.
Levinson, Paul. *New New Media.* Boston: Pearson, 2009.
Levmore, Saul, and Martha C. Nussbaum, eds. *The Offensive Internet: Speech, Privacy, and Reputation.* Cambridge, MA.: Harvard University Press, 2010.
Levy, Steven. *Crypto: How the Code Rebels Beat the Government: Saving Privacy in the Digital Age.* New York: Viking, 2001.
Lewis, Stewart. "Reputation and Corporate Responsibility." *Journal of Communication Management* 7, no. 4 (2003): 356–94.
Licklider, J. C. R., and Robert W. Taylor. "The Computer as a Communication Device." *Science and Technology* 76, no. 2 (1968): 20–41.
Lipschultz, Jeremy Harris. *Broadcast and Internet Indecency: Defining Free Speech.* New York: Routledge, 2008.
Lumby, Catharine, Lelia Green, and John Hartley. *Untangling the Net: The Scope of Content Caught By Mandatory Internet Filtering.* Sydney: Australian Law Reform Commission, 2011. http://www.alrc.gov.au/sites/default/files/pdfs/ci_2522_1_green_w_attachments.pdf .
Macedo, Stephen. *Deliberative Politics: Essays on Democracy and Disagreement.* New York: Oxford University Press, 1999.
Madelin, Robert. "The Evolving Social Responsibilities of Internet Corporate Actors: Pointers Past and Present." *Philosophy and Technology* 24, no. 4 (2011): 455–61.
Markus, Hazel Rose, Carol D. Ryff, Alana L. Conner, Eden K. Pudberry, and Katherine L. Barnette. "Themes and Variations in American Understandings of Responsibility." In *Caring and Doing for Others: Social Responsibility in the Domains of Family, Work, and Community,* edited by Alice S. Rossi, 349–99. Chicago: University of Chicago Press, 2001.
Marsden, Christopher T. *Net Neutrality: Towards a Co-Regulatory Solution.* London: Bloomsbury, 2010.
Mathiason, John. *Internet Governance: The New Frontier of Global Institutions.* Abingdon, UK: Routledge, 2009.
Mayer-Schönberger, Viktor. "The Shape of Governance: Analyzing the World of Internet Regulation." *Virginia Journal of International Law* 43 (2003): 605–73.
McLuhan, Marshall. *Understanding Media: The Extensions of Man.* Cambridge, MA: MIT Press, 1994.
McQuail, Denis. *Media Accountability and Freedom of Publication.* New York: Oxford University Press, 2003.
Meiklejohn, Alexander. *Political Freedom.* New York: Oxford University Press, 1965.
Mill, John Stuart. *Utilitarianism, Liberty, and Representative Government.* London: J. M. Dent, 1948.
Miller, David. "Deliberative Democracy and Social Choice." *Political Studies* 40, suppl. 1 (1992): 54–67.
Moreno-Riaño, Gerson, ed. *Tolerance in the Twenty-First Century: Prospects and Challenges.* Lanham, MD: Lexington Books, 2006.

Mossberger, Karen, Caroline J. Tolbert, and Ramona S. McNeal. *Digital Citizenship: The Internet, Society, and Participation.* Cambridge, MA: MIT Press, 2007.

Mueller, Milton L. *Networks and States: The Global Politics of Internet Governance.* Cambridge, MA: MIT Press, 2010.

Murthy, Dhiraj. *Twitter: Social Communication in the Twitter Age.* Cambridge: Polity, 2013.

Myskja, Bjørn K. "The Categorical Imperative and the Ethics of Trust." *Ethics and Information Technology* 10, no. 4 (2008): 213–20.

National Research Council. *Global Networks and Local Values: A Comparative Look at Germany and the United States.* Washington, DC: National Academies Press, 2001.

Nelson, Michael R. "A Response to Responsibility of and Trust in ISPs by Raphael Cohen-Almagor." *Knowledge, Technology, and Policy* 23, no. 3 (2010): 403–7.

Nelson, Theodor Holm. *Literary Machines.* Watertown, MA: Mindful Press, 1982.

Nino, Carlos Santiago. *The Constitution of Deliberative Democracy.* New Haven, CT: Yale University Press, 1996.

Nissenbaum, Helen. "Hackers and the Contested Ontology of Cyberspace." In *The Impact of the Internet on Our Moral Lives,* edited by Robert J. Cavalier, 139–60. Albany: State University of New York Press, 2005.

——— "Securing Trust Online: Wisdom or Oxymoron." *Boston University Law Review* 81, no. 3 (2001): 635–64. http://www.nyu.edu/projects/nissenbaum/papers/securing trust.pdf .

Nussbaum, Martha C. *Love's Knowledge: Essays on Philosophy and Literature.* New York: Oxford University Press, 1990.

Pagallo, Ugo. "ISPs and Rowdy Web Sites before the Law: Should We Change Today's Safe Harbour Clauses?" *Philosophy and Technology* 24, no. 4 (2011): 419–36.

Paré, Daniel. "The Digital Divide: Why the 'The' is Misleading." In *Human Rights in the Digital Age,* edited by Mathias Klang and Andrew Murray, 85–98. London: GlassHouse, 2005.

Parkinson, John, and Jane Mansbridge, eds. *Deliberative Systems: Deliberative Democracy at the Large Scale.* Cambridge: Cambridge University Press, 2012.

Passerin d'Entrèves, Maurizio. *Democracy as Public Deliberation: New Perspectives.* Piscataway, NJ: Transaction, 2006.

Pavlik, John Vernon. *Media in the Digital Age.* New York: Columbia University Press, 2008.

Peters, William. *A Class Divided: Then and Now.* New Haven, CT: Yale University Press, 1971.

Pettit, Philip. "Trust, Reliance, and the Internet." *Analyse and Kritik* 26, no. 1 (2004): 108–21.

Pittaro, Michael L. "Cyber Stalking: An Analysis of Online Harassment and Intimidation." *International Journal of Cyber Criminology* 1, no. 2 (2007): 180–97.

Ploug, Thomas. *Ethics in Cyberspace: How Cyberspace May Influence Interpersonal Interaction.* Dordrecht, Netherlands: Springer, 2009.

Polder-Verkiel, Saskia E. "Online Responsibility: Bad Samaritanism and the Influence of Internet Mediation." *Science and Engineering Ethics* 18, no. 1 (2012): 117–41.

Popper, Karl R. *The Open Society and Its Enemies.* London: Routledge and Kegan Paul, 1962.

"Toleration and Intellectual Responsibility." In *On Toleration*, edited by Susan Mendus and David Edwards, 17–34. Oxford: Clarendon Press, 1987.

Porter, Michael E., and Mark R. Kramer. "Strategy and Society: The Link between Competitive Advantage and Corporate Social Responsibility." *Harvard Business Review*, December 2006, 78–92.

Post, David G. "Against 'Against Cyberanarchy.'" *Berkeley Technology Law Journal* 17 (2002): 1365–87.

"What Larry Doesn't Get: Code, Law, and Liberty in Cyberspace." *Stanford Law Review* 52 (2000): 1439–59.

Price, Monroe E., and Stefaan G. Verhulst. "The Concept of Self-Regulation and the Internet." In *Protecting Our Children on the Internet: Towards a New Culture of Responsibility*, edited by Jens Waltermann, and Marcel Machill, 133–98. Gütersloh, Germany: Bertelsmann Foundation, 2000.

Rawls, John. *The Law of Peoples*. Cambridge, MA: Harvard University Press, 2002.

A Theory of Justice. Boston: Belknap Press, 2005.

Raz, Joseph. *The Morality of Freedom*. Oxford: Clarendon Press, 1986.

Value, Respect, and Attachment. Cambridge: Cambridge University Press, 2001.

Reidenberg, Joel R. "Technology and Internet Jurisdiction." *University of Pennsylvania Law Review* 153 (2005): 1951–74.

"Yahoo and Democracy on the Internet." *Jurimetrics* 42 (2002): 261–80.

Resnick, David. "Tolerance and the Internet." In *Tolerance in the Twenty-First Century: Prospects and Challenges*, edited by Gerson Moreno-Riaño, 213–20. Lanham, MD: Lexington Books, 2006.

Rheingold, Howard. "The Emerging Wireless Internet Will Both Improve and Degrade Human Life." In *The Future of the Internet*, edited by Tom Head, 19–32. Farmington Hills, MI: Greenhaven Press, 2005.

Rorty, Richard. "Justice as a Larger Loyalty." *Ethical Perspectives* 4, no. 3 (1997): 139–51.

Rossi, Alice S., ed. *Caring and Doing for Others: Social Responsibility in the Domains of Family, Work, and Community*. Chicago: University of Chicago Press, 2001.

Rotenberg, Marc. "Fair Information Practices and the Architecture of Privacy: (What Larry Doesn't Get)." *Stanford Technology Law Review* 1 (2001). https://journals.law.stanford.edu/sites/default/files/stanford-technology-law-review/online/rotenberg-fair-info-practices.pdf .

Rothman, Jennifer E. "Freedom of Speech and True Threats." *Harvard Journal of Law and Public Policy* 25, no. 1 (2001): 283–367.

Ryan, Johnny. "The Essence of the 'Net: A History of the Protocols That Hold the Network Together." Ars Technica, March 8, 2011. http://arstechnica.com/tech-policy/news/2011/03/the-essence-of-the-net.ars/ .

Sacco, Dena, Rebecca Argudin, James Maguire, and Kelly Tallon. "Sexting: Youth Practices and Legal Implications." Berkman Research Publication 2010–8, Berkman Center for Research and Technology, Harvard University, Cambridge, MA, June 22, 2010.

Sadurski, Wojciech. *Freedom of Speech and Its Limits*. Dordrecht, Netherlands: Kluwer, 1999.

Salus, Peter H. *Casting the Net: From ARPANET to Internet and Beyond*. Reading, MA: Addison-Wesley, 1995.

Saunders, Kevin W. *Saving Our Children from the First Amendment*. New York: New York University Press, 2003.

Schachter, Madeleine, and Joel Kurtzberg. *Law of Internet Speech*. Durham, NC: Carolina Academic Press, 2008.

Schauer, Frederick. "The Cost of Communicative Tolerance." In *Liberal Democracy and the Limits of Tolerance: Essays in Honor and Memory of Yitzhak Rabin*, edited by Raphael Cohen-Almagor, 28–42. Ann Arbor: University of Michigan Press, 2000.

Schneider, Gary P., and Jessica Evans. *New Perspectives on the Internet: Comprehensive*. 6th ed. Boston: Thomson, 2007.

New Perspectives on the Internet: Introductory. Boston: Thomson, 2007.

Schuler, Douglas, and Peter Day. *Shaping the Network Society: The New Role of Civil Society in Cyberspace*. Cambridge, MA: MIT Press, 2004.

Shachtman, Noah. *Pirates of the ISPs: Tactics for Turning Online Crooks into International Pariahs*. Washington, DC: Brookings Institution, 2011.

Sher, George. *Who Knew? Responsibility without Awareness*. New York: Oxford University Press, 2009.

Shirky, Clay. *Here Comes Everybody: The Power of Organizing with Organizations*. London: Penguin, 2008.

Simpson, Thomas W. "e-Trust and Reputation." *Ethics and Information Technology* 13, no. 1 (2011): 29–38.

Slevin, James. *The Internet and Society*. Oxford: Polity, 2000.

Smith, Bradford L. "The Third Industrial Revolution: Policymaking for the Internet." *Columbia Science and Technology Law Review* 3 (2001–2002): 1. http://www.columbia.edu/cu/stlr/html/Archive/ .

Smith, Ron F. *Groping for Ethics in Journalism*. Ames: Iowa State Press, 2003.

Spinello, Richard A. *Cyberethics: Morality and Law in Cyberspace*. Sudbury, MA: Jones & Bartlett, 2000.

Stanley, Janet. "The Internet Can Be Dangerous for Children." In *Does the Internet Benefit Society?*, edited by Cindy Mur, 95–101. Farmington Hills, MI: Greenhaven, 2005.

Szoka, Berin, and Adam Marcus, eds. *The Next Digital Decade: Essays on the Future of the Internet*. Washington, DC: TechFreedom, 2010.

Tarbox, Katherine. *Katie.com: My Story*. New York: E. P. Dutton, 2000.

Tavani, Herman T. *Ethics and Technology: Controversies, Questions, and Strategies for Ethical Computing*. Hoboken, NJ: Wiley, 2011.

"The Impact of the Internet on Our Moral Condition: Do We Need a New Framework of Ethics?" In *The Impact of the Internet on Our Moral Lives*, edited by Robert J. Cavalier, 215–38. Albany: State University of New York Press, 2005.

Tavani, Herman T., and Frances S. Grodzinsky. "Cyberstalking, Personal Privacy, and Moral Responsibility." *Ethics and Information Technology* 4, no. 2 (2002): 123–32.

Tengblad, Stefan, and Claes Ohlsson. "The Framing of Corporate Social Responsibility and the Globalization of National Business Systems." *Journal of Business Ethics* 93, no. 4 (2010): 653–69.

Thierer, Adam, and Clyde Wayne Crews Jr., eds. *Who Rules the Net? Internet Governance and Jurisdiction*. Washington, DC: Cato Institute, 2003.

Thornburgh, Dick, and Herbert S. Lin. eds. *Youth, Pornography, and the Internet*. Washington, DC: National Academies Press, 2002.

Travis, Hannibal, ed. *Cyberspace Law: Censorship and Regulation of the Internet.* London: Routledge, 2013.

Turkle, Sherry. *Life on the Screen: Identity in the Age of the Internet.* New York: Simon & Schuster, 1995.

Uslaner, Eric M. "Trust, Civic Engagement, and the Internet." *Political Communication* 21, no. 2 (2004): 223–42.

van den Hoven, Jeroen, and John Weckert, eds. *Moral Philosophy and Information Technology.* Cambridge: Cambridge University Press, 2008.

van Dijk, Jan. *The Network Society.* London: Sage, 2012.

Vedder, Anton H. "Accountability of Internet Access and Service Providers: Strict Liability Entering Ethics." *Ethics and Information Technology* 3, no. 1 (2001): 67–74.

Vogel, David. *The Market for Virtue: The Potential and Limits of Corporate Social Responsibility.* Washington, DC: Brookings Institution, 2005.

Wagner, Ben. "The Politics of Internet Filtering: The United Kingdom and Germany in a Comparative Perspective." *Politics* 34, no. 1 (2014): 58–71.

Waisbord, Silvio. *Reinventing Professionalism.* Cambridge: Polity, 2013.

Wallace, Jonathan, and Mark Mangan. *Sex, Laws, and Cyberspace.* New York: Henry Holt, 1996.

Waltermann, Jens, and Marcel Machill, eds. *Protecting Our Children on the Internet: Towards a New Culture of Responsibility.* Gütersloh, Germany: Bertelsmann Foundation, 2000.

Walther, Joseph B., Brandon Van Der Heide, Sang-Yeon Kim, David Westerman, and Stephanie Tom Tong. "The Role of Friends' Appearance and Behavior on Evaluations of Individuals on Facebook: Are We Known by the Company We Keep?" *Human Communication Research* 34, no. 1 (2008): 28–49.

Werther, William B., and David B. Chandler. *Strategic Corporate Social Responsibility: Stakeholders in a Global Environment.* Thousand Oaks, CA: Sage, 2010.

Wu, Tim. *The Master Switch: The Rise and Fall of Information Empires.* New York: Knopf, 2010.

Zhao, Shanyang, Sherry Grasmuck, and Jason Martin. "Identity Construction on Facebook: Digital Empowerment in Anchored Relationships." *Computers in Human Behavior* 24, no. 5 (2008): 1816–36. http://astro.temple.edu/~bzhao001/ Identity%20Construction%20on%20Facebook.pdf .

Zimmermann, Philip R. *The Official PGP User's Guide.* Cambridge, MA: MIT Press, 1995.

Zittrain, Jonathan L. "Beware the Cyber Cops." *Forbes Magazine*, July 8, 2002.

The Future of the Internet – And How to Stop It. New Haven, CT: Yale University Press, 2008.

"The Generative Internet." *Harvard Law Review* 119, no. 7 (2006): 1974–2040.

"Without a Net." Legal Affairs, January–February 2006. http://www.legalaffairs.org/ issues/January-February-2006/feature_zittrain_janfeb06.msp .

CHILD PORNOGRAPHY, CRIME, AND CYBERBULLYING

Akdeniz, Yaman. "Computer Pornography: A Comparative Study of the U.S. and U.K. Obscenity Laws and Child Pornography Laws in Relation to the Internet." *International Review of Law, Computers, and Technology* 10, no. 2 (1996): 235–61.

"Governance of Pornography and Child Pornography on the Global Internet: A Multi-layered Approach." In *Law and the Internet: Regulating Cyberspace*, edited by Lilian Edwards and Charlotte Waelde, 223–41. Oxford: Hart, 1997.

Internet Child Pornography and the Law: National and International Responses. Aldershot, UK: Ashgate, 2008.

"The Regulation of Pornography and Child Pornography on the Internet." *Journal of Information, Law, and Technology* 1 (1997). http://www2.warwick.ac.uk/fac/soc/law/elj/jilt/1997_1/akdenizi .

Sex on the Net: The Dilemma of Policing Cyberspace. London: South Street, 1999.

Bartow, Ann. "Internet Defamation as Profit Center: The Monetization of Online Harassment." *Harvard Journal of Law and Gender* 32, no. 2 (2009): 384–429.

Berg, Terrence. "The Internet Facilitates Crime." In *Does the Internet Benefit Society?*, edited by Cindy Mur, 35–39. Farmington Hills, MI: Greenhaven, 2005.

Berlan, Elise D., Heather L. Corliss, Alison E. Field, Elizabeth Goodman, and S. Bryn Austin. "Sexual Orientation and Bullying among Adolescents in the Growing Up Today Study." *Journal of Adolescent Health* 46, no. 4 (2010): 366–71.

Bibby, Peter C., ed. *Organized Abuse: The Current Debate.* Brookfield, VT: Ashgate, 1996.

Carr, Indira, ed. *Computer Crime.* Surrey, UK: Ashgate, 2009.

Cavalier, Robert J., ed. *The Impact of the Internet on Our Moral Lives.* Albany: State University of New York Press, 2005.

Child Exploitation and Online Protection Centre (CEOP). "Annual Review: 2009–10." London: CEOP, 2010.

"Strategic Threat Assessment: Child Trafficking in the U.K." London: CEOP, April 2009.

Understanding Online Social Network Services and Risks to Youth. London: CEOP, November 2006.

Cohen-Almagor, Raphael. "Online Child Sex Offenders: Challenges and Counter-measures." *Howard Journal of Criminal Justice* 52, no. 2 (2013): 190–215.

Cohen-Almagor, Raphael, and Sharon Haleva-Amir. "Bloody Wednesday in Dawson College: The Story of Kimveer Gill, or Why Should We Monitor Certain Websites to Prevent Murder." *Studies in Ethics, Law and Technology* 2, no. 3 (2008): article 1. http://works.bepress.com/raphael_cohen_almagor/1 .

"Why Monitor Violent Websites? A Justification." *Beijing Law Journal* 3, no. 2 (2012): 64–71.

Connors, Paul G. "Internet Sexual Crime Legislation in Michigan." Legislative Research Division, Michigan Legislative Service Bureau, Lansing, 2002.

Daniels, Peggy, ed. *Policing the Internet.* Farmington Hills, MI: Greenhaven, 2007.

Dwyer, Susan. "Enter Here – At Your Own Risk: The Moral Dangers of Cyberporn." In *The Impact of the Internet on Our Moral Lives*, edited by Robert J. Cavalier, 69–94. Albany: State University of New York Press, 2005.

Edwards, Lilian, and Charlotte Waelde, eds. *Law and the Internet.* Oxford: Hart, 1997.

Eneman, Marie. "The New Face of Child Pornography." In *Human Rights in the Digital Age*, edited by Mathias Klang and Andrew Murray, 27–40. London: GlassHouse, 2005.

Europol. *The Internet Organised Crime Threat Assessment (iOCTA).* Den Haag, Netherlands: European Police Office, 2014.

Ferraro, Monique Mattei, Eoghan Casey, and Michael McGrath. *Investigating Child Exploitation and Pornography: The Internet, the Law, and Forensic Science.* Burlington, MA: Elsevier, 2005.

Flores, J. Robert. "Internet Pornography Should Be Censored." In *Censorship: Opposing Viewpoints*, edited by Andrea C. Nakaya, 79–83. Farmington Hills, MI: Greenhaven Press, 2005.

Friedman, Lauri S., ed. *The Internet: Introducing Issues with Opposing Viewpoints.* Farmington Hills, MI: Greenhaven, 2008.

Gerson, Ruth, and Nancy Rappaport. "Cyber Cruelty: Understanding and Preventing the New Bullying." *Adolescent Psychiatry* 1, no. 1 (2011): 67–71.

Gillespie, Alisdair. *Child Exploitation and Communication Technologies.* Dorset, UK: Russell House, 2008.

Goldstein, Seth. *The Sexual Exploitation of Children: A Practical Guide to the Assessment, Investigation, and Intervention.* Boca Raton, FL: CRC Press, 2009.

Good, R. Stephanie. *Exposed: The Harrowing Story of a Mother's Undercover Work with the FBI to Save Children from Internet Sex Predators.* Nashville, TN: Thomas Nelson, 2007.

Graham, William R., Jr. "Uncovering and Eliminating Child Pornography Rings on the Internet: Issues Regarding and Avenues Facilitating Law Enforcement's Access to 'Wonderland.'" *Law Review of Michigan State University, Detroit College of Law* 2 (2000): 457–84.

Haas, Larry. *Safeguarding Your Child Online: How to Deal with Stranger Contact and Pornography on the Internet.* Philadelphia: Xlibris, 2000.

Henderson, Harry. *Internet Predators.* New York: Facts on File, 2005.

Hiber, Amanda, ed. *Child Pornography.* San Diego, CA: Greenhaven Press, 2009.

Hinduja, Sameer, and Justin W. Patchin. *Bullying beyond the Schoolyard: Preventing and Responding to Cyberbullying.* Thousand Oaks, CA: Sage, 2009.

"Bullying, Cyberbullying, and Suicide." *Archives of Suicide Research* 14, no. 3, (2010): 206–21.

"Offline Consequences of Online Victimization: School Violence and Delinquency." *Journal of School Violence* 6, no. 3 (2007): 89–112.

Information Resources Management Association. *Cyber Crime: Concepts, Methodologies, Tools, and Applications.* 3 vols. Hershey, PA: IGI Global, 2012.

Jenkins, Philip. *Beyond Tolerance: Child Pornography on the Internet.* New York: New York University Press, 2001.

"How Europe Discovered Its Sex Offender Crisis." In *How Claims Spread: Studies in the Cross-National Diffusion of Social Problems Claims*, edited by Joel Best, 147–68. Hawthorne, NY: Aldine de Gruyter, 2001.

Intimate Enemies: Moral Panics in Contemporary Great Britain. Hawthorne, NY: Aldine de Gruyter, 1992.

Jewkes, Yvonne, and Majid Yar, eds. *Handbook of Internet Crime.* Portland, OR: Willan, 2010.

Kowalski, Robin M., and Susan P. Limber. "Electronic Bullying among Middle School Students." *Journal of Adolescent Health* 41, no. 6 (2007): S22–30.

Kowalski, Robin, Susan Limber, and Patricia W. Agatston. *Cyberbullying: Bullying in the Digital Age.* Malden, MA: Blackwell, 2008.

Krinsky, Charles, ed. *Moral Panics over Contemporary Children and Youth*. Farnham, UK: Ashgate, 2008.

Lipschultz, Jeremy Harris. *Broadcast and Internet Indecency: Defining Free Speech*. New York: Routledge, 2008.

Lipton, Jacqueline D. "Combating Cyber-Victimization." *Berkeley Technology Law Journal* 26, no. 2 (2011): 1103–56.

Livingstone, Sonia, and Leslie Haddon, eds. *Kids Online: Opportunities and Risks for Children*. Bristol, UK: Policy Press, 2009.

Livingstone, Sonia, Leslie Haddon, and Anke Görzig, eds. *Children, Risk, and Safety Online: Research and Policy Challenges in Comparative Perspective*. Bristol, UK: Policy Press, 2012.

Livingstone, Sonia, Leslie Haddon, Anke Görzig, and Kjartan Ólafsson. *Risks and Safety on the Internet: The Perspective of European Children – Full Findings and Policy Implications from the EU Kids Online Survey*. London: EU Kids Online Network, 2011.

Lobe, Bojana, Sonia Livingstone, Kjartan Ólafsson, and Hana Vodeb. *Cross-National Comparison of Risks and Safety on the Internet*. London: EU Kids Online, 2011.

Mansell, Robin, and Brian S. Collins, eds. *Trust and Crime in Information Societies*. Cheltenham, UK: Edward Elgar, 2005.

McAlinden, Anne-Marie. *"Grooming" and the Sexual Abuse of Children: Institutional, Internet, and Familial Dimensions*. Oxford: Oxford University Press, 2012.

McLaughlin, James F. "Characteristics of a Fictitious Child Victim: Turning a Sex Offender's Dreams into His Worst Nightmare." *International Journal of Communications Law and Policy* 9 (2004): 1–27. http://www.ijclp.net/files/ijclp_web-doc_6-cy-2004.pdf .

McMahon, Elaine M., Udo Reulbach, Helen Keeley, Ivan J. Perry, and Ella Arensman. "Bullying Victimisation, Self-Harm, and Associated Factors in Irish Adolescent Boys." *Social Science and Medicine* 71, no. 7 (2010): 1300–1307.

Meyer, Anneke. *The Child at Risk: Pedophiles, Media Responses, and Public Opinion*. Manchester, UK: Manchester University Press, 2007.

National Policing Improvement Agency. *Investigating Child Abuse and Safeguarding Children*. Wyboston, UK: Association of Chief Police Officers, 2009.

National Research Council. *Global Networks and Local Values: A Comparative Look at Germany and the United States*. Washington, DC: National Academies Press, 2001.

O'Neill, Brian, Elisabeth Staksrud, and Sharon McLaughlin, eds. *Towards a Better Internet for Children? Policy Pillars, Players, and Paradoxes*. Göteborg, Germany: Nordicom, 2013.

Ost, Suzanne. *Child Pornography and Sexual Grooming: Legal and Societal Responses*. Cambridge: Cambridge University Press, 2009.

Patchin, Justin W., and Sameer Hinduja. "Trends in Online Social Networking: Adolescent Use of MySpace over Time." *New Media and Society* 12, no. 2 (2010): 197–216.

Quayle, Ethel, and Max Taylor. "Child Pornography and the Internet: Perpetuating a Cycle of Abuse." *Deviant Behavior* 23, no. 4 (2002): 331–62.

Saunders, Kevin W. *Saving Our Children from the First Amendment*. New York: New York University Press, 2003.

Shariff, Shahreen. *Confronting Cyber Bullying: What Schools Need to Know to Control Misconduct and Avoid Legal Consequences*. Cambridge: Cambridge University Press, 2009.

Singer, P. W., and Allan Friedman, *Cybersecurity and Cyberwar: What Everyone Needs to Know*. New York: Oxford University Press, 2014.

Stanley, Janet. "The Internet Can Be Dangerous for Children." In *Does the Internet Benefit Society?*, edited by Cindy Mur, 95–101. Farmington Hills, MI: Greenhaven, 2005.

Talbot, David. "The Private Sector Should Fight Cybercrime." In *The Internet: Introducing Issues with Opposing Viewpoints*, edited by Lauri S. Friedman, 81–86. Farmington Hills, MI: Greenhaven, 2008.

Tarbox, Katherine. *Katie.com: My Story*. New York: E. P. Dutton, 2000.

Taylor, Max, Gemma Holland, and Ethel Quayle. "Typology of Pedophile Picture Collections." *Police Journal* 74, no. 2 (2001): 97–107.

Taylor, Max, and Ethel Quayle. *Child Pornography: An Internet Crime*. Hove, UK: Brunner-Routledge, 2003.

Thornburgh, Dick, and Herbert S. Lin, eds. *Youth, Pornography, and the Internet*. Washington, DC: National Academies Press, 2002.

U.S. Department of Justice, Office for Victims of Crime. "Internet Threats to Children Are Increasing." In *Does the Internet Increase the Risk of Crime?*, edited by Lisa Yount, 10–18. Farmington Hills, MI: Greenhaven, 2006.

von Feilitzen, Cecilia, and Johanna Stenersen, eds. *Young People, Media, and Health: Risks and Rights*. Göteborg, Germany: Nordicom, 2014.

White, Amy E. *Virtually Obscene: The Case for an Uncensored Internet*. Jefferson, NC: McFarland, 2006.

Williams, Kirk R., and Nancy G. Guerra. "Prevalence and Predictors of Internet Bullying." *Journal of Adolescent Health* 41, no. 6 (2007): S14–21.

Wilson, Clay. "Botnets, Cybercrime, and Cyberterrorism: Vulnerabilities and Policy Issues for Congress." CRS Report for Congress, Congressional Research Service, Washington, DC, January 29, 2008.

Wolak, Janis. *Child Sexual Exploitation on the Internet*. Oxford: Wiley, 2010.

Wolak, Janis, David Finkelhor, and Kimberly J. Mitchell. "Child Pornography Possessors: Trends in Offender and Case Characteristics." *Sexual Abuse* 23, no. 1 (2011): 22–42.

Wolak, Janis, David Finkelhor, Kimberly J. Mitchell, and Michele L. Ybarra. "Online 'Predators' and Their Victims: Myths, Realities, and Implications for Prevention and Treatment." *American Psychologist* 63, no. 2 (2008): 111–28.

Wright, Kevin B., and Lynn M. Webb, eds. *Computer-Mediated Communication in Personal Relationships*. New York: Peter Lang, 2011.

Wyre, Ray, and Tim Tate. *The Murder of Childhood: Inside the Mind of One of Britain's Most Notorious Child Murderers*. London: Penguin, 1995.

Ybarra, Michele L., Marie Diener-West, and Philip J. Leaf. "Examining the Overlap in Internet Harassment and School Bullying: Implications for School Intervention." *Journal of Adolescent Health* 41, no. 6 (2007): S42–50.

Ybarra, Michele L., Dorothy L. Espelage, and Kimberly J. Mitchell. "The Co-occurrence of Internet Harassment and Unwanted Sexual Solicitation

Victimization and Perpetration: Associations with Psychosocial Indicators." *Journal of Adolescent Health* 41, no. 6 (2007): S31–41.

Yount, Lisa, ed. *Does the Internet Increase the Risk of Crime?* Farmington Hills, MI: Greenhaven, 2006.

HATE

Akdeniz, Yaman. *Racism on the Internet.* Strasbourg, France: Council of Europe Publishing, December 2009.

Allport, Gordon W. *The Nature of Prejudice.* Cambridge, MA: Addison-Wesley, 1954.

Anti-Defamation League. "Combating Extremism in Cyberspace: The Legal Issues Affecting Internet Hate Speech." Anti-Defamation League, New York, 2000. http://archive.adl.org/civil_rights/newcyber.pdf .

"Responding to Cyberhate." Anti-Defamation League, New York, August 2010.

Barnett, Brett A. *Untangling the Web of Hate: Are Online "Hate Sites" Deserving of First Amendment Protection?* Youngstown, OH: Cambria, 2007.

Berkowitz, Leonard. "The Case for Bottling Up Rage." *Psychology Today* 7, no. 2 (1973): 24–31.

Burch, Edgar. "Comment: Censoring Hate Speech in Cyberspace – A New Debate in a New America." *North Carolina Journal of Law and Technology* 3, no. 1 (2001): 175–92.

Citron, Danielle Keats, and Helen Norton. "Intermediaries and Hate Speech: Fostering Digital Citizenship for Our Information Age." *Boston University Law Review* 91, no. 4 (2011): 1435–84.

Cohen-Almagor, Raphael. "Countering Hate on the Internet." *Annual Review of Law and Ethics* 22 (2014): 431–43.

"Countering Hate on the Internet: A Rejoinder." *Amsterdam Law Forum* 2, no. 2 (2010): 125–32.

"Ethical Considerations in Media Coverage of Hate Speech in Canada." *Review of Constitutional Studies* 6, no. 1 (2001): 79–100.

"Fighting Hate and Bigotry on the Internet." *Policy and Internet* 3, no. 3 (2011): 1–26.

"Holocaust Denial Is a Form of Hate Speech." *Amsterdam Law Forum* 2, no. 1 (2009): 33–42.

"Is Law Appropriate to Regulate Hateful and Racist Speech? The Israeli Experience." *Israel Studies Review* 27, no. 2 (2012): 41–64.

ed. *Liberal Democracy and the Limits of Tolerance: Essays in Honor and Memory of Yitzhak Rabin.* Ann Arbor: University of Michigan Press, 2000.

"Regulating Hate and Racial Speech in Israel." *Cardozo Journal of International and Comparative Law* 17 (2009): 101–10.

Daniels, Jessie. *Cyber Racism: White Supremacy Online and the New Attack on Civil Rights.* Lanham, MD: Rowman & Littlefield, 2009.

Delgado, Richard, and Jean Stefancic. *Understanding Words That Wound.* Boulder, CO: Westview, 2004.

Eberwine, Eric T. "Sound and Fury Signifying Nothing: Jürgen Büssow's Battle against Hate-Speech on the Internet." *New York Law Review* 49 (2004): 353–410.

Ehrlich, Howard J. *Campus Ethnoviolence and the Policy Options*. Baltimore: National Institute against Prejudice and Violence, 1990.

Foxman, Abraham H., and Christopher Wolf. *Viral Hate: Containing Its Spread on the Internet*. New York: Palgrave Macmillan, 2013.

Goldsmith, Jack L., and Tim Wu. *Who Controls the Internet? Illusions of a Borderless World*. New York: Oxford University Press, 2006.

Gruen, Madeleine. "Innovative Recruitment and Indoctrination Tactics by Extremists: Video Games, Hip-Hop, and the World Wide Web." In *The Making of a Terrorist: Recruitment, Training, and Root Causes*, edited by James J. F. Forest, 11–22. Westport, CT: Praeger, 2006.

"White Ethnonationalist and Political Islamist Methods of Fund-Raising and Propaganda on the Internet." In *The Changing Face of Terrorism*, edited by Rohan Gunaratna, 127–45. Singapore: Marshall Cavendish, 2004.

Haney, Craig, Curtis Banks, and Philip Zimbardo. "Interpersonal Dynamics in a Simulated Prison." *International Journal of Criminology and Penology* 1 (1973): 69–97.

Hare, Ivan, and James Weinstein, eds. *Extreme Speech and Democracy*. Oxford: Oxford University Press, 2009.

Henthoff, Nat. *Free Speech for Me – But Not for Thee: How the American Left and Right Relentlessly Censor Each Other*. New York: Harper Collins, 1992.

Hoffman, David S. *The Web of Hate: Extremists Exploit the Internet*. New York: Anti-Defamation League, 1996.

Kessler, Jordan. *Poisoning the Web: Hatred Online*. New York: Anti-Defamation League, 1999.

Lawrence, Frederick M. "The Hate Crime Project and Its Limitations: Evaluating the Societal Gains and Risk in Bias Crime Law Enforcement." Public Law Research Paper 216, George Washington University Law School, Washington, DC, 2006. http://papers.ssrn.com/sol3/papers.cfm?abstract_id=921923 .

Le Menestrel, Marc, Mark Hunter, and Henri-Claude de Bettignies. "Internet E-ethics in Confrontation with an Activists' Agenda: Yahoo! on Trial." *Journal of Business Ethics*, 39, nos. 1–2 (2002): 135–44.

Maitra, Ishani, and Mary Kate McGowan. "On Racist Hate Speech and the Scope of a Free Speech Principle." *Canadian Journal of Law and Jurisprudence* 23, no. 2 (2010): 343–72.

Marriott, Michel. "Rising Tide: Sites Born of Hate." *New York Times*, March 18, 1999, G1.

Matsuda, Mari J., Charles R. Lawrence III, Richard Delgado, and Kimberly W. Crenshaw, eds. *Words That Wound: Critical Race Theory, Assaultive Speech, and the First Amendment*. Boulder, CO: Westview Press, 1993.

Moon, Richard. *The Constitutional Protection of Freedom of Expression*. Toronto, ON: University of Toronto Press 2000.

"Report to the Canadian Human Rights Commission Concerning Section 13 of the Canadian Human Rights Act and the Regulation of Hate Speech on the Internet." Canadian Human Rights Commission, Ottawa, October 2008. http://www.safs.ca/moonreport.pdf .

National Research Council. *Global Networks and Local Values: A Comparative Look at Germany and the United States*. Washington, DC: National Academies Press, 2001.

Newman, Stephen L. "American and Canadian Perspectives on Hate Speech and the Limits of Free Expression." In *Constitutional Politics in Canada and the United States*, edited by Stephen L. Newman, 153–73. Albany: State University of New York Press, 2004.

"Should Hate Speech Be Allowed on the Internet? A Reply to Raphael Cohen-Almagor." *Amsterdam Law Forum* 2, no. 2 (2010): 119–23.

Peters, William. *A Class Divided: Then and Now*. New Haven, CT: Yale University Press, 1971.

Phillips Marsh, Elizabeth. "Purveyors of Hate on the Internet: Are We Ready for Hate Spam?" *Georgia State University Law Review* 17, no. 2 (2000): 379–407.

Roversi, Antonio. *Hate on the Net: Extremist Sites, Neo-fascism On-line, Electronic Jihad*. Aldershot, UK: Ashgate, 2008.

Southern Poverty Law Center. *Active Hate Sites on the Internet in the Year 2001: Intelligence Report* 101 (2002): 40–43.

Spinello, Richard A. *Cyberethics: Morality and Law in Cyberspace*. Sudbury, MA: Jones & Bartlett, 2000.

Stone, Deborah. "To Hate, Click Here: Antisemitism on the Internet." Special Report 38, B'nai B'rith Anti-Defamation Commission, Melbourne, Australia, August 2008.

Tiven, Lorraine. *Hate on the Internet: A Response Guide for Educators and Families*. Washington, DC: Anti-Defamation League, December 2003.

Tsesis, Alexander. "Prohibiting Incitement on the Internet." *Virginia Journal of Law and Technology* 7, no. 2 (2002): 1–41. http://vjolt.net/vol7/issue2/v7i2_a05-Tsesis. pdf .

Vick, Douglas W. "Regulating Hatred." In *Human Rights in the Digital Age*, edited by Mathias Klang and Andrew Murray, 41–54. London: GlassHouse, 2005.

Waldron, Jeremy. *The Harm in Hate Speech*. Cambridge, MA: Harvard University Press, 2012.

"Is Dignity the Foundation of Human Rights?" Public Law Research Paper 12–73, New York University School of Law, New York, January 3, 2013.

Wallace, Jonathan, and Mark Mangan. *Sex, Laws, and Cyberspace*. New York: Henry Holt, 1996.

Watt, Horatia Muir. "Yahoo! Cyber-Collision of Cultures: Who Regulates?" *Michigan Journal of International Law* 24, no. 3 (2003): 673–96.

Wistrich, Robert S. *A Lethal Obsession: Anti-semitism from Antiquity to the Global Jihad*. New York: Random House, 2010.

Wolf, Christopher. "Needed: Diagnostic Tools to Gauge the Full Effect of Online Anti-semitism and Hate." Presented at the Organization for Security and Co-operation in Europe meeting on the Relationship between Racist, Xenophobic, and Anti-semitic Propaganda on the Internet and Hate Crimes. Paris, June 16, 2004.

TERRORISM

Altheide, David L. *Terrorism and the Politics of Fear*. Lanham, MD: AltaMira Press, 2006.

Terror Post 9/11 and the Media. New York: Peter Lang, 2009.

Anonymous. *Al Qaeda Manual*. http://www.fas.org/irp/world/para/manualpart1_1.pdf .

Anonymous [Rita Katz]. *Terrorist Hunter*. New York: HarperCollins, 2003.

Atwan, Abdel Bari. *After Bin Laden: Al Qaeda, the Next Generation*. London: Saqi, 2012.

The Secret History of al Qaeda. Berkeley: University of California Press, 2006.

Bobbitt, Philip. *Terror and Consent: The Wars of the Twenty-First Century*. New York: Knopf, 2008.

Brachman, Jarret M. *Global Jihadism: Theory and Practice*. London: Routledge, 2009.

Bunt, Gary R. *Islam in the Digital Age: E-Jihad, Online Fatwas, and Cyber Islamic Environments*. London: Pluto Press, 2003.

Calvert, John. *Sayyid Qutb and the Origins of Radical Islamism*. New York: Columbia University Press, 2010.

Clarke, Richard A. *Against All Enemies: Inside America's War on Terror*. New York: Free Press, 2004.

Cohen-Almagor, Raphael. "In Internet's Way: Radical, Terrorist Islamists on the Free Highway." *International Journal of Cyber Warfare and Terrorism* 2, no. 3, (2012): 39–58.

"The Terrorists' Best Ally: The Quebec Media Coverage of the FLQ Crisis in October 1970." *Canadian Journal of Communication* 25, no. 2 (2000): 251–84.

Combs, Cindy C. "The Media as a Showcase for Terrorism." In *Teaching Terror: Strategic and Tactical Learning in the Terrorist World*, edited by James J. F. Forest, 133–54. Lanham, MD: Rowman & Littlefield, 2006.

Conway, Maura. "Terrorist 'Use' of the Internet and Fighting Back." *Information and Security* 19 (2006): 9–30.

Coolsaet, Rik, ed. *Jihadi Terrorism and the Radicalisation Challenge: European and American Experiences*. Surrey, UK: Ashgate, 2011.

Daniels, Peggy, ed. *Policing the Internet*. Farmington Hills, MI: Greenhaven, 2007.

Denning, Dorothy E. "Terror's Web: How the Internet Is Transforming Terrorism." In *Handbook of Internet Crime*, edited by Yvonne Jewkes and Majid Yar, 194–213. Portland, OR: Willan, 2010.

"A View of Cyberterrorism Five Years Later." In *Internet Security: Hacking, Counterhacking, and Society*, edited by Kenneth E. Himma, 123–40. Sudbury, MA: Jones & Bartlett, 2007.

Forest, James J. F., ed. *Teaching Terror: Strategic and Tactical Learning in the Terrorist World*. Lanham, MD: Rowman & Littlefield, 2006.

"Training Camps and Other Centers of Learning." In *Teaching Terror: Strategic and Tactical Learning in the Terrorist World*, edited by James J. F. Forest, 69–109. Lanham, MD: Rowman & Littlefield, 2006.

Gable, Gerry, and Paul Jackson. *Lone Wolves: Myth or Reality*. London: Searchlight, 2011.

Geltzer, Joshua Alexander. *U.S. Counter-terrorism Strategy and al-Qaeda: Signalling and the Terrorist World-View*. New York: Routledge, 2009.

Giroux, Henry A. *Beyond the Spectacle of Terrorism: Global Uncertainty and the Challenge of the New Media*. Boulder, CO: Paradigm, 2006.

Golumbic, Martin C. *Fighting Terror Online: The Convergence of Security, Technology, and the Law*. New York: Springer, 2008.

Green, Joshua. "Cyberterrorism Is Not a Major Threat." In *Does the Internet Increase the Risk of Crime?*, edited by Lisa Yount, 56–65. Farmington Hills, MI: Greenhaven, 2006.

"The Threat of Cyberterrorism Is Greatly Exaggerated." In *Does the Internet Benefit Society?*, edited by Cindy Mur, 40–49. Farmington Hills, MI: Greenhaven, 2005.

Gruen, Madeleine. "Innovative Recruitment and Indoctrination Tactics by Extremists: Video Games, Hip-Hop, and the World Wide Web." In *The Making of a Terrorist: Recruitment, Training, and Root Causes*, edited by James J. F. Forest, 11–22. Westport, CT: Praeger, 2006.

Gunaratna, Rohan, ed. *The Changing Face of Terrorism*. Singapore: Marshall Cavendish, 2004.

Hewitt, Steve. *The British War on Terror: Terrorism and Counter-terrorism on the Home Front since 9–11*. London: Continuum, 2008.

Hoffman, Bruce. "Internet Terror Recruitment and Tradecraft: How Can We Address an Evolving Tool While Protecting Free Speech?" Statement before the House Committee on Homeland Security, Subcommittee on Intelligence, Information Sharing and Terrorism Risk Assessment, Washington, DC, May 26, 2010.

Jenkins, Brian Michael. "No Path to Glory: Deterring Homegrown Terrorism." Statement before the House Committee on Homeland Security, Subcommittee on Intelligence, Information Sharing and Terrorism Risk Assessment, Washington, DC, May 26, 2010.

Would-Be Warriors: Incidents of Jihadist Terrorist Radicalization in the United States since September 11, 2001. Santa Monica, CA: RAND, 2010.

Kenney, Michael. "How Terrorists Learn." In *Teaching Terror: Strategic and Tactical Learning in the Terrorist World*, edited by James J. F. Forest, 33–51. Lanham, MD: Rowman & Littlefield, 2006.

Lappin, Yaakov. *Virtual Caliphate: Exposing the Islamist State on the Internet*. Washington, DC: Potomac Books, 2011.

Martin, Gus. *Essentials of Terrorism: Concepts and Controversies*. Thousand Oaks, CA: Sage, 2011.

Meijer, Roel, ed. *Global Salafism: Islam's New Religious Movement*. New York: Columbia University Press, 2009.

Meleagrou-Hitchens, Alexander. "New English-Language al-Qaeda Explosive Manual Released Online." *ICSR Insight*, International Centre for the Study of Radicalization, King's College, London, December 31, 2010.

Morris, John B. "Free Speech and Online Intermediaries in an Age of Terror Recruitment." Statement before the House Committee on Homeland Security, Washington, DC, May 26, 2010.

National Security Program. "Countering Online Radicalization in America." Bipartisan Policy Center, Washington, DC, December 2012.

"Jihadist Terrorism and Other Unconventional Threats." Bipartisan Policy Center, Washington, DC, September 2014, http://bipartisanpolicy.org/sites/default/files/BPC%20HSP%202014%20Jihadist%20Terrorism%20and%20Other%20Unconventional%20Threats%20September%202014.pdf .

Perl, Raphael F. "Terrorist Use of the Internet: Threat, Issues, and Options for International Co-operation." Remarks before the Second International Forum on Information Security, Garmisch-Partenkirchen, Germany, April 7–10, 2008.

Rheingold, Howard. "The Emerging Wireless Internet Will Both Improve and Degrade Human Life." In *The Future of the Internet*, edited by Tom Head, 19–32. Farmington Hills, MI: Greenhaven Press, 2005.

Rogan, Hanna. "Jihadism Online: A Study of How al-Qaida and Radical Islamist Groups Use the Internet for Terrorist Purposes." Norwegian Defence Research Establishment, Kjeller, Norway, 2006.

Rollins, John, Liana Sun Wyler, and Seth Rosen. "International Terrorism and Transnational Crime: Security Threats, U.S. Policy, and Considerations for Congress." CRS Report for Congress, Congressional Research Service, Washington, DC, March 18, 2010.

Romero, Anthony D. "Internet Terror Recruitment and Tradecraft: How Can We Address an Evolving Tool While Protecting Free Speech?" Statement before the House Committee on Homeland Security, Washington, May 26, 2010.

Roversi, Antonio. *Hate on the Net: Extremist Sites, Neo-fascism On-line, Electronic Jihad*. Aldershot, UK: Ashgate, 2008.

Sageman, Marc. *Leaderless Jihad: Terror Networks in the Twenty-First Century*. Philadelphia: University of Pennsylvania Press, 2008.

Understanding Terror Networks. Philadelphia: University of Pennsylvania Press, 2004.

Schmid, Alex P., ed. *The Routledge Handbook of Terrorism Research: Research, Theories, and Concepts*. London: Routledge, 2011.

Sieber, Ulrich, and Phillip W. Brunst. *Cyberterrorism: The Use of the Internet for Terrorist Purposes*. Strasbourg, France: Council of Europe, 2008.

United Nations Office on Drugs and Crime. *The Use of the Internet for Terrorist Purposes*. New York: United Nations, 2012.

Wade, Marianne, and Almir Mljevic, eds. *A War on Terror? The European Stance on a New Threat, Changing Laws, and Human Rights Implications*. New York: Springer, 2011.

Wallace, Jonathan, and Mark Mangan. *Sex, Laws, and Cyberspace*. New York: Henry Holt, 1996.

Weimann, Gabriel. "New Terrorism and New Media." Woodrow Wilson International Center for Scholars, Washington, DC, 2014.

"Terrorist Dot Com: Using the Internet for Terrorist Recruitment and Mobilization." In *The Making of a Terrorist: Recruitment, Training, and Root Causes*, edited by James J. F. Forest, 53–65. Westport, CT: Praeger, 2006.

Terror on the Internet: The New Arena, the New Challenges. Washington, DC: U.S. Institute of Peace Press, 2006.

"Virtual Training Camps: Terrorists' Use of the Internet." In *Teaching Terror: Strategic and Tactical Learning in the Terrorist World*, edited by James J. F. Forest, 110–32. Lanham, MD: Rowman & Littlefield, 2006.

Weimann, Gabriel, and Katharina Von Knop. "Applying the Notion of Noise to Countering Online Terrorism." *Studies in Conflict and Terrorism* 31, no. 10 (2008): 883–902.

Wiktorowicz, Quintan. *Global Jihad: Understanding September 11*. Falls Church, VA: Sound Room Publishers, 2002.

"The New Global Threat: Transnational Salafis and Jihad." *Middle East Policy* 8, no. 4 (2001): 18–38, http://groups.colgate.edu/aarislam/wiktorow.htm .

Radical Islam Rising: Muslim Extremism in the West. Lanham, MD: Rowman & Littlefield, 2005.

Wright, Lawrence. *The Looming Tower: Al-Qaeda and the Road to 9/11.* New York: Knopf, 2006.

Index

The letter *t* following a page number denotes a table. The letter *f* following a page number denotes a figure.

Optus, 197
Orchid Club, 295–96
Organization for Security and Co-operation in Europe (OSCE), 288, 303
outcome measures, 51
OwnWhatYouThink.com, 90

packet-level filtering, 41
packet radio, 23–24
packet sniffers, 42–43
packet switching, 17, 21, 22–23, 31, 35–36, 154
Page, Larry, 28
Panama, 27
parental responsibility, 97–98
particularism, 311–12
password crackers, 185
Pecard, Christophe, 236
Pedophile Liberation Front, 191–92
pedophiles/pedophilia: advertising campaigns against, 255; defined, 191–92; on MySpace, 42; in search results, 223; task forces combating, 293–94; wrongful depiction of teachers as, 145. *See also* online child sex offenders
peer-to-peer (P2P) file sharing, 39–40, 193, 226
PeoplesDirt.com, 88
performance standards, 51
personal identity, 68–69, 71. *See also* anonymity
Peru, 26
Pew Internet Project, 103
PGP (Pretty Good Privacy), 26, 45–46, 181–82
Philippines, 27, 299, 300
Philosophy of Information, The (Floridi), 148
PhotoDNA, 226
Pierce, William, 157
"Place of Dark Desires" (webpage), 115
Planet, the (Web-hosting company), 156
Platform for Internet Content Selection (PICS), 41
pluralism, 10, 172–73
Plutarch, 43
Poland, 25–26, 139n145, 298
political neutrality principle, 173
Popper, Karl, 232
pornography/adult websites, 98–99, 145, 225. *See also* child pornography
Porter, Michael, 150
Portugal, 26, 296
Postal 2 (video game), 120–21

Post and Telecommunications Authority (Norway), 168
Postel, Jonathan B., 23
Pricewert (Web-hosting company), 156
"Priority of Right and Ideas of the Good, The" (Rawls), 174–75
privacy: importance of for adolescents, 113; on social networking sites, 200; tools for, 13, 43–47; violations of, 250–52. *See also* anonymity; security
procedural neutrality, 174–75
promotional approach, 15, 174–76, 214, 216, 223, 318
"Proposal for an International Convention on Cyber Crime and Terrorism, A," 276–77
Protection from Harassment Act (1997, UK), 263
Protect Our Children Act (2008, US), 198
Protocol for Web Description Resources (POWDER), 41, 262
PSINET, 27
Public Order Act (1936, UK), 233
Public Order Act (1986, UK), 234

Race Relations Act (1965, UK), 233–34
Racial Hatred Act (1995, Australia), 246
racism: defined, 148, 246; European legislation against, 232–35, 279; freedom of speech and, 61; on hate sites, 157; organizations against, 301–5; search engine results and, 221–23; state responsibility and, 231; white supremacists, 207–13, 215, 221. *See also* antisemitism; hate speech
Radin, David, 249–50
RAND Corporation, 20–21
rape, 86, 115, 125–26, 201, 210, 223
ratings. *See* reputation systems; self-regulation
Ravi, Dharun, 109
Rawls, John, 9, 174–75
Raz, Joseph, 174
reader responsibility: cyberstalking and, 142–45; importance of, 13, 14; murders and, 14, 116, 120, 126–28, 130–6; suicides and, 137–40
recognition respect, 8
Reding, Viviane, 304
Red Lake High School, 130
Reedy, Janice, 275
Reedy, Thomas, 275
Register.com, 188
relativism, 10, 189